Lecture Notes in Mathematics

Edited by A. Dold and B. Eckmann

957

Differential Equations

Proceedings of the 1st Latin American School of Differential Equations, Held at São Paulo, Brazil, June 29–July 17, 1981

Edited by D.G. de Figueiredo and C.S. Hönig

Springer-Verlag
Berlin Heidelberg New York 1982

Editor
Djairo Guedes de Figueiredo
Department of Mathematics, University of Brasilia
70910 Brasilia DF, Brazil

Chaim Samuel Hönig
Institute of Mathematics and Statistics
University of São Paulo
01000 São Paulo SP, Brazil

AMS Subject Classifications (1980): 34 D 05, 35 J 65, 45 D 05, 46 E 35, 47 E 05, 47 F 05, 47 H 99, 49 B 40, 49 G 05

ISBN 3-540-11951-5 Springer-Verlag Berlin Heidelberg New York
ISBN 0-387-11951-5 Springer-Verlag New York Heidelberg Berlin

Printing and binding: Beltz Offsetdruck, Hemsbach/Bergstr.
2146/3140-543210

Foreword

These are the proceedings of the first Latin American School of Differential Equations (ELAED 1 for short) held at the University of São Paulo during the period from June 29 to July 17, 1981. The School has been started as a consequence of a joint project of Brazilian and Mexican mathematicians, but hopefully it will evolve towards a larger program in Latin America with the active participation of other countries.

This first meeting received financial support from the Conselho Nacional do Desenvolvimento Científico e Tecnológico (CNPq, Brazil), Consejo Nacional de Ciencia y Tecnologia (CONACYT, Mexico), Sociedade Brasileira de Matemática, Fundação de Amparo à Pesquisa do Estado de São Paulo (FAPESP, Brazil), Universidade de São Paulo, Instituto Politecnico Nacional de Mexico, Universidad Nacional Autonoma de Mexico.

There were 6 courses, which were delivered by Alfonso Castro B. (Reduction Methods via Minimax), Djairo G. de Figueiredo (Positive Solutions of Semilinear Elliptic Problems), Jorge Ize (Introduction to Bifurcation Theory), Onésimo Hernandez (Introduction to Stochastic Differential Equations), Pedro Nowosad (Quanta and Geometry) and Paul Rabinowitz (The Mountain Pass Theorem: Theme and Variations), 15 invited lectures and a number of research announcements. The number of participants exceeded 80.

The organizational committee was composed by Chaim Samuel Hönig (USP, Brazil), Djairo Guedes de Figueiredo (UnB, Brazil) and Samuel Gitler (IPN, Mexico).

We would like to acknowledge the warm hospitality offered to all the participants during the meeting by the faculty and staff of the Instituto de Matemática e Estatística da Universidade de São Paulo. Our thanks to them, to the supporting agencies and to the lecturers that have kindly answered our call for their manuscripts. A word of acknowledgement to José Pereira dos Santos for his excellent typing of the text, and to Springer Verlag for including these proceedings in their well known collection Lecture Notes in Mathematics.

Brasília

May, 1982 Djairo Guedes de Figueiredo

LIST OF CONTRIBUTORS

ALFONSO CASTRO B., Departamento de Matemáticas, Centro de Investigación del IPN, Apartado Postal 14740, México 14, D.F. México

DJAIRO GUEDES DE FIGUEIREDO, Universidade de Brasília, Departamento de Matemática, Brasília, DF - Brazil

SAUL HAHN-GOLDBERG, Centro de Investigación del IPN, Departamento de Matemáticas, Apartado Postal 14-740, México 14, D.F.

DANIEL B. HENRY, Instituto de Matemática e Estatística, Universidade de São Paulo, São Paulo, SP - Brazil

CHAIM SAMUEL HÖNIG, Instituto de Matemática e Estatística, Universidade de São Paulo, São Paulo, SP - Brazil

A. F. IZÉ, Universidade de São Paulo, Instituto de Ciências Matemáticas de São Carlos, Departamento de Matemática, São Paulo, SP - Brazil

JORGE IZE, IIMAS - Universidad Nacional Autonoma de Mexico, Apartado Postal 20-726 Mexico DF - Mexico

JORGE LEWOWICZ, Universidad Simon Bolivar, Departamento de Matemática, Caracas, Venezuela

P. S. MILOJEVIĆ, Universidade Federal de Minas Gerais, Department of Mathematics - ICEx, Belo Horizonte, MG - Brazil

GUSTAVO PERLA MENZALA, Instituto de Matemática, Universidade Federal do Rio de Janeiro, Rio de Janeiro, RJ - Brazil

PAUL H. RABINOWITZ, Department of Mathematics, University of Wisconsin, Madison, Wisconsin, USA

LARRY L. SCHUMAKER, Department of Mathematics and Center for Approximation Theory, Texas A&M University, College Station, Texas 77843 USA

J. SOTOMAYOR, Instituto de Matemática Pura e Aplicada, Rio de Janeiro, RJ - Brazil

CONTENTS

REDUCTION METHODS VIA MINIMAX

Alfonso Castro B.

Departamento de Matemáticas
Centro de Investigación del IPN
Apartado Postal 14740
México 14, D.F. México

1. Introduction

Let H be a real Hilbert space and $J:H \to \mathbb{R}$ a functional of class C^1. That is, there exists a continuous function $\nabla J:H \to H$ such that for $x,y \in H$

$$\lim_{t\to 0} \frac{J(x+ty)-J(x)}{t} = \langle \nabla J(x), y \rangle$$

where $\langle \, , \, \rangle$ is the inner product in H.

In this note we consider the existence of critical points of J, which are points $u \in H$ such that $\nabla J(u) = 0$. The particular kind of functional J that we study have the property that there exist closed subspaces X and Y with $H = X \oplus Y$ and such that the existence of critical points of J is equivalent to the existence of critical points of a new functional $\tilde{J}:X \to \mathbb{R}$. The functional $\tilde{J}$ is given in the form $\tilde{J}(x) = J(x+r(x))$, where $r:X \to Y$ is a continuous function defined via a "minimax characterization"; that is, some functional takes a minimum, maximum or minimax value at $r(x)$ (see Lemmas 1 and 3).

In section 2 the reader will find the basic abstract tools which will be used throughout the applications. As applications we present the existence of solutions for Hammerstein integral equations, periodic solutions of the forced pendulum equation and solutions to a nonlinear Dirichlet problem.

2. Reduction Lemmas

Lemma 1. *Let* X *and* Y *be closed subspaces of a real Hilbert space* H *such that* $H = X \oplus Y$. *Let* $J:H \to \mathbb{R}$ *be a functional of class* C^1. *If there exists an increasing function* $\phi:(0, \infty) \to (0, \infty)$ *such that* $\phi(t) \to \infty$ *as* $t \to \infty$ *and*

$$\langle \nabla J(x+y) - \nabla J(x+y_1), y-y_1 \rangle \geq \|y-y_1\| \phi(\|y-y_1\|) \tag{2.1}$$

for all $x \in X$, $y,y_1 \in Y$, $y \neq y_1$, *then:*

i) *there exists a continuous function* $r:X \to Y$ *such that* $J(x+r(x)) = \min\{J(x+y);\ y \in Y\}$; *moreover,* $r(x)$ *is the only critical point of the functional* $J_x:Y \to \mathbb{R}$, $y \to J(x+y)$.

ii) *the function* $\tilde{J}:X \to R$, $x \to J(x+r(x))$ *is of class* C^1 *and* $\langle \nabla\tilde{J}(x), x_1 \rangle = \langle J(x+r(x)), x_1 \rangle$ *for all* $x,x_1 \in X$.

Proof: From (2.1) and the assumption that ϕ takes only positive values it follows that J_x has at most one critical point. Also from (2.1) we have

$$\begin{aligned} J_x(y) &= J_x(0) + \int_0^1 \langle \nabla J_x(sy), y \rangle ds \\ &\geq J_x(0) - \|\nabla J_x(0)\| \|y\| + \int_0^1 s\|y\| \phi(\|sy\|) ds. \end{aligned} \tag{2.2}$$

Since we are assuming that $\phi(t) \to \infty$ as $t \to \infty$, there exists $R > 0$ such that $\phi(t) \geq 2(\|\nabla J_x(0)\| + 1)$ for $t > R$. Hence, for $\|y\| \geq 2R$ we have $J_x(y) \geq J_x(0) + 3\|y\| \to \infty$ as $\|y\| \to \infty$. Therefore, in order to prove that J_x has a unique point of minimum it is sufficient to show that J_x is convex. Let $y_1, y_2 \in Y$, $\psi(t) = J_x(y_1 + t(y_2 - y_1))$ and $0 < \alpha < \beta < 1$. Since J is of class C^1, ψ is of class C^1. From (2.1) we have

$$\psi'(\beta)-\psi'(\alpha) = \langle \nabla J(x+y_1+\beta(y_2-y_1))-\nabla J(x+y_1+\alpha(y_2-y_1)), y_2-y_1\rangle$$

$$\geq \|(\beta-\alpha)(y_2-y_1)\|\phi(\|(\beta-\alpha)(y_2-y_1)\|)/(\beta-\alpha) > 0.$$

Hence ψ is convex, which implies that J_x is convex. Consequently we have proved that J_x has a unique point of minimum which we denote by $r(x)$.

Now we show that $r(x)$ is continuous. If not, let $\delta > 0$ and $\{x_n\}$ be a sequence converging to some $x \in X$ such that

$$\|r(x_n) - r(x)\| \geq 2\delta. \tag{2.3}$$

Since ∇J is a continuous function and $\langle \nabla J(x+r(x)), y\rangle = 0$ for all $y \in Y$, we have for n sufficiently large

$$\|P^*(\nabla J(x_n+r(x)))\| < \phi(\delta), \tag{2.4}$$

where P^* denotes the adjoint of the operator $P(x+y) = y$ for $x \in X$, $y \in Y$. Thus from (2.1) we obtain

$$\|P^*(\nabla J(x_n+r(x)))\|\|r(x_n)-r(x)\| \geq \langle -\nabla J(x_n+r(x)), r(x_n)-r(x)\rangle$$

$$\geq \langle \nabla J(x_n+r(x_n)) - \nabla J(x_n+r(x)), r(x_n)-r(x)\rangle$$

$$\geq \|r(x_n) - r(x)\|\phi(2\delta).$$

Since this inequality contradicts (2.4), we have proved that r is continuous. This completes the proof of part i).

Let $t > 0$ and $h \in X$. From the fact that J_x attains its minimum at $r(x)$ we have

$$((\tilde{J}(x+th)-\tilde{J}(x))/t) = ((J(x+th+r(x+th))-J(x+r(x))$$

$$\leq ((J(x+th+r(x))-J(x+r(x)))/t)$$

$$= \int_0^1 <\nabla J(x+r(x)+sth),\ h>ds. \qquad (2.5)$$

In a similar manner it can be proved that

$$((\tilde{J}(x+th)-\tilde{J}(x))/t) \geq \int_0^1 <\nabla J(x+r(x+th) + sth),\ h>ds. \qquad (2.6)$$

From (2.5), (2.6) and the continuity of ∇J and r we obtain

$$<\nabla\tilde{J}(x),\ h> = <\nabla J(x+r(x)),\ h>.$$

Hence $\tilde{J}$ has continuous directional derivatives. Therefore $\tilde{J}$ is of class C^1 and the Lemma has been proved.

Remarks. 1. In most of the applications the function ϕ has the form $\phi(t) = mt$.

2. It is not necessary that ϕ be increasing, double checking the proof we see that it is sufficient to assume that $\inf\{\phi(t);\ t \geq \eta\} > 0$ for $\eta > 0$.

3. It is easy to verify that if (2.1) is replaced by "J_x has a unique point of minimum $r(x)$ and the function r is bounded in bounded sets, and ∇J is continuous in the weak topology of H", then the conclusions of Lemma 1 still hold.

4. The function $\tilde{J}$ inherits the degree of differentiability of J. For example we have

Lemma 2. *If in addition to the hypotheses of Lemma* 1 *we assume that* J *is of class* C^2 *then* $\tilde{J}$ *is of class* C^2.

Proof. See [13, Theorem 4].

The condition (2.1) says that the graph of J_x looks like a "parabola". There are cases in which the graph of J_x looks like a "global saddle point". In the following Lemma we consider this situation.

Lemma 3. *Let* X, Y *and* Z *be closed subspaces of* H *such that* $H = X \oplus Y \oplus Z$. *Let* $J: H \to \mathbb{R}$ *be a functional of class* C^1. *If there exists an increasing function* $\phi: (0, \infty) \to (0, \infty)$ *such that* $\phi(t) \to \infty$ *as* $t \to \infty$ *and*

$$-\|z_1 - z_2\| \phi(\|z_1 - z_2\|) \geq <\nabla J(x+y+z_1) - \nabla J(x+y+z_2),\ z_1 - z_2>, \tag{2.6}$$

$$<\nabla J(x+y+z) - \nabla J(x+y_1+z),\ y-y_1> \geq \|y-y_1\| \phi(\|y-y_1\|) \tag{2.7}$$

for all $x \in X$, $y, y_1 \in Y$, $y \neq y_1$, $z, z_1 \in Z$, $z \neq z_1$, *then:*

i) *there exists a continuous function* $r: X \to Y \oplus Z$ *such that* $J(x+r(x)) = \max\{\min\{J(x+y+z);\ y \in Y\};\ z \in Z\} = \min\{\max\{J(x+y+z);\ z \in Z\};\ y \in Y\}$.

ii) *the function* $\tilde{J}: X \to R$, $x \to J(x+r(x))$ *is of class* C^1 *and* $<J(x), x_1> = <\nabla J(x+r(x)), x_1>$ *for* $x, x_1 \in X$.

Proof: From (2.7) and Lemma 1 it follows that there exists a continuous function $\psi: X \oplus Z \to Y$ such that $J(x+z+\psi(x+z)) = \min\{J(x+z+y);\ y \in Y\}$, $J_1(x+z) \equiv J(x+z+\psi(x+z))$ is of class C^1 and $<\nabla J_1(x+z), u> = <\nabla J(x+z+\psi(x+z)), u>$ for $u \in X \oplus Z$.

From the definition of ψ, (2.6) and the fundamental theorem of calculus we infer

$$\begin{aligned} J_1(x+z) &= J(x+z+\psi(x+z)) \leq J(x+z) \\ &= J(x) + <\nabla J(x), z> + \int_0^1 <\nabla J(x+tz) - \nabla J(x), z> dt \\ &\leq J(x) + \|\nabla J(x)\| \|z\| - \int_0^1 \|z\| \phi(\|tz\|) dt. \end{aligned}$$

Arguing as in the paragraph following (2.2) we see that

$$J_1(x+z) \to -\infty \quad \text{as} \quad \|z\| \to \infty \tag{2.8}$$

Now we fix $x \in X$. Let $\{z_n\}$ be a sequence in Z such that $\lim_n J_1(z_n) = \sup\{J_1(x+z);\ z \in Z\}$. From (2.8) we see that $\{z_n\}$ is bounded. Therefore, without loss of generality, we can assume that $\{z_n\}$ converges weakly to some $z_o \in Z$. By the definition of ψ we have

$$\begin{aligned} J_1(x+z_n) &\leq J(x+z_n + \psi(x+z_n)) \\ &\leq J(x+z_n+\psi(x+z_o)). \end{aligned} \qquad (2.9)$$

From (2.6) it follows that the function $z \to J(x+z+\psi(x+z_o))$ is concave, therefore (see [20])

$$\begin{aligned} \limsup_n J(x+z_n+\psi(x+z_o)) &\leq J(x+z_o+\psi(x+z_o)) \\ &= J_1(x+z_o). \end{aligned} \qquad (2.10)$$

Hence, from (2.9) and (2.10) we see that J_1 attains its maximum value at z_o. Thus $r(x) = z_o + \psi(x+z_o)$ satisfies

$$J(x+r(x)) = \max\{\min\{J(x+y+z);\ y \in Y\};\ z \in Z\}. \qquad (2.11)$$

Now we show that $r(x)$ is the only critical point of $J_x : Y+Z \to R$, $y+z \to J(x+y+z)$. Suppose y_1+z_1 and y_2+z_2 are two different critical points of J_x and say $y_1 \neq y_2$. Since, from (2.7), the function $y \to J_x(z_1+y)$ is convex and y_1 is a critical point of this function we have $J_x(z_1+y_1) < J(z_1+y_2)$. From (2.6) we see that the function $z \to J_x(y_2+z)$ is concave. Hence, since z_2 is a critical point of this function, we have $J(z_1+y_2) \leq J(y_2+z_2)$. From the last two inequalities we have $J_x(y_1+z_1) < J_x(y_2+z_2)$. In a similar manner it can be proved that $J_x(y_2+z_2) < J_x(y_1+z_1)$. Since the last two inequalities contradict each other, we have proved that J_x has a unique critical point. Consequently $r(x)$ is the unique critical point of J_x.

Another way to obtain a critical point for J_x is like this.

For each $y \in Y$ the function $z \to J_x(y+z)$ has a unique point of maximum $\rho(x+y)$. Repeating the analysis we did on the function J_1 it is shown that the function $y \to J_x(y+\rho(x+y))$ attains a minimum value at some point y_0. Therefore we have

$$\begin{aligned} J_x(y_0+\rho(x+y_0)) &= J(x+y_0+\rho(x+y_0)) = \min\left\{J(x+y+\rho(x+y));\ y \in Y\right\} \\ &= \min\left\{\max\left\{J(x+y+z);\ z \in Z\right\};\ y \in Y\right\}. \end{aligned} \tag{2.12}$$

Combining (2.11) with (2.12) and using the fact that $r(x)$ is the unique critical point of J_x we have $r(x) = y_0 + \rho(x+y_0)$; thus we have proved the variational characterization of the conclusion i). The proof of the continuity of r is essentially identical to the proof of the continuity of the function r in Lemma 1. For this reason we leave this as an exercise.

We prove now that $\tilde{J}$ is differentiable. From (2.12) we see that for $t > 0$, $x,h \in X$ we have

$$\begin{aligned} \frac{\tilde{J}(x+th)-\tilde{J}(x)}{t} &= \frac{J(x+th+r_1(x+th)+r_2(x+th)) - J(x+r_1(x) + r_2(x))}{t} \\ &\le \frac{J(x+th+r_1(x)+r_2(x+th))-J(x+r_1(x)+r_2(x))}{t} \\ &\le \frac{J(x+th+r_1(x)+r_2(x+th))-J(x+r_1(x)+r_2(x+th))}{t} \\ &= \int_0^1 \langle\nabla J(x+sth+r_1(x)+r_2(x+th)),\ h\rangle ds. \end{aligned} \tag{2.13}$$

In the above we have written $r(u) = r_1(u) + r_2(u)$ with $r_1(x) \in Y$ and $r_2(u) \in Z$. In the first inequality we have used that $J(x+th+r(x+th)) = \min\{J(x+th+y+r_2(x+th));\ y \in Y\}$. Similar argument justifies the next inequality. We leave it to the reader to show that this type of analysis also leads to

$$\frac{\tilde{J}(x+th)-\tilde{J}(x)}{t} \ge \int_0^1 \langle\nabla J(x+sth+r_1(x+th)+r_2(x)),h\rangle ds. \tag{2.14}$$

Clearly (2.13) and (2.14) imply that J is of class C^1 and

$$<\nabla\tilde{J}(x), h> = <\nabla J(x+r_1(x)+r_2(x)), h> \tag{2.15}$$

and the Lemma is proven.

We invite the reader to state for Lemma 3 analogs of Remarks 1-4 of Lemma 1. Several applications of Lemma 3 can be found in [3].

3. Applications to Hammerstein integral equations

Let $\Omega \subseteq \mathbb{R}^n$ be a bounded region and $K:\Omega\times\Omega \to \mathbb{R}$ be such that $K(x,y) = K(y,x)$ for all $x,y \in \Omega$ and $K \in L^2(\Omega\times\Omega)$. We define $k:L^2(\Omega) \to L^2(\Omega)$ by

$$k(u)(x) = \int_\Omega K(x,y)u(y)dy. \tag{3.1}$$

Since $\|k(u)\|_{L^2(\Omega)} \leq \|K\|_{L^2(\Omega\times\Omega)} \|u\|_{L^2(\Omega)}$, the operator k is continuous. From Fubini's theorem it follows that k is compact (see [16, ch. IV]) and selfadjoint.

Let $g:\Omega\times\mathbb{R} \to \mathbb{R}$, $(x,y) \to g(x,y)$ satisfy the Caratheodory condition, that is, let g be continuous in u for each x and measurable in x for each u. We consider the *Hammerstein integral equation*

$$u(x) = \int_\Omega K(x,y)g(y,u(y))dy. \tag{3.2}$$

Now we show that finding solutions to (3.2) is equivalent to finding critical points of a certain functional to be defined (see Lemma 5). Since k is compact and $L^2(\Omega)$ is separable there exists a complete orthonormal set $\{\phi_1,\phi_2,...\}$ in $L^2(\Omega)$ and a sequence of real numbers $\{\lambda_1,\lambda_2,...\}$ tending to 0 with $k\phi_i = \lambda_i\phi_i$, $i=1,2,....$ We denote by Y the closed subspace of $L^2(\Omega)$ generated by $\{\phi_i;\ \lambda_i > 0\}$. It is easy to show that if X denotes the orthogonal complement of Y then X is the closed subspace generated by $\{\phi_i;\ \lambda_i \leq 0\}$. Hence, given $y \in Y$ there exists a sequence of real numbers $\{c_i\}$ such that $c_i = 0$ if $\lambda_i \leq 0$ and $y = \sum_{i=1}^{\infty} c_i\phi_i$. For $u \in L^2(\Omega)$ with

$u = \sum_{i=1}^{\infty} d_i\phi_i$ we define $Q_1(u) = \sum_{i=1}^{\infty} a_i\phi_i$ with $a_i = d_i(\lambda_i)^{\frac{1}{2}}$ for $\lambda_i \geq 0$ and $a_i = 0$ for $\lambda_i < 0$. It is easy to verify that

$$Q_1(x) = 0 \quad \text{for} \quad x \in X, \text{ and } Q_1(Q_1(y)) = k(y) \tag{3.4}$$

for all $y \in Y$. In the same manner it can be shown that there exists a selfadjoint operator $Q:L^2(\Omega) \to L^2(\Omega)$ such that

$$Q(y) = 0 \text{ for } y \in Y, \text{ and } Q(Q(x)) = -k(x) \tag{3.5}$$

for $x \in X$. Now we can prove:

Lemma 4. *For* $w \in L^2(\Omega)$ *let* $g_1(w)(\xi) \equiv g(\xi, w(\xi))$. *If* $(x,y) \in X \times Y$ *satisfies*

$$-x(\xi)+y(\xi) = Q(g_1(Q_1(y)+Q(x))(\xi))+Q_1(g_1(Q_1(y)+Q(x))(\xi)) \tag{3.6}$$

then $u = Q(x) + Q_1(y)$ *is a solution of* (3.2). *Conversely, if* u *is a solution of* (3.2) *then there exists* $(x,y) \in X \times Y$ *which satisfies* (3.6) *and such that* $u = Q(x) + Q_1(y)$.

Proof: Let $P:L^2(\Omega) \to L^2(\Omega)$ be the orthogonal projection onto X and $P_1 = I-P$ the orthogonal projection onto Y. Let (x,y) be a solution to (3.6) and let $u = Q(x) + Q_1(y)$. Applying Q and Q_1 to (3.6) we obtain

$$Q(x) = -Q^2(g_1(u)) = k(P(g_1(u))) \tag{3.7}$$

$$Q_1(y) = k(P_1(g_1(u))) \tag{3.8}$$

respectively. Adding (3.7) and (3.8) we see that u is a solution of (3.2).

Suppose now that u is a solution to (3.2). Let $x_1 = P(u)$ and $y_1 = P(u)$. Hence $x_1 = P(k(g_1(u))) = k(P(g_1(u))) = -Q^2(P(g_1(u)))$. Thus, if we set $x = -Q(P(g_1(u)))$ then $Q(x) = x_1$. Similarly $y_1 = Q_1(y)$ with $y = Q_1(g_1(u))$. Hence $-x = +Q(P(g_1(x_1+y_1))) =$

$+Q(G_1(Q(x) + Q_1(y)))$ and $y = Q_1(g_1(Q(x) + Q_1(y)))$. Adding the last two equations we see that $(x,y) \in X\times Y$ is a solution of (3.6) and the Lemma is proved.

Lemma 5. *Suppose there exist a real number* A *and a function* $B \in L^2(\Omega)$ *such that*

$$|g(x, u)| \le A|u| + B(x) \tag{3.9}$$

for $(x,u) \in \Omega\times\mathbb{R}$. *For* $w \in L^2(\Omega)$ *let* $G_1(w)(\xi) = \int_0^{w(\xi)} g(\xi, s)ds$. *If* $J:L^2(\Omega) \to \mathbb{R}$ *is the functional defined by*

$$J(u) = \int_\Omega (((P_1u)^2 - (Pu)^2)/2)-G_1(Q(u)+Q_1(u))(\xi)d\xi, \tag{3.10}$$

then J *is of class* C^1 *and* u_o *is a critical point of* J *iff* $(P(u_o), P_1(u_o)) \in X \times Y$ *is a solution to* (3.6).

Proof: For $(x,u) \in \Omega\times\mathbb{R}$ let $G(x,y) = \int_0^u g(x,s)ds$. Since g satisfies the Caratheodory condition, G also satisfies it. From (3.9) we have $|G(\xi,u)| \le (A+1)|u| + B^2(\xi)$. Hence J is well defined for all $u \in L^2(\Omega)$. Let us show that J is of class C^1. Let $u,v \in L^2(\Omega)$. A simple computation shows that

$$\frac{J(u+tv)-J(u)}{t} = \int_\Omega (P_1(u)P_1(v)-P(u)P(v))(1-(t/2))$$
$$-\int_\Omega (G_1((Q+Q_1)(u+tv))-G_1((Q+Q_1)(u)))/t).$$

Hence, using the definition of G_1 and the fact that $(\partial G/\partial u)(x,u) = g(x,u)$ we infer that

$$\lim_{t\to 0} \frac{J(u+tv)-J(u)}{t} = \int_\Omega (P_1(u)P_1(v)-P(u)P(v))$$
$$-\int_\Omega g_1((Q+Q_1)(u))(Q+Q_1)(v).$$
$$= \int_\Omega (P_1(u)-P(u)-(Q+Q_1)(g_1((Q+Q_1)(u))))v. \tag{3.11}$$

We observe that in obtaining the last equality we have used the selfadjointness of Q and Q_1. But P, P_1, Q and Q_1 are bounded operators. Hence because of (3.9) and Nemytskii's theorem (see [20, chapter 2]) the expression $P_1(u) - P(u) - (Q+Q_1)(g_1((Q+Q_1)(u)))$ defines a continuous operator with domain $L^2(\Omega)$ and values in $L^2(\Omega)$. Hence J is of class C^1 and

$$\nabla J(u) = P_1(u)-P(u)-(Q+Q_1)(g_1((Q+Q_1)(u))). \tag{3.12}$$

From (3.12) we see that if $\nabla J(u_o) = 0$ then

$$P(u_o) = -Q(g_1((Q+Q_1)(u_o))) \tag{3.13}$$

$$P_1(u_o) = Q_1(g_1((Q+Q_1)(u_o))). \tag{3.14}$$

Adding the last two equations we see that $(P(u_o), P_1(u_o)) \in X \times Y$ is a solution to (3.6).

Conversely, if $(P(u_o), P_1(u_o))$ is a solution to (3.6) then u_o satisfies (3.13) and (3.14). Hence from (3.12) we see that $\nabla J(u_o) = 0$ and the Lemma has been proved.

Since the eigenvalues $\{\lambda_i;\ i=1,2,\dots\}$ form a sequence tending to zero we can assume that they are ordered in the following way: $\lambda_1 \geq \lambda_2 \geq \dots \geq 0 \geq \dots \geq \lambda_{-2} \geq \lambda_{-1}$. Now we can proceed to apply the results of the previous section to the equation (3.2).

Theorem 6. *If* (3.9) *holds and there exist an integer* N *and real numbers* γ, γ' *and* C *such that:*

a) $(1/\lambda_N) < \gamma \leq \gamma' < (1/\lambda_{N+1})$,

b) *for* $u,v \in \mathbb{R}$, $x \in \Omega$ $(g(x,u) - g(x,v))(u-v) \leq \gamma'(u-v)^2$,

c) $G(x,u) \geq (\gamma/2)u^2-C$ *for* $(x,u) \in \Omega\times\mathbb{R}$

then:

A) *the equation* (3.2) *has a solution,*

B) *if the hypothesis* c) *is replaced by* $(g(x,u) - g(x,v))$

$(u-v) \geq \gamma(u-v)^2$ *for all* $u,v \in \mathbb{R}$, $x \in \Omega$, *then* (3.2) *has a unique solution.*

Proof: In order to prove claim A) it is sufficient to prove that J has a critical point (see Lemma 5). Without loss of generality we can assume that $N > 0$. Let Y_1 be the closed subspace of $L^2(\Omega)$ containing $\{\phi_{N+1}, \phi_{N+2}, \ldots\}$ and X_1 be its orthogonal complement. Thus X_1 is the closed subspace generated by the Kernel of k, X and the set $\{\phi_N, \phi_{N-1}, \ldots, \phi_1\}$. Applying b) and (3.12) we have for $x \in X_1$, $y, y_1 \in Y_1$

$$\begin{aligned}<\nabla J(x+y_1)-\nabla J(x+y), y_1-y> &= \int((y_1 - y)^2 - (g_1((Q+Q_1)(x+y_1)) \\ &\quad - g_1((Q+Q_1)(x+y)))(Q+Q_1)(x+y)) \\ &\geq \int((y_1 - y)^2 - \gamma'(Q+Q_1)^2(y_1-y)). \qquad (3.15)\end{aligned}$$

From the definition of Q_1 we have that $\int_\Omega (Q_1(y))^2 \leq \lambda_{N+1} \int y^2$. Combining this inequality with (3.15) we have

$$<\nabla J(x+y_1)-\nabla J(x+y), y_1-y> \geq (1-\gamma'\lambda_{N+1})\|y_1-y\|^2_{L^2(\Omega)}. \qquad (3.16)$$

Since hypothesis a) implies that $(1-\gamma'\lambda_{N+1}) > 0$, by Lemma 1 there exists a continuous function $r: X_1 \to Y_1$ sucht that $\tilde{J}(x) \equiv J(x+r(x)) = \min\{J(x+y);\ y \in Y_1\}$. Also from Lemma 1 (claim ii)) we know that in order to prove that J has a critial point it is sufficient to prove that $\tilde{J}$ has a critical point.

For $x \in X_1$, using c) we have

$$\begin{aligned}-2\tilde{J}(x) &\geq -2J(x) \\ &= -\int_\Omega ((P_1(x))^2 - (P(x))^2 - 2G_1((Q+Q_1)(x))) \\ &\geq -\|P_1(x)\|^2_{L^2(\Omega)} + \|P(x)\|^2_{L^2(\Omega)} + \gamma\|(Q+Q_1)(x)\|^2_{L^2(\Omega)} \\ &\quad -2C\,\mathrm{meas}(\Omega)\end{aligned}$$

$$\geq (\gamma\lambda_N - 1)\|x_1\|^2_{L^2(\Omega)} + \|P(x)\|^2_{L^2(\Omega)} - 2C\ \mathrm{meas}(\Omega), \tag{3.17}$$

where x_1 denotes the orthogonal projection of x on the subspace generated by $\{\phi_1,\ldots,\phi_N\}$. From (3.17) it is clear that

$$\tilde{J}(x) \to -\infty \quad \text{as} \quad \|x\| \to \infty, \quad x \in X_1. \tag{3.18}$$

Also (3.17) implies that $\tilde{J}$ is bounded above.

Let $\{x_n\}$ be a sequence in X_1 such that $\tilde{J}(x_n) \to \sup\{\tilde{J}(x);\ x \in X_1\}$. From (3.18) we see that $\{x_n\}$ is bounded. Thus without loss of generality we can assume that $\{x_n\}$ converges weakly to $x_o \in X_1$. Let x_n' denote the orthogonal projection of x_n on the subspace generated by $\{\phi_1,\ldots,\phi_N\}$. Clearly $\{x_n'\}$ converges to x_o', where x_o' is the orthogonal projection of x_o on that subspace. Let us see that $\{r(x_n)\}$ converges to $r(x_o)$. For $y \in Y_1$ we have

$$\begin{aligned} 0 &= \lim_n \langle \nabla J(x_n + \phi(x_n)),\ y\rangle \\ &= \lim_n \int((x_n' + r(x_n))y - g_1((Q+Q_1)(x_n + r(x_n)))Q_1(y)). \end{aligned} \tag{3.19}$$

Setting, in (3.16), $x = x_n$, $y_1 = r(x_n)$ and $y = 0$ we see that $\{r(x_n)\}$ is bounded. Hence we can also assume that $\{\phi(x_n)\}$ converges weakly to some element $y_o \in Y_1$. Hence using the compactness of Q and Q_1 we obtain

$$0 = \int y_o y - g_1((Q+Q_1)(x_o + y_o)))Q_1(y) \tag{3.20}$$

for all $y \in Y_1$. Since by Lemma 1 the only critical point of $J_{x_o}(y) \equiv J(x_o + y)$ is $r(x_o)$, from (3.2) we see that $y_o = r(x_o)$. Therefore, setting in (3.16), $x = x_n$, $y_1 = r(x_n)$ and $y = r(x_o)$, we have

$$\begin{aligned} (1-\gamma'\lambda_{N+1})\|r(x_n) - r(x_o)\|^2_{L^2(\Omega)} &\leq \langle \nabla J(x_n + r(x_o)) - \nabla J(x_n + r(x_n)), \\ r(x_o) - r(x_n)\rangle &= \int((r(x_o))(r(x_o) - r(x_n)) - g_1((Q+Q_1)(x_n + r(x_o))) \\ &\qquad Q_1(r(x_o) - r(x_n))). \end{aligned} \tag{3.21}$$

Since Q_1 is compact the last expression in (3.21) tends to zero. Consequently $\{r(x_n)\}$ converges to $r(x_o)$. Hence we have

$$\sup\{\tilde{J}(x);\ x \in X_1\} = \lim \tilde{J}(x_n)$$
$$= \left(\int (r(x_n))^2 - (P(x_n))^2 - (x_n')^2 - 2G_1((Q+Q_1)(x_n+r(x_n)))\right)/2$$
$$\leq J(x_o + r(x_o)) = \tilde{J}(x_o). \qquad (3.22)$$

In obtaining the last inequality we have used that $\{r(x_n)\}$ converges to $r(x_o)$, that x_n' converges to x_o' and that $\int (P(x_o))^2 \leq \liminf \int (P(x_n))^2$. From (3.22) it follows that x_o is a point of maximum of $\tilde{J}$, which proves that J has a critical point. Thus claim A) has been proved.

Now we prove claim B). Imitating the proof of (3.15) we have

$$<\nabla J(x_1+y) - \nabla J(x+y),\ x_1-x \leq (1-\gamma\lambda_N)\|x_1 - x\|^2_{L^2(\Omega)},$$

for $x, x_1 \in X_1$ and $y \in Y_1$. Hence using Lemma 3 with $X = \{0\}$, $Z = X_1$, $Y = Y_1$ we infer that J has a unique critical point and, therefore, that (3.2) has a unique solution.

<u>Remark</u>. If in addition to a), b), c) we assume that $g(x,0) = 0$, $(\partial g/\partial u)(x,0) < (1/\lambda_N)$ for all $x \in \Omega$, $\phi_N \in L^\infty(\Omega)$ and $r(s\phi_N) \in L^\infty(\Omega)$ for $|s|$ small, then (3.2) has a nontrivial solution. The proof of this follows the following pattern. From the proof of the claim A) of Theorem 5 we know that (3.2) has a solution which comes from a point of maximum of $\tilde{J}$. The additional hypothesis permits showing that for some small s, $\tilde{J}(s\phi_1) > 0 = \tilde{J}(0)$. Hence such a point of maximum is not 0.

4. <u>On a result by A. Ambrosetti and G. Prodi</u>

In this section we present a version of a classical result

proved in [4]. The proof of the next theorem is essentially taken from [6].

Let Ω be a bounded region in $\mathbb{R}^n$. Let (λ_i, ϕ_i) be the ith eigenvalue-eigenfunction pair of the problem

$$\Delta u + \lambda u = 0 \quad \text{in} \quad \Omega$$

$$u = 0 \quad \text{in} \quad \partial\Omega,$$

where Δ denotes the Laplacian operator $\partial^2/\partial x_1^2 + \ldots + \partial^2/\partial x_n^2$. We assume $\int_\Omega \phi_i^2 = 1$ for all $i=1,2,\ldots$.

We consider the problem

$$\Delta u + g(u) = \rho\phi_1 + h \quad \text{in} \quad \Omega, \tag{4.1}$$

$$u = 0 \quad \text{in} \quad \partial\Omega, \tag{4.2}$$

where ρ is a real parameter, $h \in L^2(\Omega)$, $\int_\Omega h\phi_1 = 0$ and g is a continuous function satisfying

(I) $\lim_{x\to-\infty} g(x)/x = \mu < \lambda_1$

(II) $\lim_{x\to\infty} g(x)/x = \nu \in (\lambda_1, \lambda_2)$

(III) $(g(u) - g(v))/(u-v) \leq \gamma < \lambda_2$ if $u \neq v$.

Let H denote the Sobolev space of square integrable functions in Ω vanishing on $\partial\Omega$ and having generalized first order partial derivatives in $L^2(\Omega)$ (see [1, ch. III]). In order to state our next theorem we recall that $u \in H$ is a *weak solution* to (4.1)-(4.2) iff u is a critical point of the functional $J:H \to \mathbb{R}$ defined by

$$J(u) = \int_\Omega (\|\nabla u\|^2/2 - G(u) + \rho\phi_1 u + hu)$$

where $G(u) = \int_0^u g(s)ds$.

Theorem 7. *If* g *and* h *are as above then there exists* $\rho(h)$ *such that the problem* (4.1)-(4.2) *has*

(A) *at least two solutions for* $\rho > \rho(h)$

(B) *at least one solution for* $\rho = \rho(h)$

(C) *no solution for* $\rho < \rho(h)$.

Further, if $\{h_n\}$ *converges weakly to* h *in* $L^2(\Omega)$ *then* $\rho(h_n) \to \rho(h)$. *If, in addition,* (IV) g *is strictly convex, then* (A) *and* (B) *are valid with "at least" replaced by "precisely".*

Proof: By choosing $\varepsilon < \lambda_1 - \mu$, C large and considering the equation

$$\Delta u + (\lambda_1-\varepsilon)u + g(u)-(\lambda_1-\varepsilon)u+C) = \rho\phi_1+h_1$$

it is clear that there is no loss of generality in assuming $\mu < 0$ and $g \geq 0$. We will also take $\phi_1 \geq 0$.

Let X denote the subspace generated by ϕ_1 and let Y be its orthogonal complement in H. Since Δ is selfadjoint it is easy to see that Y is the closure of the subspace generated by $\{\phi_2, \phi_3, \ldots\}$. We leave it to the reader to prove that the functional J is of class C^1. From the hypothesis (III) follows it that for $x \in X$, $y, y_1 \in Y$

$$\langle \nabla J(x+y_1)-\nabla J(x+y), y_1-y\rangle \geq (1-\gamma/\lambda_2)\|\nabla(y_1 - y)\|^2_{L^2(\Omega)}. \tag{4.3}$$

Since X is one dimensional, from (4.3) and Lemma 1 we see that for each real number t there exists $r(t) \in Y$ such that $J(t\phi_1+r(t)) = \min\{J(t\phi_1+y);\ y \in Y\}$. We wish to show that $d(\tilde{J}(t))/dt \to -\infty$ as $|t| \to \infty$. Note that, by claim ii) of Lemma 1,

$$\frac{d\tilde{J}(t)}{dt} = \langle \nabla J(t\phi_1+r(t)), \phi_1\rangle = t\int_\Omega |\nabla\phi_1|^2+\rho-\int_\Omega g(t\phi_1+r(t))\phi_1. \tag{4.4}$$

Since g is nonnegative, $\tilde{J}'(t) \to -\infty$ as $t \to -\infty$. Also

$$J(t\phi_1+r(t)) \leq J(t\phi_1) = t^2\int_\Omega |\nabla\phi_1|^2/2 - \int_\Omega G(t\phi_1) + t\rho. \tag{4.5}$$

From hypothesis (II) we see that for any $\varepsilon > 0$ there exists C such that for all $t > 0$, $2G(t) \geq (\nu-\varepsilon)t^2 - 2C$. Hence, from (4.5) we obtain

$$J(t\phi_1+r(t)) \leq (t^2\lambda_1-(\nu-\varepsilon)t^2)/2+C \to -\infty \quad \text{as} \quad t \to \infty.$$

Hence, if $\tilde{J}'(t)$ does not tend to $-\infty$ as $t \to \infty$, there must exist a sequence $t_n \to \infty$ such that $J'(t_n)$ is bounded. From the inequality in (3.4), the definition of J, and (III) we have

$$\int_\Omega \|\nabla r(t_n)\|^2 \leq \gamma\int_\Omega (t_n\phi_1+r(t_n))^2 + g(0)\int_\Omega (t_n\phi_1+r(t_n))$$

$$\leq \gamma t_n^2 + \gamma\int_\Omega (r(t_n))^2+|g(0)|(t_n\int_\Omega \phi_1+(\text{meas } \Omega)\ \frac{1}{\sqrt{\lambda_2}}\ \|\nabla r(t_n)\|^2_{L^2(\Omega)}).$$

This implies that $\{\|\nabla r(t_n)\|_{L^2(\Omega)}/t_n\}$ is bounded and by taking subsequence, we can suppose that $\{\phi(t_n)/t_n\}$ converges weakly in Y to ψ say. Replacing t by t_n in (4.4), dividing by t_n and taking limit as $n \to \infty$ give

$$0 = \int |\nabla\phi_1|^2 - g_1(\phi_1+\psi)\phi_1 \tag{4.6}$$

where $g_1(s) = \mu s$ if $s \leq 0$, $g_1(s) = \nu s$ if $s > 0$. Notice that $\lim_{n\to\infty} \langle\nabla J(t_n\phi_1 + r(t_n))/t_n, y\rangle = \lim_n 0 = 0$ for all $y \in Y$, i.e.,

$$0 = \lim_n \int_\Omega (\nabla r(t_n)\cdot\nabla y-g(t_n\phi_1+r(t_n))y+hy)/t_n$$

$$= \int_\Omega \nabla\psi\cdot\nabla y - g_1(\phi_1+\psi)y.$$

But putting $y = \psi$ gives

$$\lambda_2\|\psi\|^2_{L^2(\Omega)} \leq \|\nabla\psi\|^2_{L^2(\Omega)} = \int g_1(\phi_1+\psi)\psi \leq \nu\int\psi^2$$

which implies that $\psi = 0$. Now (4.6) becomes $0 = \lambda_1-\nu$, a contradiction. Thus $\tilde{J}'(t) \to -\infty$ as $|t| \to \infty$, which implies the existence of $\rho(h)$ satisfying (A), (B) and (C). Observe that this also implies that the set of zeros of $\tilde{J}'$ is bounded.

The fact $\rho(h)$ depends continuously on h will follow by showing that $r(t)$ depends continuously on h. Let $h_1, h_2 \in L^2(\Omega)$ with $\int_\Omega \phi_1 h_1 = \int_\Omega \phi_1 h_2 = 0$ and for fixed $t \in \mathbb{R}$ let ψ_1 and ψ_2 be the $r(t)$ corresponding to replacing h in J by h_1 and h_2 respectively. Hence

$$0 = \int_\Omega (\nabla\psi_1 \cdot \nabla(\psi_1 - \psi_2) - (g(t\phi_1+\psi_1)-h_1)(\psi_1-\psi_2))$$

and

$$0 = \int_\Omega (\nabla\psi_2 \cdot \nabla(\psi_1 - \psi_2) - (g(t\phi_1+\psi_2)-h_2)(\psi_1-\psi_2)).$$

Subtraction gives

$$\begin{aligned} 0 &= \|\nabla(\psi_1-\psi_2)\|^2 - \int_\Omega (g(t\phi_1+\psi_1)-g(t\phi_1+\psi_2))(\psi_1-\psi_2) \\ &\quad + \int_\Omega (h_1-h_2)(\psi_1-\psi_2) \\ &\geq \|\nabla(\psi_1-\psi_2)\|^2 - \gamma\|\psi_1-\psi_2\|^2 - \|h_1-h_2\|\|\psi_1-\psi_2\| \end{aligned} \tag{4.7}$$

where all the above norms are in $L^2(\Omega)$. From (4.7) we have

$$\|\nabla(\psi_1-\psi_2)\| \leq \sqrt{\lambda_2}/(\lambda_2-\gamma)\|h_1 - h_2\|. \tag{4.8}$$

Thus, for fixed t, the mapping $h \to r(t)$ is globally Lipschitzian from $L^2(\Omega)$ to H. Now (4.8) shows that if $h_n \to h$ weakly, and if ψ_n, ψ denote the corresponding $r(t)$'s, then $\{\psi_n\}$ is bounded in H. Hence by the Sobolev embedding Theorem ([1, pp. 97]) we can assume that $\psi_n \to \psi$ in $L^2(\Omega)$. Finally (4.7) shows that $\psi_n \to \psi$ in H. Thus we have proved that $\rho(h)$ is continuous from the weak topology in $L^2(\Omega)$ to the reals.

The proofs of the implications of the hypothesis (IV) do not depend on the abstract developements of section 2, so we rather refer the reader to [6] for these details.

5. Periodic solutions of the forced pendulum equation

Let $g:\mathbb{R} \to R$ be a continuous T-periodic function such that $(g(u) - g(v))(u-v) < (u-v)^2$ for $u \neq v$. Let $p_0:\mathbb{R} \to \mathbb{R}$ be a continuous 2π-periodic function such that $\int_0^{2\pi} p_0 = 0$. In [9], using the methods of section 2 as we applied them in the previous two sections it was proved that the 2π-periodic solvability of the *pendulum equation*

$$x''(t) + g(x(t)) = p(t)=p_0(t)+p_1 \tag{5.1}$$

where p_1 is a constant, is given by:

Theorem 8: *If* p_0, p_1 *and* g *are as above, then there exist real numbers* $d(p_0)$, $D(p_0) \in [\min\{g(s): s \in \mathbb{R}\}, \max\{g(s): s \in \mathbb{R}\}]$ *such that the equation* (5.1) *has a solution iff* $d(p_0) \leq p_1 \leq D(p_0)$. *The functions* d *and* D *are continuous in the weak topology of* $L^2[0, 2\pi]$, *i.e., if* $\{p_0^n\}$ *is a sequence converging weakly to* p_0 *in* $L^2(\Omega)$ *then* $d(p_0^n) \to d(p_0)$, $D(p_0^n) \to D(p_0)$. *Moreover,* $d(p_0) \leq \left(\int_0^T g(s)ds\right)/T \leq D(p_0)$. *If, in addition,* $\{x: g(x) = \min\{g(s): s \in \mathbb{R}\}\}$ *and* $\{x: g(x) = \max\{g(s): s \in \mathbb{R}\}\}$ *are discrete then* $d(p_0) = 0$ *or* $D(p_0) = 0$ *iff* $p_0 = 0$.

For details of the proof we refer the reader to [9].

REFERENCES

[1] R. Adams, *Sobolev Spaces*, Academic Press, New York, San Francisco, London (1975).

[2] S. Ahmad, A.C. Lazer and J. Paul, *Elementary critical point theory and perturbations of elliptic boundary value problems at resonance*, Indiana Univ. Math. J. 25 (1976), 933-944.

[3] H. Amann, *Saddle points and multiple solutions of differential equations*, Math. Z. (1979), 127-166.

[4] A. Ambrosetti, and G. Prodi, *On the inversion of some differentiable mappings with singularities between Banach spaces*, Annali Math. Pura Appl., 93 (1972) 231-247.

[5] P. Bates and A. Castro, *Existence and uniqueness for a variational hyperbolic system without resonance*, Nonlinear Analysis, Theory, Methods and Applications, 4 (1980), 1151-1156.

[6] P. Bates and A. Castro, *Necessary and sufficient conditions for existence of solutions to equations with noninvertible linear part*, to appear in volume XV (numbers 1-2) of the Revista Colombiana de Matemáticas.

[7] A. Castro, *Hammerstein integral equations with indefinite Kernel*, Internat. J. Math. and Math. Sci., 1 (1978), 187-201.

[8] A. Castro, *A two point boundary value problem with jumping nonlinearities*, Proc. Amer. Math. Soc., 79 (1980), 207-211.

[9] A. Castro, *Periodic solutions of the forced pendulum equation*, Differential Equations, Editors S. Ahmad, M. Keener and A.C. Lazer, Academic Press (1980), 149-160.

[10] A. Castro, *Existence of infinitely many solutions for a class of superlinear problems*, (preprint).

[11] A. Castro, *Métodos variacionales y análisis funcional no lineal*, edited by the Universidad Pedagógica y Tecnológ ca de Colombia and the Sociedad Colombiana de Matemáticas (1980).

[12] A. Castro and A. C. Lazer, *Applications of a maxmin principle*, Rev. Colombiana Mat., 10 (1976), 141-149.

[13] A. Castro and A. C. Lazer, *Critical point theory and the number of solutions of a nonlinear Dirichlet problem*. Annali Mat. Pura Appl., CXX (1979), 113-137.

[14] E. M. Landesman and A. C. Lazer, *Nonlinear perturbations of linear elliptic equations at resonance*, J. Math. Mech., 19 (1970), 609-623.

[15] A. C. Lazer, E. M. Landesman and D. Meyers, *On saddle point problems in the calculus of variations, the Ritz algorithm, and monotone convergence*, J. Math. Anal. Appl., 52 (1975), 594-614.

[16] L. A. Lusternik and V. J. Sobolev, *Functional analysis*, Hindustan Publishing Corpn. (India), Delhi (1961).

[17] L. Nirenberg, *Variational and topological methods in nonlinear problems*, Bull. A. Math. Soc., 4 (1981), 267-302.

[18] P. Rabinowitz, *Periodic solutions of Hamiltonian systems*, Comm. Pure Appl. Math., 31 (1978), 157-184.

[19] H. L. Royden, *Real analysis*, McMillan Publishing Co., New York (1968).

[20] M. Vainberg, *Variational methods for the study of nonlinear operators*, Holden Day, San Francisco (1964).

ON MULTIPLE SOLUTIONS OF NONLINEAR ELLIPTIC EQUATIONS WITH ODD NONLINEARITIES

Alfonso Castro B.

Departamento de Matemáticas
Centro de Investigación del IPN
Apartado Postal 14740
México 14, D.F. México

and

J. V. A. Gonçalves

Universidade de Brasília
Departamento de Matemática
Brasília, DF - Brazil

1. Introduction

Let $\Omega \subset \mathbb{R}^N$ be a bounded domain with smooth boundary $\partial\Omega$. Let Δ be the Laplacian and $\lambda_1 < \lambda_2 \leq \ldots, \lambda_j \to +\infty$, the sequence of eigenvalues of the boundary value problem

$$\begin{cases} \Delta u + \lambda u = 0 & \text{in } \Omega \\ u = 0 & \text{on } \partial\Omega \end{cases}$$

with respective eigenfunctions denoted by $\phi_1, \phi_2, \ldots$. It is well known that λ_1 is simple, positive and ϕ_1 can be chosen positive in Ω.

In this paper we stablish results on multiplicity of solutions for the boundary value problem

$$\begin{cases} \Delta u + \alpha u + f(u) = 0 & \text{in } \Omega \\ u = 0 & \text{on } \partial\Omega \end{cases} \tag{1}$$

where $\alpha \in \mathbb{R}$ and $f: \mathbb{R} \to \mathbb{R}$ is an odd continuous function. The existence of multiple solutions has been studied by many authors, under various conditions on f, see e.g. Hempel [1], Ambrosetti [2,3], Rabinowitz [4,5], Castro-Lazer [7], Ambrosetti-Mancini [6], Thews [11,12]. We shall be concerned here with the behavior of f both at infinity and at the origin, i.e., we shall explore the condition

$$\limsup_{s \to +\infty} f(s) < 0 \tag{2}$$

and the positions of both α and the limit

$$\ell \equiv \liminf_{s \to 0^+} \frac{f(s)}{s}$$

with respect to the eigenvalues of $-\Delta$. A polinomial growth condition is also required, namely, for all $s \in \mathbb{R}$

$$|f(s)| \leq a|s|^\sigma + b \tag{3}$$

with $a,b,\sigma \in [0,+\infty)$ and $1 \leq \sigma < \frac{N+2}{N-2}$ if $N > 2$. Our main result is as follows

<u>Theorem 1</u>. *Assume* $f: \mathbb{R} \to \mathbb{R}$ *is an odd continuous function satisfying* (2)-(3) *and suppose* $\alpha \leq \lambda_1$. *Then, problem* (1) *has*

(i) *at least* $2j+1$ *solutions if* $\ell > \lambda_j - \alpha$

(ii) *infinitely many solutions if* $\ell = +\infty$.

In order to prove Theorem 1 we associate to (1) the family of problems

$$\begin{cases} \Delta u + \alpha u + f_n(u) = 0 & \text{in } \Omega \\ u = 0 & \text{on } \partial\Omega \end{cases} \tag{1$_n$}$$

by truncating the function f conveniently. Then we obtain an L^∞ - a priori bound for the solutions of $(1)_n$, independent of n. Let

$$\lambda_{k-p} < \lambda_{k-p+1} = \dots = \lambda_k < \lambda_{k+1}$$

$k \geq 1$, $p \geq 1$, i.e., we assume λ_k has multiplicity p. The following result will be applied to solve $(1)_n$.

Theorem 2. *Assume* $f: \mathbb{R} \to \mathbb{R}$ *is an odd bounded and continuous function satisfying* (2). *Suppose in addition that* $\alpha = \lambda_k$ *or* $\alpha < \lambda_1$ *in* (1). *Then problem* (1) *has*

(i) *at least* $2(j-k+p)+1$ *solutions if* $\ell > \lambda_j - \alpha$, $j \geq k$

(ii) *infinitely many solutions if* $\ell = +\infty$

Our next result corresponds to the case in that α is between two consecutive eigenvalues of $-\Delta$ and f is sublinear in the sense that

$$\lim_{s\to\infty} \frac{f(s)}{s} = 0. \tag{4}$$

Theorem 3. *Suppose* $f: \mathbb{R} \to \mathbb{R}$ *is an odd continuous function satisfying* (4) *and the following inequality*

$$(f(s)-f(t))(s-t) \leq \gamma(s-t)^2 \tag{5}$$

for all $s,t \in \mathbb{R}$ *and some constant* γ. *Let* $\alpha \in (\lambda_k, \lambda_{k+1})$. *Then problem* (1) *has at least* $2|j-k| + 1$ *solutions provided either* $\ell > \lambda_j > \lambda_k$ *or*

$$\limsup_{s\to 0} \frac{f(s)}{s} < \lambda_j < \lambda_k.$$

Theorems 2 and 3 are in fact an exploration of a result due to Clark [10], (cf. section 2), concerning the existence of critical points for even C^1-functionals. In the proof of Theorem 3 we apply Clark's result in connection with reduction arguments. Our Theorem 1 improves a result by Ambrosetti-Mancini [6], (Cf. Th. 5.7) where it is assumed that $f \in C^1$ and f' is bounded from above. Theorems 2 and 3 are related to results by Rabinowitz [4], Thews [11,12] and Castro-Lazer [7]. Our results remain true for higher order operators.

2. The Abstract Framework and Notations

Let Σ denote the collection of closed, symmetric (with respect to the origin), subsets of $E\backslash\{0\}$, where E is a real Banach space. The genus $\gamma(A)$ of an element $A \in \Sigma$ is defined to be the least integer $j \geq 0$ such that there is an odd $\phi \in C^0(A, R^j\backslash\{0\})$. For the properties of genus see e.g. Rabinowitz [5] or Castro [8]. A C^1-functional $J: E \to \mathbb{R}$ satisfies $(PS)^-$, provided any sequence $u_n \in E$ for which Ju_n is bounded from below, $Ju_n < 0$ and $J'u_n \to 0$ has a convergent subsequence. If J is even we define

$$i_1(J) = \lim_{a \to 0^-} \gamma(J_a)$$

and

$$i_2(J) = \lim_{a \to -\infty} \gamma(J_a)$$

where $J_a = \{u \in E \mid J(u) \leq a\}$. The following theorem is a specialization of a result due to Clark [10] and follows from Clark's version of the Ljusternik-Schnirelman Theory. For a proof see also Castro [8].

Theorem 4 (Clark [10]). *Suppose* $J: E \to \mathbb{R}$ *is an even* C^1-*functional with* $J(0) = 0$, *satisfying* $(PS)^-$. *Then,* J *has at least* $2m$ *critical points* $u \in E$ *with* $J(u) < 0$ *provided* $m \geq 1$ *is an integer such that* $i_1(J) - i_2(J) \geq m$.

In what follows we shall take E to be the Sobolev space H_0^1 whose norm and inner product are given by $|u|_1 = \int|\nabla u|^2$ and $\langle u,v\rangle_1 = \int \nabla u\cdot\nabla v$ respectively. (All integrals here are taken over Ω). We recall the following inequalities

$$\int|\nabla u|^2 \leq \lambda_j\int u^2, \quad u \in \langle\phi_1,\ldots,\phi_j\rangle \tag{6}$$

$$\int|\nabla v|^2 \geq \lambda_{j+1}\int v^2, \quad v \in \langle\phi_1,\ldots,\phi_j\rangle^{\perp}. \tag{7}$$

Now, if $f\colon \mathbb{R} \to \mathbb{R}$ is an odd continuous function satisfying (3) it follows that $J\colon H_0^1 \to \mathbb{R}$ given by

$$J(u) = \frac{1}{2}\int|\nabla u|^2 - \alpha u^2 - \int F(u) \tag{8}$$

where

$$F(z) = \int_0^z f(s)ds, \quad z \in \mathbb{R}$$

is a well defined even C^1-functional with $J(0) = 0$. In fact,

$$\langle\nabla J(u),v\rangle_1 = J'(u)\cdot v = \int \nabla u.\nabla v - \alpha uv - f(u)v, \qquad u,v \in H_0^1.$$

3. Proofs

We associate with the odd, continuous function f a sequence of odd, bounded and continuous functions as follows: for an integer $n \geq 1$,

$$f_n(s) = \begin{cases} -n \ , & f(s) < -n \\ f(s), & |f(s)| \leq n \\ n \ , & f(s) > n \end{cases}$$

It is immediate that f_n satisfies (2)-(3) with the same constants a,b,σ. Now, it follows, by applying the Linear Elliptic Theory, that a weak solution u of $(1)_n$ belongs to $W^{2,p}$ $2 \leq p < +\infty$. We denote by J_n the energy functional associated with f_n as in (8).

Lemma 5. *Let* $f: \mathbb{R} \to \mathbb{R}$ *be an odd continuous function satisfying* (2)-(3) *and suppose* $\alpha \leq \lambda_1$ *in* (1). *Then, there exists a constant* C, *independent of* n, *such that*

$$|u|_\infty \leq C$$

for all solutions u *of* $(1)_n$ *satisfying* $J_n(u) \leq 0$.

Proof of Lemma 5. Let u be a solution of $(1)_n$. Taking the L^2-inner product with u in $(1)_n$ we find

$$\int u f_n(u) \geq 0. \tag{9}$$

Now, it follows from (2) that

$$f_n(s) \leq C, \quad s \geq 0$$

for some constant C. (We shall use C to denote various constants independent of n). Hence,

$$s f_n(s) \leq C|s|, \quad s \in \mathbb{R}. \tag{10}$$

Consequently,

$$s f_n(s) \leq -|s f_n(s)| + 2C|s|, \quad s \in \mathbb{R}$$

and this together with (9) yeld

$$\int |u||f_n(u)| \leq C|u|_{L^2}. \tag{11}$$

The remaining part of the proof is divided into three steps.

1) Assume that $|u_n|_{L^2} \to \infty$ with

$$\Delta u_n + \alpha u_n + f_n(u_n) = 0 \text{ in } \Omega \tag{12}$$

and $J_n(u_n) \leq 0$. We write $u_n = t_n \phi_1 + \omega_n$, ω_n orthogonal to ϕ_1. It follows from (2) that

$$F_n(z) \leq C, \quad z \in \mathbb{R}.$$

Therefore,

$$\frac{1}{2}\int|\nabla u_n|^2 - \alpha u_n^2 \le \int F_n(u_n) \le C.$$

Hence

$$\left\{1 - \frac{\alpha}{\lambda_2}\right\}|\omega_n|_1^2 \le C$$

and $|\omega_n|_1$ is bounded. Then, $|t_n| \to \infty$ and

$$v_n = \frac{u_n}{|u_n|_{L^2}} \xrightarrow{L^2} t^*\phi_1$$

with $t^* = \pm 1$. Suppose $t^* = 1$. By passing to subsequences, if necessary, we may assume that

$$v_n \to \phi_1, \quad u_n \to +\infty \quad \text{and} \quad |v_n| \le h \qquad \text{a.e. in } \Omega$$

where h is some function in L^2. On the other hand, we find from (2) and (10) that

$$\limsup f_n(u_n) < 0 \quad \text{a.e. in } \Omega$$

and

$$v_n f_n(u_n) \le C|v_n| \qquad \text{a.e. in } \Omega$$

respectively. Then, by applying Fatou's Theorem in (9) and using the inequality

$$\liminf v_n(C-f_n(u_n)) \ge \phi_1 \liminf(C-f_n(u_n)) \quad \text{a.e. in } \Omega$$

we get

$$0 \le \limsup \int v_n f_n(u_n) \le \int \limsup v_n f_n(u_n) \le \int \phi_1 \limsup f_n(u_n)$$

which is a contradiction. In the case $t^* = -1$ we obtain

$$-v_n \to \phi_1 \quad \text{and} \quad -u_n \to +\infty \qquad \text{a.e. in } \Omega$$

so that the earlier reasoning applies, since f is odd. Assume now, that $|u_m|_{L^2} \to \infty$, where u_m is a solution of $(1)_{n_0}$ for some $n_0 \ge 1$. We arrive at a contradiction by a reasoning similar to the above one.

This proves that the solutions u of $(1)_n$ such that $J_n(u) \leq 0$ are bounded in L^2.

2) We find from (11) and step 1 that

$$\int |\nabla u|^2 - \alpha u^2 \leq \int |u||f_n(u)| \leq C|u|_{L^2}.$$

Therefore,

$$|u|_1 \leq C, \quad (C \text{ independent of } n)$$

for all solutions u of $(1)_n$ such that $J_n(u) \leq 0$.

3) We shall use now a bootstrap argument to obtain $|u|_\infty \leq C$. If $N = 2$, we get from Sobolev's Imbedding Theorem and step 2 that $|u|_{L^p} \leq C$, $2 \leq p < +\infty$. Then, we apply the a-priori estimates of Elliptic Theory to $(1)_n$ and use the late inequality together with (3) to get $|u|_{W^{2,p}} \leq C$, $2 \leq p < +\infty$. Thus, $|u|_\infty \leq C$. In the case $N > 2$, Sobolev's Imbedding Theorem and step 2 give $|u|_{L^{2^*}} \leq C$, $\frac{1}{2^*} = \frac{1}{2} - \frac{1}{N}$. Let $p_1 = 2^*|\sigma$. Then, using (3) and the a-priori estimates for elliptic operators we get $|u|_{W^{2,p_1}} \leq C$. Next, one studies two cases: $2 \geq \frac{N}{p_1}$ and $2 < \frac{N}{p_1}$. After repeating the argument a finite number of times one arrives at $|u|_\infty \leq C$. This proves Lemma 5.

<u>Proof of Theorem 2</u>. Let $\alpha = \lambda_k$ $k \geq 1$, $Y = \langle\phi_1,\ldots,\phi_{k-p}\rangle$, $N = \langle\phi_{k-p+1},\ldots,\phi_k\rangle$, $W = \langle\phi_{k+1},\ldots\rangle$ and $u_n = y_n + v_n + \omega_n$ with $y_n \in V$, $v_n \in N$ and $w_n \in W$. Assume $|u_n|_1 \to \infty$ and $J(u_n) \to 0$. Then, (6)-(7) and the boundedness of f yeld

$$\langle\nabla J(u_n),\omega_n\rangle_1 = \int |\nabla\omega_n|^2 - \lambda_k\omega_n^2 - f(y_n+v_n+\omega_n)\omega_n \geq \left(1 - \frac{\lambda_k}{\lambda_{k+1}}\right)|\omega_n|_1^2 - C|\omega_n|_1.$$

Therefore, $|\omega_n|_1$ is bounded. Similarly, we find, by computing

$\langle \nabla J(u_n), y_n \rangle_1$, that $|y_n|_1$ is also bounded. So, $|v_n|_1 \to \infty$. On the other hand,

$$J(u_n) = \frac{1}{2}\int |\nabla\omega_n|^2 - \lambda_k\omega_n^2 + \frac{1}{2}\int |\nabla y_n|^2 - \lambda_k y_n^2 - \int F(y_n + v_n + \omega_n)$$

$$\geq -C - \int F(v_n).$$

Now, it follows from (2) that $F(z) \to -\infty$ as $|z| \to \infty$. Consequently, (Cf. Rabinowitz [4])

$$\int F(v_n) \to -\infty.$$

Hence, $J(u_n) \to +\infty$ so that every sequence u_n such that $-C \leq J(u_n) < 0$ and $\nabla J(u_n) \to 0$ is necessarily bounded. It is easily seen that

$$\nabla J(u) = u - Ku, \quad u \in H_0^1$$

where K is a compact mapping in H_0^1. This proves $(PS)^-$. We shall apply Clark's result. It suffices to show that $i_1(J)-i_2(J) \geq j-k+p$. We find from $\ell > \lambda_j - \lambda_k$ $j \geq k$, that

$$F(z) > \frac{\eta}{2} z^2, \quad |z| \leq \varepsilon$$

for any $\varepsilon > 0$ and some $\eta > \lambda_j - \lambda_k$. Let $u \in \langle\phi_1,\ldots,\phi_k,\ldots,\phi_j\rangle$, $u = u_1 + u_2$ with $u_1 \in \langle\phi_1,\ldots,\phi_k\rangle$ and $u_2 \in \langle\phi_{k+1},\ldots,\phi_j\rangle$. Then, for $\varepsilon^* > 0$ properly chosen and $|u|_1 = \varepsilon^*$,

$$J(u) = \frac{1}{2}\int |\nabla u_1|^2 - \lambda_k u_1^2 + \frac{1}{2}\int |\nabla u_2|^2 - \lambda_k u_2^2 - \int F(u)$$

$$\leq \frac{1}{2}\left(1 - \frac{\lambda_k}{\lambda_j}\right)|u_2|_1^2 - \frac{\eta}{2\lambda_j}|u|_1^2.$$

Therefore, $\sup_{u \in S_{\varepsilon^*}} J(u) < 0$, where S_{ε^*} is the sphere of radius ε^* in $\langle\phi_1,\ldots,\phi_j\rangle$. Consequently, (we recall that $\gamma(S_{\varepsilon^*}) = j$), $i_1(J) \geq j$. On the other hand, for $u = v+\omega \in N \oplus W$

$$J(u) = \frac{1}{2}\int |\nabla\omega|^2 - \lambda_k\omega^2 - \int F(v+\omega) \geq \frac{1}{2}\left(1 - \frac{\lambda_k}{\lambda_{k+1}}\right)|\omega|_1^2 - C|\omega|_1 - \int F(v)$$

$$\geq -C.$$

Let $a < -C$. Then $J_a \cap (N \oplus W) = \emptyset$ so that $\gamma(J_a) \leq k-p$. Thus, $i_2(J) \leq k-p$. In the case $\ell = +\infty$ we find that $i_1(J) \geq j$ for any $j \geq k$. The case $\alpha < \lambda_1$ is handled similarly by observing that J is bounded from below in H_0^1. Theorem 2 is proved.

Proof of Theorem 1. We apply Theorem 2 to solve $(1)_n$ with $\alpha \leq \lambda_1$ and observe that the solutions u obtained via Th. 2 satisfy $J_n(u) \leq 0$. Consequently, $|u|_\infty \leq C$ and from (3),

$$|f(u(x))| \leq C, \quad x \in \bar{\Omega}.$$

Let $n_1 > C$. Then, the solutions of $(1)_{n_1}$ obtained via Th. 2 are also solutions of (1). This proves Theorem 1.

Proof of Theorem 3. It follows from (4) that

$$|f(s)| \leq \varepsilon|s| + C_\varepsilon \qquad \varepsilon > 0, \quad s \in \mathbb{R}.$$

Suppose $u_n = v_n + \omega_n$ with $v_n \in \langle\phi_1,\ldots,\phi_k\rangle$ and $\omega_n \in W \equiv \langle\phi_{k+1},\ldots\rangle$ satisfies:

$$\nabla J(u_n) \to 0 \quad \text{and} \quad |Ju_n| \leq C.$$

Then,

$$|\langle\nabla J(v_n+\omega_n), \quad v_n-\omega_n\rangle_1| \geq$$

$$\geq \left|\int|\nabla v_n|^2 - \alpha v_n^2 - \int|\nabla\omega_n|^2 - \alpha\omega_n^2\right| - \int|f(v_n + \omega_n)||v_n - \omega_n|$$

$$\geq \left(\frac{\alpha}{\lambda_k} - 1\right)|v_n|_1^2 + \left(1 - \frac{\alpha}{\lambda_{k+1}}\right)|\omega_n|_1^2 - \varepsilon|v_n+\omega_n|_1^2 - C_\varepsilon|v_n+\omega_n|_1.$$

On the other hand,

$$|\langle\nabla J(v_n+\omega_n), v_n-\omega_n|_1 \leq \varepsilon|v_n+\omega_n|_1.$$

Therefore, u_n is bounded, so that J satisfies $(PS)^-$. Now, let $m > k+1$ be such that $\alpha+\gamma < \lambda_{m+1}$, $X = \langle\phi_1,\ldots,\phi_m\rangle$ and Y the orthogonal complement of X in H_0^1. By well known results on the reduction method (Cf. Castro [9]) we get, by applying (5), that there exists a continuous mapping ϕ: X → Y such that

$$J(v+\phi(v)) = \min_{\omega\in Y} J(v+\omega), \quad v \in X.$$

Moreover, $u \in H_0^1$ is a critical point of J iff $u = v_0 + \phi(v_0)$ with v_0 a critical point of the functional $\tilde{J}$: X → ℝ given by

$$\tilde{J}(v) \equiv J(v+\phi(v)), \quad v \in X.$$

On the other hand, it is easily seen that $\tilde{J}$ satisfies $(PS)^-$ once J satisfies it. Thus, according to Clark's Theorem it suffices to show that $i_1(\tilde{J}) - i_2(\tilde{J}) \geq |j-k|$. We consider only the case $j > k$. The proof of the case $j < k$ is the same, replacing $\tilde{J}$ by $-\tilde{J}$. Now, it follows by using the condition $\ell > \lambda_j > \lambda_k$, as in the proof of Theorem 2, that

$$\tilde{J}(v) \leq J(v) < 0 \quad v \in \langle\phi_1,\ldots,\phi_j\rangle, \quad |v|_1 = \varepsilon^*$$

with $\varepsilon^* > 0$ properly chosen. Consequently,

$$i_1(\tilde{J}) \geq j.$$

Next, we show that

$$i_2(\tilde{J}) \leq k.$$

From (4) it follows that for any $\varepsilon > 0$ there exists $C_\varepsilon > 0$ such that

$$F(z) \leq \frac{\varepsilon}{2} z^2 + C_\varepsilon|z|, \quad z \in \mathbb{R}.$$

Let $\varepsilon > 0$ be such that $\alpha+\varepsilon < \lambda_{k+1}$. Hence, if $\omega \in W$, we find

$$J(\omega) \geq \frac{1}{2}\int|\nabla\omega|^2 - (\alpha+\varepsilon)\omega^2 - C_\varepsilon|\omega|_1$$
$$\geq -C_\varepsilon.$$

Thus,

$$\tilde{J}(v_1) \geq -C_\varepsilon, \quad v_1 \in \langle \phi_{k+1}, \ldots, \phi_m \rangle.$$

This implies, as in the proof of Theorem 2, that for $a < -C_\varepsilon$

$$\gamma(\tilde{J}_a) \leq k.$$

Hence, $i_2(\tilde{J}) \leq k$ and Theorem 3 is proved.

REFERENCES

[1] Hempel, A. J. Multiple Solutions for a Class of Nonlinear Boundary Value Problems. Indiana Univ. Math. J. 20 (1971) 983-996.

[2] Ambrosetti, A. Existenza de Infinite Soluzione per Problemi Nonlineari in Assenza di Parametro. Atti Acad. Naz. Lincei 52 (1972) 660-667.

[3] ______ On the Existence of Multiple Solutions for a Class of Nonlinear Boundary Value Problems. Rend. Sem. Mat. Univ. Padova 49 (1973) 195-204.

[4] Rabinowitz, P. Some Minimax Theorems and Applications to Nonlinear Partial Diff. Equations. In "Nonlinear Analysis". Academic Press (1978) 161-177.

[5] ______ Variational Methods for Nonlinear Eigenvalue Problems. Proc. Symp. on Eigenvalue Nonlinear Problems. Varena, Italy (1974) 141-195.

[6] Ambrosetti, A. & Mancini, G. Theorems of Existence and Multiplicity for Nonlinear Elliptic Problems with Noninvertible Linear Part. Ann. Sc. Norm. Sup. Pisa 5 (1978) 15-28.

[7] Castro, A. & Lazer, A. Critical Point Theory and the Number of Solutions of a Nonlinear Dirichlet Problem. Ann. Mat. Pura App. (1979) 113-137.

[8] Castro, A. Mētodos Variacionales y Analysis Funcional no Lineal. X Coloquio Colombiano de Matemāticas (1980).

[9] ______ Hammerstein Integral Equations with Indefinite Kernel. Internat. J. Math. and Math. Sci. 1 (1978) 187-201.

[10] Clark, D. A Variant of the Ljusternik-Schnirelman Theory. Indiana Univ. Math. J 22 (1972) 65-74.

[11] Thews, K. Multiple Solutions for Elliptic Boundary Value Problems with Odd Nonlinearities Math. Z. 163 (1972) 163-175.

[12] ______ Nontrivial Solutions of Elliptic Equations at Resonance. Proc. Royal Soc. Edinburgh 85 A (1980) 119-129.

[13] de Figueiredo, D.G., Lions, P-L. & Nussbaum, R. A priori Estimates for Positive Solutions of Semilinear Elliptic Equations (To appear).

[14] Brezis, H. & Nirenberg, L. Characterizations of the Ranges of Some Nonlinear Operators and Applications to Boundary Value Problems. Ann. Sc. Norm. Sup. Pisa 5 (1978) 225-326.

POSITIVE SOLUTIONS OF SEMILINEAR ELLIPTIC PROBLEMS

Djairo Guedes de Figueiredo

Universidade de Brasília
Departamento de Matemática
Brasília, DF - Brazil

Introduction

The past twenty years have witnessed a tremendous development of the theory of boundary value problems for semilinear elliptic equations - positive solutions. We do not intend to survey it in this set of lecture notes which were rather prepared to be an introductory course on the subset for presentation at the Latin American School of Differential Equations held at the University of São Paulo, July 1981. For example we have nothing on bifurcation problems or existence of branches of positive solutions; these questions will be eventually discussed by other lecturers in this conference. Extensive references may be seen in Krasnosel'skii's book [1] and in the paper by Amann [2] for the literature up to 1975. In the present work we concentrate in a single aspect of the theory: the existence of positive solutions of the Dirichlet problem. Our aim here is threefold. First we develop in detail the theory of eigenvalue problems with weight, which has shown to be an important tool in treating these problems; see Ambrosetti [3], Ambrosetti-Mancini [4], Berestycki [5], Kato-Hess [6], among others. The possibility of dealing with indefinite weights enables us to obtain more general results, see Section 2. Second we show that sublinear problems can be studied efficiently via the method of monotone iteration; this line of research was intensively pursued by Herbert Keller, Donald Cohen and others. In the present form is essentially due to the work of Herbert Amann and David Sattinger; see reference [2]. Third we study some special cases of superlinear problems. These are much harder problems due to the difficulty in obtaining a priori bounds. Some progress has been done by the work of Ambrosetti-Rabinowitz [7], Brezis-Turner [8], Crandall-Rabinowitz [9], Nussbaum [10], Puel [11], Turner [12]. There is still a lot to be done on this subject and many questions require a more complete study, and we hope that these lectures will motivate some students to take them up. For the reader's convenience we have divided the present work in chapters and sections as listed next.

Chapter 1

LINEAR EIGENVALUE PROBLEMS WITH AN INDEFINITE WEIGHT

1.1 SPECTRAL ANALYSIS OF COMPACT SYMMETRIC OPERATORS

Let $T: H \to H$ be a compact symmetric linear operator in a Hilbert space H. [Inner product: $(\ , \ )$. Norm $\|\cdot\|$]. A linear operator T on a Hilbert space is said to be compact if it is bounded and takes bounded sets into relatively compact sets. Or equivalently, if it takes weakly convergent sequences into strongly convergent ones. Most of the results presented in this section are quite standard and may be found in books on functional analysis. However they are included here for the reader's convenience. Since they are presented in their logical order, an eventual reader may consider their omitted proofs as exercises.

Lemma 1.1. *If*

$$\lambda_1 \equiv \sup[(Tx, x): \|x\| = 1] > 0 \tag{1}$$

then there exists $\phi_1 \in H$, *with* $\|\phi_1\| = 1$, *such that*

$$(T\phi_1, \phi_1) = \lambda_1, \quad T\phi_1 = \lambda_1\phi_1 \tag{2}$$

Proof. Since $(Tx, x) \leq \|T\|$ for $\|x\| = 1$, we have $\lambda_1 < \infty$. Let us take $\|x_n\| = 1$ such that $(Tx_n, x_n) \to \lambda_1$. Passing to subsequences we may assume that $x_n \rightharpoonup \phi_1$ [$\rightharpoonup$ designates weak convergence, while $\longrightarrow$ means strong convergence] and $Tx_n \to T\phi_1$, in view of the compactness of T. So we obtain the first part of (2). For the second part we introduce the operator $A = T-\lambda_1$, which is seen to satisfy: $(A\phi_1, \phi_1) = 0$ and $(Ax, x) \geq 0$, for all $x \in H$. Writing down the last inequality with $x = \phi_1 + ty$, where $t \in R$ and $y \in H$ are arbitrary, we obtain $t^2(Ay, y) + 2t(A\phi_1, y) \geq 0$. Now letting

$t = \theta(A\phi_1, y)$ we obtain $\theta(A\phi_1, y)^2[2 + \theta(Ay, y)] \geq 0$. We then remark that this last inequality cannot hold when θ is small and changes sign, unless $(A\phi_1, y) = 0$. So the proof is complete.

A similar argument proves the following result.

Lemma 1.2. *If*

$$\lambda_{-1} \equiv \inf[(Tx, x):\|x\| = 1] < 0 \tag{3}$$

then there exists $\phi_{-1} \in H$, $\|\phi_{-1}\| = 1$, *such that*

$$(T\phi_{-1}, \phi_{-1}) = \lambda_{-1}, \quad T\phi_{-1} = \lambda_{-1}\phi_{-1} \tag{4}$$

Remark. Observe that λ_1 defined by (2) is the largest eigenvalue of T. Similarly λ_{-1} is the smallest one. This procedure can be repeated to obtain other eigenvalues of T, considering its restriction to the orthogonal complement of $\mathbb{R}\phi_1$. This possibility rests on the following easily proved fact: "Let $V \subset H$ be a subspace of H such that $T(V) \subset V$. Then $T: V^\perp \to V^\perp$ is compact and symmetric". So we obtain

Proposition 1.3. *If*

$$\lambda_n \equiv \sup\left\{(Tx, x):\|x\| = 1, \ x \perp \phi_1,\ldots,\phi_{n-1}\right\} > 0 \tag{5}$$

then there exists $\phi_n \in H$, *with* $\|\phi_n\| = 1$, *such that*

$$(T\phi_n, \phi_n) = \lambda_n \qquad T\phi_n = \lambda_n\phi_n.$$

A similar statement holds when

$$\lambda_{-n} \equiv \inf\left\{(Tx, x):\|x\| = 1, \ x \perp \phi_{-1},\ldots,\phi_{-(n-1)}\right\} < 0. \tag{6}$$

Before proceeding further we collect in the following lemma some well known facts about compact linear operators. Proofs are immediate.

Lemma 1.4. (a) *Eigenvectors corresponding to different eigenvalues are orthogonal.* (b) *If* λ *is an eigenvalue of* T, *then the eigenspace*

$N(T-\lambda I)$ *has finite dimension.* (c) *The eigenvalues of* T *cannot accumulate at a point* $\lambda \neq 0$.

Next we claim that all the eigenvalues $\neq 0$ of T are obtained by the process described in Proposition 1.3. To prove that we need the following result.

Lemma 1.5. *For each* $x \in H$ *one has*

$$Tx = \sum_{i=-\infty}' \lambda_i (x, \phi_i)\phi_i. \tag{7}$$

[$\sum'$ means that $i = 0$ does not appear in the summation].

Indeed suppose that there is an eigenvalue $\lambda \neq 0$ which is different from λ_i for all i. By Lemma 1.4(a) a corresponding eigenfunction ϕ $(T\phi = \lambda\phi)$ is orthogonal to all ϕ_i. By (7) $T\phi = 0$, which is impossible. In the proof of Lemma 1.5 we need the following well known fact.

Lemma 1.6. *Let* $T: H \to H$ *be a symmetric operator in* H. *Then*

$$\sup_{\|x\|=1} |(Tx, x)| = \sup_{\|x\|=1} \|Tx\| \equiv \|T\|. \tag{8}$$

Proof of Lemma 1.5. Let

$$x_n = x - \sum_{i=-n}^{n}{}' (x, \phi_i)\phi_i. \tag{9}$$

It suffices to prove that $Tx_n \to 0$. Since $x_n \perp \phi_i$ for $i = -n,\dots,-1,1,\dots,n$ we have by Proposition 1.3 and Lemma 1.6 that

$$\|Tx_n\| \leq \max\{|\lambda_{-n}|, \lambda_n\}\|x_n\|. \tag{10}$$

So we have the conclusion because $\|x_n\|^2 = x^2 - \sum_{i=-n}^{n} (x, \phi_i)^2 \leq \|x\|^2$.

Q.E.D.

Formula (5) for the n-th positive eigenvalue presents the inconvenience of requiring the knowledge of all previous positive ones. Similar statement for λ_{-n} in (6). The result next improves this situation.

Proposition 1.7. *For each positive* n

(11) $$\lambda_n = \inf_{F_{n-1}} \sup [(Tx, x) : \|x\| = 1, x \perp F_{n-1}]$$

where the infimum is taken over all subspaces F_{n-1} *of* H *with dimension* n-1. *A similar formula holds for* λ_{-n}

(12) $$\lambda_{-n} = \sup_{F_{n-1}} \inf [(Tx, x) : \|x\| = 1, x \perp F_{n-1}].$$

Proof. Let us denote by Λ_n the right side of (11). Taking F_{n-1} = subspace generated by $\phi_1, \ldots, \phi_{n-1}$, we see by Proposition 1.4 that $\Lambda_n \le \lambda_n$. On the other hand let $h_1, \ldots, h_{n-1}$ be any system of mutually orthogonal vectors in H, and F_{n-1} the subspace generated by them. Choose a vector $\phi = \sum_{i=1}^{n} \alpha_i \phi_i$ in such a way that $(\phi, h_j) = 0$ for $j = 1, \ldots, n-1$, and normalize it, i.e. $\sum_{i=1}^{n} \alpha_i^2 = 1$. Now $\|\phi\| = 1$ and

(13) $$(T\phi, \phi) = \sum_{i=1}^{n} \alpha_i^2 \lambda_i \ge \lambda_n.$$

which implies

(14) $$\sup [(Tx, x) : \|x\| = 1, x \perp F_{n-1}] \ge \lambda_n, \quad \forall F_{n-1}.$$

and consequently $\Lambda_n \ge \lambda_n$.

Q.E.D.

Formula (11) has still a set back since the sup (Tx, x) has to be taken over x in a subspace of infinite dimension. This situation is improved by the following result.

Proposition 1.8. *For each* n *positive*

(15) $$\lambda_n = \sup_{F_n} \inf [(Tx, x) : \|x\| = 1, x \in F_n]$$

where the supremum is taken over all subspaces F_n *of* H *with dimension* n. *A similar statement holds for* λ_{-n}

(16) $$\lambda_{-n} = \inf_{F_n} \sup [(Tx, x) : \|x\| = 1, x \in F_n].$$

Proof. Let us denote by Γ_n the right side of (15). Taking F_n the subspace generated by $\phi_1, \ldots, \phi_n$ one has for all $x = \sum_{i=1}^{n} \alpha_i \phi_i \in F_n$, $\|x\|^2 = \sum \alpha_i^2 = 1$, that

(17) $$(Tx, x) = \sum_{i=1}^{n} \alpha_i^2 \lambda_i \geq \lambda_n.$$

So $\Gamma_n \geq \lambda_n$. Conversely given any subspace F_n of dimension n, choose a $x \in F_n$ such that $x \perp \phi_1, \ldots, \phi_{n-1}$. By Proposition 1.3. $(Tx, x) \leq \lambda_n$. Consequently

$$\inf_{x \in F_n} (Tx, x) \leq \lambda_n, \quad \forall F_n,$$

which implies $\Gamma_n \leq \lambda_n$. Q.E.D.

Remark. Observe that all the inf's a sup's in the above characterizations of λ_n and λ_{-n} are actually assumed. So they could be replaced by min's and max's.

1.2 THE EIGENVALUE PROBLEM WITH AN INDEFINITE WEIGHT: THE VARIATIONAL FORMULATION

Let us consider the eigenvalue problem

(18) $$Lu = \mu m u \quad \text{in } \Omega, \quad u = 0 \quad \text{on } \partial\Omega,$$

where

$$Lu = - \sum_{i,j=1}^{N} \frac{\partial}{\partial x_j}\left(a_{ij}(x) \frac{\partial u}{\partial x_i}\right) + a_o(x) u$$

is a strongly elliptic operator in a bounded domain Ω of R^N, i.e., there is a constant $c_0 > 0$ such that

(19) $$\sum_{i,j=1}^{N} a_{ij}(x)\xi_i\xi_j \geq c_0|\xi|^2, \quad \forall\, x \in \bar{\Omega}, \quad \forall\, \xi \in R^N.$$

We assume $a_{ij} = a_{ji}$ and $a_0(x) \geq 0$ in Ω. The weight function $m:\bar{\Omega} \to R$ is assumed to be in $L^r(\Omega)$ with $r > N/2$. [In particular, this is the case if m is continuous in $\bar{\Omega}$. However we should allow discontinuous functions m viewing future applications]. We emphasize that m *could change sign* in Ω. One assumes $a_{ij} \in L^\infty(\Omega)$ and $a_0 \in L^{N/2}(\Omega)$.

Let us recall Poincaré's inequality

$$\|u\|_{L^2} \leq c\|u\|_{H^1}, \qquad \forall\, u \in H_0^1(\Omega),$$

which implies that the inner products

(20) $$\int \nabla u \cdot \nabla v \qquad \text{and} \qquad \int [\nabla u \cdot \nabla v + uv]$$

are equivalent in H_0^1.

Problem (18) is studied here in its variational formulation:

(21) $$\begin{cases} a[u, v] = \mu \int muv, & \forall\, v \in H_0^1, \\ u \in H_0^1 \end{cases}$$

where $a: H_0^1 \times H_0^1 \to R$ is the bilinear form defined by

$$a[u, v] = \int \left[\sum a_{ij}(x) \frac{\partial u}{\partial x_i} \frac{\partial v}{\partial x_j} + a_0(x)uv\right] dx.$$

Since $H_0^1 \subset L^{2^*}$, where $\frac{1}{2^*} = \frac{1}{2} - \frac{1}{N}$, if $N \geq 3$, we see, using the hypotheses on a_0 and m, that expressions like $\int muv$ and $\int a_0 uv$, for $u,v \in H_0^1$ are well defined. The case of $N \leq 2$ can be treated similarly and it is even easier. The above bilinear form defines in H_0^1 an inner-product equivalent to the original one [in (20)] since

$$a[u, v] = a[v, u], \ |a(u, v)| \le c\|u\|_{H^1}\|v\|_{H_1}, \ a[u, u] \ge c\|u\|^2_{H_1}.$$

Let us denote this inner-product by $(\cdot, \cdot)_a$ and the corresponding norm by $\|\cdot\|_a$.

For fixed $u \in H^1_o$, the map $v \longmapsto \int muv$ is a bounded linear functional in H^1_o. So by the Riesz-Frêchet representation theorem, there is an element in H^1_o, denote it by Tu, such that

$$(22) \qquad (Tu, v)_a = \int muv.$$

Clearly $T: H^1_o \to H^1_o$ is linear symmetric and bounded. Moreover T is compact; indeed let (u_n) be a bounded sequence in H^1_o. Passing to a subsequence we may assume that $u_n \rightharpoonup u$ in H^1_o, $u_n \to u$ in L^s, for $s < 2^*$, using the Rellich-Kondrachov theorem on compact imbeddings of Sobolev spaces. Use (22) with u replaced by $u_n - u$ and v by $Tu_n - Tu$, and apply Hölder's inequality with $\frac{1}{s} = 1 - \frac{1}{r} - \frac{1}{2^*}$ to get

$$\|Tu_n - Tu\|^2_a \le \|m\|_{L^r}\|u_n - u\|_{L^s}\|Tu_n - Tu\|_{L^{2^*}} \Longrightarrow \|Tu_n - Tu\|_a \le c\|u_n - u\|_{L^s}$$

which implies that $Tu_n \to Tu$ in H^1_o.

Problem (21) may be rewritten as

$$(u, v)_a = \mu(Tu, v)_a \quad \forall\, v \in H^1_o \Longleftrightarrow Tu = \frac{1}{\mu} u.$$

So we can apply the theory developed in the previous section to describe the eigenvalues and eigenvectors (in H^1_o) of problem (18).

<u>Proposition 1.10</u>. *The eigenvalue problem* (21) *has a double sequence of eigenvalues*

$$\dots \le \mu_{-2} \le \mu_{-1} < 0 < \mu_1 \le \mu_2 \le \dots$$

whose variational characterizations are

$$(23) \quad \frac{1}{\mu_n} = \sup_{F_n} \inf \left[\int mu^2 : \|u\|_a = 1,\ u \in F_n\right], \ \frac{1}{\mu_{-n}} = \inf_{F_n} \sup \left[\int mu^2 : \|u\|_a = 1,\ u \in F_n\right]$$

where F_n *varies over all n-dimensional subspaces of* H_0^1. *The corresponding eigenfunctions* ϕ_n *are such that*

(24) $$a[\phi_n, v] = \mu_n \int m\phi_n v, \quad \forall\, v \in H_0^1$$

and

(25) $$a[\phi_n, \phi_n] = 1 \qquad \frac{1}{\mu_n} = \int m\phi_n^2.$$

By the results of the previous section the eigenvalues above do not accumulate, except at $+\infty$ or $-\infty$, and the situation is completely described by the following result.

<u>Proposition 1.11</u>. *Let* $\Omega_\pm = \{x \in \Omega: m(x) \gtrless 0\}$. *Then*

(a) $|\Omega_+| = 0 \Longrightarrow$ *there is no positive* μ_n.

(b) $|\Omega_-| = 0 \Longrightarrow$ *there is no negative* μ_{-n}

(c) $|\Omega_+| > 0 \Longrightarrow$ *there is a sequence of positive* $\mu_n \to +\infty$

(d) $|\Omega_-| > 0 \Longrightarrow$ *there is a sequence of negative* $\mu_{-n} \to -\infty$.

[$|\cdot|$ denotes the Lebesgue measure of the set].

<u>Proof</u>. (a) and (b) follow readily from Proposition 1.4. To prove (c), let $B_1,\ldots,B_n$ be a set pairwise disjoint balls in Ω such that the sets $B_j \cap \Omega_+$ have positive measure. Let $u_1,\ldots,u_n$ be C_0^∞ functions with supports in the corresponding balls and such that $\int mu_j^2 > 0$. [u_j can be obtained as an $L^{2r'}$-approximation to the characteristic function of the set $B_j \cap \Omega_+$, where $1/r' + 1/r = 1$]. Let us normalize them so as to have $\int mu_j^2 = 1$ for $j = 1,\ldots,n$. We claim that $\mu_n > 0$. Let F_n be the subspace generated by the mutually orthogonal functions u_j. So for $u = \sum\alpha_j u_j \in F_n$ we have

$$\int mu^2 = \sum\alpha_j^2 \int mu_j^2 = \sum\alpha_j^2$$

and

$$a[u, u] = \sum\alpha_j^2 a[u_j, u_j] \le c\sum\alpha_j^2$$

where $c = \max_{j=1,\ldots,n} a[u_j, u_j]$. Hence

$$\int mu^2 \geq \frac{1}{c} a[u, u], \quad \forall u \in F_n.$$

The result follows from this and (23). A similar proof for (d).

Q.E.D.

The next two results explain how $\mu_n = \mu_n(m)$ varies as a function of m. Both of them follow readily from Proposition 1.10.

Proposition 1.12 A. *Let* $m, \hat{m}: \bar{\Omega} \to R$ *be* L^r *functions, with* $r > N/2$, *such that* $m(x) \leq \hat{m}(x)$ *for* $x \in \Omega$. *Let us suppose that for a given* n, $n = \pm 1, \pm 2, \ldots$, *the eigenvalues* $\mu_n(m)$ *and* $\mu_n(\hat{m})$ *exist. Then*

(26) $$\mu_n(m) \geq \mu_n(\hat{m}).$$

Moreover, if $m(x) < \hat{m}(x)$ *on a subset of positive measure in* Ω, *then one has strict inequality in* (26).

Proposition 1.12 B. $\mu_n(m)$ *is a continuous function of* m, *in the norm of* $L^{N/2}(\Omega)$. *In other words, if* $m_j \in L^r(\Omega)$ *converges in the* $L^{N/2}$ - *norm to* $m \in L^r(\Omega)$, *then* $\mu_n(m_j)$

Remark. If $m(x) \equiv 1$ in Ω we use the following notation $\mu_j(1) = \lambda_j$, $j = 1,2,\ldots$. Observe that in this special case there are no λ_j with negative j. It is easy to see that

$$\mu_j(\alpha) = \lambda_j \alpha^{-1} \quad \text{for} \quad \alpha > 0.$$

This relation will be used extensively later on for values of α being equal to eigenvalues λ_k.

1.3 A THEOREM OF THE KREIN-RUTMAN TYPE

Theorem 1.13. *Let* $m: \Omega \to R$ *be an* L^r-*function, with* $r > N/2$, *(not necessarily positive). Suppose that* $m > 0$ *on a subset of* Ω *with positive measure. Then the first positive eigenvalue* μ_1 *of* (21) *is simple, and* ϕ_1 *can be taken* > 0 *in* Ω. *A similar statement holds if* $m < 0$ *on a subset of* Ω *with positive measure.*

Remark. The above result is due to Manes-Micheletti [13]. A similar result for non self-adjoint problems with m continuous has been proved recently by Hess-Kato [6].

Proof. First step. Let $\omega \in H_0^1$ be an eigenfunction corresponding to the first positive eigenvalue μ_1. At this stage we do not know whether or not some other μ_j's are equal to μ_1, and consequently ω may be a linear combination of ϕ_j's. In any case ω is a solution of

$$a[\omega, v] = \mu_1 \int m\omega v, \quad \forall\, v \in H_0^1. \tag{27}$$

We first claim that ω does not change sign in Ω. Indeed, suppose it does, and let $\omega^+ = \max(\omega, 0)$ and $\omega^- = \min(\omega, 0)$. We know that ω^+ and ω^- are also in $H_0^1(\Omega)$, see Stampacchia [14]. So

$$\int m\omega^2 = \int m(\omega^+)^2 + \int m(\omega^-)^2 \equiv \alpha_1 + \alpha_2$$

$$a[\omega, \omega] = a[\omega^+, \omega^+] + a[\omega^-, \omega^-] \equiv \beta_1 + \beta_2, \ \beta_1, \beta_2 > 0.$$

A simple arithmetic shows that either

$$\text{(i)} \quad \frac{\alpha_1 + \alpha_2}{\beta_1 + \beta_2} = \frac{\alpha_1}{\beta_1} = \frac{\alpha_2}{\beta_2}, \quad \text{or} \quad \text{(ii)} \quad \frac{\alpha_1 + \alpha_2}{\beta_1 + \beta_2} < \max\left(\frac{\alpha_1}{\beta_1}, \frac{\alpha_2}{\beta_2}\right).$$

Applying this to

$$\frac{1}{\mu_1} = \frac{\int m\omega^2}{a[\omega, \omega]} = \frac{\int m(\omega^+)^2 + \int m(\omega^-)^2}{a[\omega^+, \omega^+] + a[\omega^-, \omega^-]}$$

we conclude that ω^+ and ω^- are also eigenfunctions corresponding to μ_1. So they are solutions of (27) and accordingly to a result of Stampacchia [15, Corollary 8.1 p. 238] $\omega^+ > 0$ in Ω(a.e.) and $\omega^- < 0$ in Ω(a.e.), which is impossible. The fact that ϕ_1 can be taken > 0 in Ω it follows from the afore-mentionned result by Stampacchia.

Second step. We now prove that the geometric multiplicity of μ_1 is 1. Indeed, let ω_1 and ω_2 be eigenfunctions corresponding to μ_1. By the previous step we know that for each $\alpha \in R$ the eigenfunction $\omega_1 + \alpha\omega_2$ has a definite sign. So the sets

$$A = \left\{\alpha \in \mathbb{R}: \omega_1 + \alpha\omega_2 \geq 0\right\} \quad \text{and} \quad B = \left\{\alpha \in \mathbb{R}: \omega_1 + \alpha\omega_2 \leq 0\right\}$$

are non-empty, closed and $A \cup B = \mathbb{R}$. Consequently there is $\tilde{\alpha} \in A \cap B$. Then $\omega_1 + \tilde{\alpha}\omega_2 = 0$, which says that ω_1 and ω_2 are linearly dependent.

Third step. The algebraic multiplicity of μ_1 is 1, i.e., μ_1 is a simple eigenvalue. Indeed, referring to Section 3, our claim is to prove that $N(I-\mu_1T)^2 = N(I-\mu_1T)$. Let $u \in N(I-\mu_1T)^2$. Then $u-\mu_1Tu = t\phi_1$ for some $t \in \mathbb{R}$. Taking inner product with ϕ_1: $(u-\mu_1Tu, \phi_1)_a = (u, \phi_1-\mu_1T\phi_1)_a = 0 = t(\phi_1, \phi_1)_a$, which implies $t = 0$. So $u \in N(I-\mu_1T)$ and the proof is complete.

Q.E.D.

1.4 A MAXIMUM PRINCIPLE. Let us start with some considerations about the Dirichlet problem

$$\text{(28)} \qquad Lu = h \text{ in } \Omega, \quad u = 0 \text{ on } \partial\Omega$$

where the operator L is the one introduced in Section 3. We shall discuss the variational formulation of (28) namely

$$\text{(29)} \qquad \begin{cases} a[u, v] = \int hv, & \forall\, v \in H_o^1 \\ u \in H_o^1(\Omega) \end{cases}$$

For the needs of the present discussion we assume that $h \in L^\sigma$ where $\sigma = 2N/(N+2)$. Under this assumption it follows from the Sobolev imbedding theorem that the map $v \longmapsto \int hv$ is a bounded linear functional in H_o^1. So by the Riesz representation theorem there exists $\tilde{h} \in H_o^1$ such that $\int hv = (\tilde{h}, v)_a$, $\forall\ v \in H_o^1$. Such an $\tilde{h}$ is the unique solution of (29) [recall that the bilinear form a is coercive in H_o^1]. In this way we have defined a bounded linear mapping $S: L^\sigma \to H_o^1$ by $S(h) = \tilde{h}$. Applying Theorem 8.1 [16, p. 168] we see that $\tilde{h} \geq 0$ whenever $h \geq 0$. Observe that this statement is a maximum principle and it has a crucial assumption: $a_o(x) \geq 0$ in Ω. The next result shows how this hypothesis can be relaxed somehow.

Theorem 1.14. *Let* $u \in H_o^1(\Omega)$ *be a solution of*

$$(30) \qquad Lu - \lambda mu = h \quad \textit{in } \Omega, \quad u = 0 \quad \textit{on } \partial\Omega$$

where $h \in L^\sigma(\Omega)$ *and* $h \geq 0$ *in* Ω. *Suppose that* $m \in L^\infty(\Omega)$ *is positive on a subset of* Ω *with positive measure. Assume also that* $0 \leq \lambda < \mu_1(m)$. *Then* $u \geq 0$ *in* Ω. *Moreover, if* $h > 0$ *in a set of positive measure then* $u > 0$ *in* Ω.

Proof. a) (30) is to be understood in the weak form

$$(31) \qquad a[u, v] - \lambda \int muv = \int hv, \quad \forall\ v \in H_o^1.$$

Without loss of generality we may assume that $|m(x)| < 1$ for $x \in \Omega$ and that fact will be used later. (31) may be rewritten as

$$(32) \qquad a_\lambda[u, v] = \lambda \int (m+1)uv + \int hv, \quad \forall\ v \in H_o^1$$

where $a_\lambda[u, v] = a[u, v] + \lambda \int uv$.

b) Given $\lambda \geq 0$ let us denote by $\mu_{1,\lambda}$ the first positive eigenvalue of

$$\begin{cases} a_\lambda[u, v] = \mu \int (m+1)uv, & \forall\ v \in H_o^1 \\ u \in H_o^1 \end{cases}$$

We claim that $0 \leq \lambda < \mu_1(m)$ implies that $\lambda < \mu_{1,\lambda}$. [To simplify the notation we set: $\alpha = \mu_{1,\lambda}$] Indeed, suppose by contradiction that $\alpha \leq \lambda$. Then

$$\frac{1}{\alpha} = \sup \frac{\int (m+1)v^2}{a[v, v]+\lambda\int v^2} \le \sup \frac{\int (m+1)v^2}{a[v, v]+\alpha\int v^2}.$$

Hence, for any given $\varepsilon > 0$ there is a $v \in H_o^1$ such that

$$\frac{1}{\alpha} - \varepsilon \le \frac{\int (m+1)v^2}{a[v, v]+\alpha\int v^2} \Longrightarrow \left(\frac{1}{\alpha} - \varepsilon\right) - \alpha\varepsilon \frac{\int v^2}{a[v, v]} \le \frac{\int mv^2}{a[v, v]}$$

By the variational characterization of the eigenvalues we get

$$\left(\frac{1}{\alpha} - \varepsilon\right) - \alpha\varepsilon \frac{1}{\mu_1(1)} \le \frac{1}{\mu_1(m)}, \quad \forall\ \varepsilon > 0,$$

which implies $\mu_1(m) \le \alpha \le \lambda < \mu_1(m)$: impossible!

c) In analogy with the mapping S defined in the beginning of the present section let us denote by S_λ the operator associated with the bilinear form a_λ. Then (32) may be written as

$$u = \lambda S_\lambda((m+1)u) + S_\lambda h.$$

Let $W: H_o^1 \to H_o^1$ be the operator defined by $W(u) = \lambda S_\lambda((m+1)u)$. This operator is compact in view of the compact imbedding $H_o^1 \subset L^\sigma$. By part b) we see that the spectral radius of W is less than 1. So

$$(33) \qquad u = (I-W)^{-1}S_\lambda h = \sum_{j=0}^{\infty} W^j(S_\lambda h)$$

and the positivity of u follows from the positivity of S_λ and W. The last statement follows readily from the strong maximum principle.

Q.E.D.

The result of the preceding theorem is sharp in the sense problem (30) cannot have a positive solution if $\lambda > \mu_1(m)$. In fact, we have

Proposition 1.15. *Let* $u \in C^1(\bar{\Omega})$ *be a solution of* (30), *where* $h \in L^\sigma(\Omega)$ *and* $h \ge 0$ *in* Ω. *Suppose that* $m \in L^\infty(\Omega)$ *is positive on a set of positive measure, and* $\lambda \geq \mu_1(m)$. *Then* $u > 0$ [*i.e.*, $u(x) \geq 0$ *in* Ω *and* $u(x') > 0$ *at some point* $x' \in \Omega$] *implies* $\lambda = \mu_1(m)$, $h = 0$ *and* $u = t\phi_1$, *where* ϕ_1 *is the eigenfunction of* (21) *corresponding to* $\mu_1(m)$.

Remark. If we have a solution $u \in H_0^1$ of (30) and the coefficients of L have the appropriate regularity for the use of the L^p-theory for the Dirichlet problem we can conclude that in fact $u \in C^1(\bar{\Omega})$.

In the proof of Proposition 1.15 we use the following form of the strong maximum principle, whose proof by its turn can be made adapting the argument in [16, p. 33] and using Theorem 8.19 in [16, p. 188].

Proposition 1.16. *Let* $u \in C^1(\bar{\Omega})$ *and assume that* $a[u, \psi] \leq 0$ *for all* $\psi \in H_0^1(\Omega)$, $\psi \geq 0$. *Suppose that* $\partial\Omega$ *satisfies the interior sphere condition at* $x_0 \in \partial\Omega$. *Then the outward normal derivative* $\frac{\partial u}{\partial \nu}(x_0)$ *is* > 0, *provided* u *is not constant in* Ω, *and* $u(x_0) \geq 0$. *The latter restriction is not necessary when* $a_0(x) \equiv 0$.

Proof of Proposition 1.15. Equation (30) may be written as (32), from which follows $a_\lambda[u, v] \geq 0$ for all $0 \leq v \in H_0^1$. Proposition 1.16 then implies that $u > t\phi_1$ for some $t > 0$. Let $t_0 = \sup\{t \geq 0: u \geq t\phi_1\}$. One has

$$a[t_0\phi_1, v] + \lambda\int t_0\phi_1 v = \mu_1(m)\int m t_0\phi_1 v + \lambda\int t_0\phi_1 v \leq \lambda\int (m+1)t_0\phi_1 v, \qquad \forall\ 0 \leq v \in H_0^1.$$

Subracting this expression from (32) and denoting by $\omega = u - t_0\phi_1$ one has

$$a[\omega, v] + \lambda\int \omega v \geq \lambda\int (m+1)\omega v + \int hv, \quad \forall\ 0 \leq v \in H_0^1.$$

Applying again Proposition 1.16 we obtain $\frac{\partial \omega}{\partial \nu} < 0$. But this implies that for $t' > 0$ sufficiently small $\omega > t'\phi_1 \implies u > (t_0 + t')\phi_1$, contradicting the maximality of t_0.

Q.E.D.

The conclusion of Theorem 1.14 can be strengthened if we assume regularity on the coefficients of L to apply the L^p theory of the Dirichlet problem.

Theorem 1.17. *Same assumptions of Theorem 1.14 plus the following: (i) regularity of the coefficients of* L, *(ii)* $h \in L^p(\Omega)$ *with* $p > N$. *Then the solution* $u \in C^{1,\alpha}(\bar{\Omega})$ *and* $\frac{\partial u}{\partial \nu} < 0$ *on* $\partial\Omega$ *if* $h > 0$ *in a set of positive measure in* Ω.

Proof. a) By a standard bootstrap argument we conclude that $u \in C^{1,\alpha}(\bar{\Omega})$. b) If $h \in L^p(\Omega)$ with $p > N$ then Sh (see beginning of this Section 5) belongs to $C^{1,\alpha}(\bar{\Omega})$ and Proposition 1.16 can be applied to conclude that $\frac{\partial}{\partial \nu}(Sh) < 0$ on $\partial\Omega$, if $h \geq 0$. c) Using part b) we conclude that the operator W (see proof of Theorem 1.14) has a similar property. Finally the result follows using (33).

Q.E.D.

Chapter 2

SUBLINEAR ELLIPTIC PROBLEMS

2.1 EXISTENCE OF POSITIVE SOLUTIONS. Let us start posing the following problem. Let $g:\bar{\Omega}\times R^+ \to R$ be a C^1 function. Suppose

(1) $$\liminf_{s \downarrow 0} \frac{g(x, s)}{s} > \lambda_1$$

and

(2) $$\limsup_{s \to +\infty} \frac{g(x, s)}{s} < \lambda_1$$

Prove that the Dirichlet problem

(3) $$-\Delta u = g(x, u) \quad \text{in } \Omega, \ u = 0 \text{ on } \partial\Omega$$

has a positive solution $u \in C^{2,\alpha}(\Omega)$.

We prove a result which includes the problem above as a special case. Besides the theory of linear eigenvalue problems with an indefinite weight function developed in the previous chapter, the other main tool used here is the method of sub and supersolutions, which we expound next.

Let L be an elliptic operator as in Section 2 of the previous chapter. We make all the assumptions on Ω and on the coefficients of L in order to be able to apply Schauder's theory and the L^p theory of the Dirichlet problem. Let $g:\bar{\Omega}\times R^+ \to R$ be a C^α function (i.e. a Hölder continuous function with exponent $0 < \alpha < 1$). We consider the Dirichlet problem

(4) $$Lu = g(x, u) \text{ in } \Omega, \quad u = 0 \text{ on } \partial\Omega.$$

A function $\bar{u} \in C^2(\bar{\Omega})$ is said to be a *supersolution* of (4) if

$$L\bar{u} \geq g(x, \bar{u}) \text{ in } \Omega, \quad \bar{u} \geq 0 \text{ on } \partial\Omega.$$

A function $\underline{u} \in C^2(\bar{\Omega})$ is said to be a *subsolution* of (4) if

$$L\underline{u} \leq g(x, \underline{u}) \text{ in } \Omega, \quad \underline{u} \leq 0 \text{ on } \partial\Omega.$$

The result below is due to Amann [17] and [2], see also Sattinger [18] and [19], and a proof is included for completeness.

Theorem 2.1. *Suppose that problem* (4) *has a subsolution* $\underline{u}$ *and a supersolution* $\bar{u}$, *with* $\underline{u} \leq \bar{u}$. *Assume that* $g:\bar{\Omega}\times R^+ \to R$ *is a* C^α *function and that there is a constant* $k \geq 0$ *such that*

(5) $$g(x, s_1) - g(x, s_2) \geq -k(s_1 - s_2)$$

for all $x \in \bar{\Omega}$ *and* $s_1 > s_2$ *with* $|s_1|, |s_2| \leq \max(\|\underline{u}\|_{C^0}, \|\bar{u}\|_{C^0})$. *Then problem* (4) *has solutions* $U, V \in C^{2,\alpha}(\bar{\Omega})$ *such that* $\underline{u} \leq U \leq V \leq \bar{u}$. *These solutions are obtained by an iteration scheme (see the Proof). Moreover any solution* u *of* (4) *with* $\underline{u} \leq u \leq \bar{u}$ *is such that* $U \leq u \leq V$. [*It is not stated that* $U \neq V$].

Proof. By the Schauder theory, for each given $u \in C^{2,\alpha}(\bar{\Omega})$ there is a unique $\omega \in C^{2,\alpha}(\bar{\Omega})$ such that

(6) $\quad L\omega + k\omega = g(x, u) + ku \text{ in } \Omega, \quad \omega = 0 \text{ on } \partial\Omega.$

This defines a mapping $T: C^{2,\alpha}(\bar{\Omega}) \to C^{2,\alpha}(\bar{\Omega})$ which is monotone in the interval $[\underline{u}, \bar{u}]$, that is, if $\underline{u} \le u_1, u_2 \le \bar{u}$, then

(7) $$u_1 \le u_2 \Longrightarrow Tu_1 \le Tu_2$$

Indeed, writing (6) with $u = u_1$ and $u = u_2$ and subtracting one from the other we obtain

$$L(\omega_1 - \omega_2) + k(\omega_1 - \omega_2) = g(x, u_1) - g(x, u_2) + k(u_1 - u_2).$$

Hence using (5) and the maximum principle we obtain (7). For the construction of solutions U and V proceed ty iteration as follows

$$u_n = Tu_{n-1}, \quad u_o = \underline{u} \quad ; \quad v_n = Tv_{n-1}, \quad v_o = \bar{u}.$$

We claim that

$$\underline{u} = u_o \le u_1 \le u_2 \le \dots \le v_2 \le v_1 \le v_o = \bar{u}, \text{ in } \Omega.$$

The proof of the monotonicity of the sequences (u_n) and (v_n) is done by induction. First observe that

$$L(\underline{u} - u_1) + k(\underline{u} - u_1) = L\underline{u} + k\underline{u} - [g(x, \underline{u}) + k\underline{u}] \le 0, \text{ in } \Omega$$

and

$$\underline{u} - u_1 = \underline{u} \le 0 \quad \text{on} \quad \partial\Omega.$$

This implies by the maximum principle that $\underline{u} - u_1 \le 0$ in $\bar{\Omega}$. A similar argument gives $v_1 \le \bar{u}$. The monotonicity of T gives the rest. So there are functions U and V such that

(8) $$u_n \to U \quad \text{and} \quad v_n \to V \quad \text{pointwisely.}$$

By the Lebesgue convergence theorem, it follows that U and V are in $L^p(\Omega)$ and the convergence in (8) is in the L^p sense for any $p \ge 1$. Using the a priori estimates for solutions of linear elliptic equations in the Sobolev spaces we get

$$\|u_n - u_m\|_{W^{2,p}} \le c\left\{\|g(x, u_{n-1}) - g(x, u_{m-1})\|_{L^p} + \|u_{n-1} - u_{m-1}\|_{L^p}\right\}$$

where C is a constant independent of n and m. A similar expression for the sequence (v_n). So the convergence in (8) is in the norm of $W^{2,p}(\Omega)$. Next we use the Sobolev imbedding theorem (take $p > N$) to conclude that U and V are in $C^{1,\alpha}(\bar{\Omega})$ and the convergence in (8) is in the norm of $C^{1,\alpha}(\Omega)$. By the Schauder estimates it follows that U and V are in $C^{2,\alpha}(\bar{\Omega})$ and the convergence in (8) is in the norm of $C^{2,\alpha}(\bar{\Omega})$. So U and V are solutions of (4). The remaining assertions of the theorem follow easily from (7).

Q.E.D.

Now let us consider the Dirichlet problem (4) with a C^α function $g:\bar{\Omega}\times R^+ \to R$ which satisfies the following three conditions:

(c_0) There is a continuous function $g_0:\bar{\Omega} \to R$, with $g_0(x) \geq 0$, and $s_0 > 0$ such that

$$g(x, s) \geq g_0(x)s, \quad \forall\, x \in \bar{\Omega}, \quad \forall\, 0 < s < s_0.$$

(c_∞) There are continuous functions g_∞, $c:\bar{\Omega} \to R$, with $c(x) \geq 0$ such that

$$g(x, s) \leq g_\infty(x)s + c(x), \quad \forall\, x \in \bar{\Omega}, \quad \forall\, s \geq 0.$$

(nd) For every real number $M > 0$ there is a corresponding $k \geq 0$ such that the function $s \longmapsto g(x, s) + ks$ is nondecreasing for $|s| \leq M$.

Theorem 2.2. *Suppose that* g *satisfies the hypotheses just stated. Assume also that*

$$\mu_1(g_0) < 1 \tag{9}$$

and

$$\mu_1(g_\infty) > 1 \tag{10}$$

[The latter assumption is made only if $g_\infty(x_0) > 0$ *for some* $x_0 \in \Omega$. *Otherwise it should be dropped, and the remaining assumptions on* g

will suffice]. *Then the Dirichlet problem* (4) *has a positive solution.*

Remarks. 1) Conditions (c_0) and (9) are satisfied, for instance, if $\lim_{s\downarrow 0} g(x, s)s^{-1} = +\infty$. In particular if $g(x, 0) > 0$.

2) If $g(x, 0) = 0$, then $u \equiv 0$ is a trivial solution.

3) The function g could assume negative values. If there is an $s_0 > 0$ such that $g(x, s_0) \leq 0$ for all $x \in \Omega$, then (c_∞) and (10) are superfluous. Indeed, in this case $\bar{u} \equiv s_0$ is a supersolution of (4).

4) Hypotheses (c_0) and (9) in the previous theorem implies that $g_0(x) > 0$ on a subset of positive measure in Ω.

5) Condition (nd) holds for instance if the function g is C^1.

6) Theorem 2.2 is essentially due to Amann [2].

Examples of functions g satisfying the assumptions of Theorem 2.2: i) $g(x, s) = \lambda e^{-s}$ for any $\lambda > 0$. ii) $g(x, s) = \lambda s - f(x, s)$, where $\lambda > \mu_1(1)$ and $f:\bar{\Omega}\times R^+ \to R$ is a C^1 function such that $\lim_{s\downarrow 0} f(x, s)s^{-1} = 0$ and $\lim_{s\to\infty} f(x, s)s^{-1} = +\infty$. iii) the function g of the Problem in the beginning of the present section.

Proof of Theorem 2.2. a) Suppose that $g_\infty(x_0) > 0$ for some $x_0 \in \Omega$. Choose C^1 functions $\hat{g}_\infty, \hat{c}:\bar{\Omega} \to R$, with $\hat{c}(x) \geq 0$, such that

$$(11) \qquad g(x, s) \leq \hat{g}_\infty(x)s + \hat{c}(x), \quad \forall\, x \in \bar{\Omega}, \quad \forall\, s \geq 0$$

and $\mu_1(\hat{g}_\infty) > 1$. This is possible in view of (10), using Proposition 1.12B. Now let $v \in C^{2,\alpha}(\bar{\Omega})$ be the solution of the Dirichlet problem

$$(12) \qquad Lv = \hat{g}_\infty(x)v + \hat{c}(x) \quad \text{in } \Omega, \quad v = 0 \text{ on } \partial\Omega.$$

It follows from Theorems 1.14 and 1.17 that $v > 0$ in Ω, and $\frac{\partial v}{\partial \nu} < 0$ on $\partial\Omega$. Expressions (11) and (12) show that v is a supersolution of (4).

b) If $g_\infty(x) \leq 0$ for all $x \in \Omega$, we can proceed as in part a) above taking $\hat{g}_\infty(x) = \varepsilon$, with $0 < \varepsilon < \mu_1(1)$.

c) Choose a C^1 function $\hat{g}_0 : \bar{\Omega} \to R$ such that $\hat{g}_0(x) \geq 0$ and

(13) $$g(x, s) \geq \hat{g}_0(x)s \qquad \forall\, x \in \bar{\Omega}, \quad \forall\, 0 \leq s \leq s_0$$

and $\mu_1(\hat{g}_0) < 1$. This is a consequence of Proposition 1.12B. Now the eigenvalue problem

$$Lu = \mu\hat{g}_0(x)u \quad \text{in} \quad \Omega, \quad u = 0 \quad \text{on} \quad \partial\Omega$$

has a positive eigenfunction $\psi \in C^{2,\alpha}(\bar{\Omega})$ corresponding to the first eigenvalue $\mu_1(\hat{g}_0)$. Let $t > 0$ be chosen in such a way that $t\psi < s_0$ and $t\psi < v$. Then using (c_0) and (13) we have

$$L(t\psi) = \mu_1(\hat{g}_0)\hat{g}_0 t\psi < g(x, t\psi),$$

which shows that $t\psi$ is a subsolution of (4). So we can apply Theorem 2.1 to finish the proof.

Q.E.D.

Let us now apply Theorem 2.2 in the discussion of the nonlinear eigenvalue problem

(14) $$Lu = \lambda f(x, u) \quad \text{in} \quad \Omega \qquad u = 0 \quad \text{on} \quad \partial\Omega$$

where $f : \bar{\Omega} \times R^+ \to R$ is a C^α function satisfying condition (nd) above, and let us suppose that the limits below are continuous functions:

$$f_0(x) = \liminf_{s \downarrow 0} \frac{f(x, s)}{s}, \quad f_\infty(x) = \limsup_{s \to +\infty} \frac{f(x, s)}{s}.$$

Then the following statements [warning: these are just sufficient conditions for the solvability of (14)] are immediate consequences of Theorem 2.2.

a) If $f_0(x) \equiv +\infty$ (in particular if $f(x, 0) > 0$) and $f_\infty(x) \leq 0$, then problem (14) has a positive solution for all $\lambda > 0$.

b) If $f_0(x) \equiv +\infty$ and $f_\infty(x_0) > 0$ for some point $x_0 \in \Omega$, then problem (14) has a positive solution for all $0 < \lambda < \mu_1(1)/[\sup f_\infty(x)]$

c) If $0 < f_0(x) < +\infty$ in $\bar{\Omega}$ and $f_\infty(x) \leq 0$, then problem (14) has a positive solution for all $\lambda > \mu_1(1)/[\inf f_0(x)]$.

d) If $0 < f_0(x) < +\infty$ in $\bar{\Omega}$ and $f_\infty(x_0) > 0$ for some $x_0 \in \Omega$, then problem (14) has a solution for all $\mu_1(1)/[\inf f_0(x)] < \lambda < \mu_1(1)/[\sup f_\infty(x)]$. Of course if $\inf f_0 < \sup f_\infty$, we cannot say anything about the existence of positive solutions of (14), at least by the present method. Observe that the above bounds on λ for existence are not necessarily sharp. Compare with [20].

2.2 UNIQUENESS OF THE POSITIVE SOLUTION. In this section we establish some uniqueness results for solutions of problem (4) involving some special type of nonlinearities g. Besides other assumptions on g we always assume

(15) the functions $s \longrightarrow g(x, s)/s$, for $s > 0$ is non-increasing.

It is easy to see that (15) holds for instance when $g(x, s)$ is concave in s and $g(x, 0) \geq 0$.

<u>Lemma 2.3</u>. *Under the hypotheses of Theorem 2.2 problem (4) has a maximal positive solution.*

<u>Remark</u>. If $g(x, 0) > 0$ then problem (4) has also a minimal positive solution. This follows readily from Theorem 2.1 using the fact that $\underline{u} \equiv 0$ is a subsolution which is not a solution.

<u>Proof of Lemma 2.3</u>. We claim that any positive solution u of (4) is such that $u \leq v$ where v is the supersolution given in (12). Once this is done we use Theorem 2.1 to conclude that $u \leq V$, and then V is the maximal solution of (4). To prove the claim we use (11) to see that

$$g(x, u) = \hat{g}_\infty(x)u + d(x), \quad d(x) \le \hat{c}(x)$$

and consequently

$$L(v-u) = \hat{g}_\infty(x)(v-u) + \hat{c}(x) - d(x).$$

Since $\mu_1(\hat{g}_\infty(x)) > 1$ and $\hat{c}(x) - d(x) \ge 0$ we can use Theorem 1.14 to conclude that $v-u \ge 0$ as required. Q.E.D.

Theorem 2.4. *Assume that the hypotheses of Theorem 2.2 and (15) above hold. Suppose also that*

(16) *the function* $\gamma(x, s) = g(x, s)/s$ *is uniformly continuous for* $x \in \bar{\Omega}$, $0 < s \le s_0$, *for some* $s_0 > 0$. *Then problem (4) has a unique positive solution.*

Remark. The function γ can be extended continuously to $s = 0$ by $\gamma(x, 0) = \lim g(x_n, s_n)/s_n$, where $\{(x_n, s_n)\}$ is any sequence converging to $(x, 0)$. Hypothesis (16) is satisfied for instance if we assume $g \in C^1(\bar{\Omega}\times R^+)$ and $g(x, 0) = 0$ for all $x \in \Omega$; in this case $\gamma(x, 0) = g_s'(x, 0)$, which follows from the mean value theorem.

Proof of Theorem 2.4. Let u_1 and u_2 be two positive solutions of (4). Without loss of generality we may assume $u_1 \le u_2$, in virtue of Lemma 2.3. Then

(17) $$Lu_i = m_i(x)u_i \text{ in } \Omega \qquad u_i = 0 \text{ on } \partial\Omega$$

for $i = 1,2$, where

$$m_i(x) = \begin{cases} \dfrac{g(x, u_i(x))}{u_i(x)} & \text{in } \{x\in\Omega: u_i(x) \ne 0\} \\ \gamma(x, 0) & \text{in } \{x\in\bar{\Omega}: u_i(x) = 0\} \end{cases}$$

Clearly m_i is a continuous function in $\bar{\Omega}$ and it is positive on a subset of positive measure in Ω. Moreover in view of (15) we see that $m_1 \ge m_2$ and $u_1 \ne u_2$ implies that $m_1 > m_2$ on a subset of positive measure. It follows from Proposition 1.12A that

$\mu_1(m_1) < \mu_1(m_2)$. But from (17) and the fact that $u_i \geq 0$ we arrive to the contradictory statement that $\mu_1(m_1) = \mu_1(m_2) = 1$. Q.E.D.

Assumption (16) in Theorem 2.4 can be relaxed as follows.

Theorem 2.5. *Assume that the hypotheses of Theorem 2.2 and (15) hold. Suppose also that*

(18) *any positive solution* u *of* (4) *is such that* $u(x) > 0$ *for all* $x \in \Omega$, *and*

(19) *there are constants* $c > 0$, $s_0 > 0$ *and* $0 < \alpha < 2/N$ *such that* $g(x, s)/s \leq cs^{-\alpha}$, *for* $0 < s < s_0$.

Then problem (4) *has a unique positive solution.*

Remark. Hypothesis (18) is verified for instance if $g(x, s) \geq 0$ for all $x \in \bar{\Omega}$, $s \geq 0$. [That is a consequence of the strong maximum principle]. Observe that hypothesis (19) is weaker than (16). While (16) requires that $g(x, s)$ vanishes at $s = 0$ like const.s we see that (19) allows a behavior like const. s^β, with $0 < \beta < 1$. An analysis of the proof below shows that (18) can be weakened, being replaced by the requirement that $u(x) > 0$ in Ω (a.e.).

Proof of Theorem 2.5. Proceed as in the proof of Theorem 2.4. However, the continuity of m_i has to be replaced by the assertion that $m_i \in L^r(\Omega)$ for some $r > N/2$, and then Proposition 1.12A is used again. In view of (19) we can choose $r > N/2$ and $r < 1/\alpha$.

Q.E.D.

2.3 SOME MULTIPLICITY RESULTS FOR SUBLINEAR PROBLEMS.

In this section we show how to use the previous results to obtain some results on the existence of multiple solution for the Dirichlet problem

(20) $$Lu = \lambda u - f(x, u) \text{ in } \Omega, \quad u = 0 \text{ on } \partial\Omega,$$

where L is the second order elliptic operator introduced in Section 1.2, [we also assume sufficient smoothness on $\partial\Omega$ as well as on the coefficients of L in order to apply the Schauder theory and the L^p theory of the linear Dirichlet problem as needed] and $f:\bar{\Omega}\times R \to R$ is a C^1 function satisfying the following assumptions:

(21) $$f(x, 0) = 0 \qquad f'_s(x, 0) = 0$$

(22) the function $s \longmapsto \dfrac{f(x, s)}{s}$ is nonincreasing in the half-line $(-\infty, 0)$ and nondecreasing in $(0, \infty)$.

(23) $$\lim_{s\to+\infty} \frac{f(x, s)}{s} \text{ and } \lim_{s\to-\infty} \frac{f(x, s)}{s} \text{ are } > \lambda_2 - \lambda_1,$$

where λ_1 and λ_2 are the first two eigenvalues associated with L for the Dirichlet problem, see end of Section 1.2. We emphasize that f is *not* supposed to be *odd* in the variable s. An example of a function satisfying hypotheses (21)-(23) is $f(x, s) = g(x)s^{2n+1}$, where $g:\bar{\Omega} \to R$ is a C^1 positive function and n is some positive integer.

<u>Theorem 2.6</u>. *Let* $f:\bar{\Omega}\times R \to R$ *be a* C^1 *function satisfying assumptions* (21), (22) *and* (23). *Suppose that* $\lambda_1 < \lambda < \lambda_2$. *Then problem* (20) *has exactly three solutions.*

<u>Remark</u>. This results extends Theorem 1.7 of Ambrosetti-Mancini [4] and Theorem 6 of Berger [21], and it is due in the present generality to Berestycki [5]. We remark however that Berger's applies to elliptic operators of higher order. The proofs in the first two papers use different techniques.

In the proof of Theorem 2.6 we use the following

<u>Lemma 2.7</u>. *Let* $g:\bar{\Omega}\times R \to R$ *be a* C^1 *function such that* $g'_s(x, s) < \lambda_2$ *for* $x \in \bar{\Omega}$ *and* $s \in R$. *Then any two solutions* u_1 *and* u_2 *of the Dirichlet problem*

(24) $$Lu = g(x, u) \quad \text{in } \Omega, \quad u = 0 \quad \text{on } \partial\Omega,$$

are ordered: either $u_1 \le u_2$ *or* $u_2 \le u_1$.

<u>Proof</u>. From $L(u_1-u_2) = g(x_1u_1) - g(x_1u_2)$ in Ω, $u_1-u_2 = 0$ on $\partial\Omega$ we obtain

(25) $$L\omega = m(x)\omega \quad \text{in } \Omega, \quad \omega = 0 \quad \text{on } \partial\Omega$$

where $\omega = u_1-u_2$ and

$$m(x) = \begin{cases} \dfrac{g(x, u_1(x)) - g(x, u_2(x))}{u_1(x) - u_2(x)}, & \text{if } u_1(x) \ne u_2(x) \\ g_s'(x, u_1(x)), & \text{if } u_1(x) = u_2(x) \end{cases}$$

From (25) and the fact that $m(x) < \lambda_2$ we obtain

(26) $$1 = \mu_j(m) > \mu_j(\lambda_2) = \frac{\lambda_j}{\lambda_2}, \quad \text{for some } j \geq 1.$$

[This is the case if $m(x) > 0$ at some point. If $m(x) \leq 0$ for all $x \in \Omega$, it follows from (25) that $\omega = 0$, and so in this case: $u_1 = u_2$]. It follows from (26) that $j = 1$, and consequently ω is an eigenfunction corresponding to the first eigenvalue $\mu_1(m)$. Thus the result follows readily from Theorem 1.13.

Q.E.D.

<u>Proof of Theorem 2.6</u>. a) First we observe that (22) implies that $f_s'(x, s) \geq 0$ for all $x \in \Omega$ and all $s \in R$. Consequently $(\lambda s - f(x, s))_s' < \lambda_2$ and Lemma 2.7 tells us that any two solutions of (20) are ordered. Since $u = 0$ is a solution, then the other solutions have a definite sign in Ω; they are either ≥ 0 or ≤ 0.

b) Hypotheses (21) and (23) permit us to apply Theorem 2.2 (or its corollary in the form of the problem posed in the very beginning of Section 2.1), and conclude that there is a positive solution u_1 and a negative solution u_2. For the existence of the negative solution we look at the Dirichlet problem

$$Lv = \lambda v - \tilde{f}(x, v) \text{ in } \Omega \quad v = 0 \quad \text{on} \quad \partial\Omega,$$

where $\tilde{f}(x, v) = -f(x, -v)$.

c) Finally using theorem 2.4 we have that u_1 is the only positive solution and u_2 is the only negative solution.

Q.E.D.

Theorem 2.8. *Let* $f:\bar{\Omega}\times R \to R$ *be a* C^1*-function satisfying assumptions* (21), (22) *and*

(27)
$$\lim_{s\to\pm\infty} \frac{f(x, s)}{s} = +\infty.$$

Then, if $\lambda_2 < \lambda < \lambda_3$ *and* λ_2 *is simple, problem* (20) *has at least four solutions. Moreover if* f *is odd then problem* (20) *has at least five solutions.*

Remark. This result is due to Lazer-McKenna [22], with a slightly different proof. Assumption (27) may be weakened: suppose that those two limits are $> \lambda_2 - \lambda_1$. The proof is the same.

Proof. As in the previous theorem we already know that there are at least 3 solutions: 0, $u_1 \geq 0$ and $u_2 \leq 0$ we also know that u_1 is the only positive solution. A similar statement for u_2. The idea now is (i) to transform problem (20) in a suitable operator equation $\Phi(u) = u - T(u)$, where T is a compact operator in $C_0^{1,\alpha}(\Omega)$, (ii) to obtain an a priori bound for the solutions of $\Phi(u) = 0$, (iii) to compute the degree of Φ in some large ball in $C_0^{1,\alpha}(\bar{\Omega})$), (iv) to compute the indices of the three known solutions, (v) to use the additivity of the Leray-Schauder degree and to conclude. We solve these questions in the sequel.

Lemma 2.9 (*A priori bound on the solutions of* (20)). *Let* $a > 0$ *be such that*

(28) $\lambda s - f(x, s) < 0$ *for* $s > a$, *and* $\lambda s - f(x, s) > 0$ *for* $s < -a$.

Then $\|u\|_{L^\infty} \leq a$ *for all solutions of* (20).

Proof. Let $u_a(x) = u(x)$, if $u(x) < a$, and $u_a(x) = a$, if $u(x) \geq a$. The solutions u of (20) are in $C^{2,\alpha}(\bar{\Omega})$ and in particular in $H^1_o(\Omega)$. By a result of Stampacchia [15, p. 196] $u_a \in H^1_o(\Omega)$, and then

$$a[u, u-u_a] = \int [\lambda u - f(x, u)](u - u_a).$$

Since $a[u_a, u-u_a] = 0$ we obtain from (28) that

$$c\|u-u_a\|^2_{H^1_o} \leq a[u-u_a, u-u_a] = \int [\lambda u - f(x, u)](u - u_a) \leq 0$$

which implies $u = u_a$ and consequently $u \leq a$ in Ω. Similarly we prove that $-a \leq u$.

Q.E.D.

Modifying the nonlinearity. The above lemma allow us to replace the nonlinearity $\lambda u - f(x, u)$ by $\psi(u)[\lambda u - f(x, u)]$ where $\psi: R \to R$ is a non-negative C^∞ function with compact support and such that $\psi(s) = 1$ for $|s| \leq a$. Indeed the Dirichlet problem

(29) $\qquad Lu = \psi(u)[\lambda u - f(x, u)]$ in Ω, $u = 0$ on $\partial\Omega$

has the same solutions as problem (20). (As in Lemma 2.9 we prove that also for the solutions u of (29) one has $\|u\|_{L^\infty} \leq a$). Problem (29) has the advantage that the right side is a bounded function of u. *So we can replace problem* (20) *by the following one:*

(30) $\qquad Lu = g(x, u)$ in Ω, $u = 0$ on $\partial\Omega$

where $g: \bar{\Omega} \times R \to R$ is a bounded C^1 function satisfying the assumptions

(31) $\qquad g(x, 0) = 0, \quad g'_s(x, 0) = \lambda \quad \lambda_2 < \lambda < \lambda_3$

(32) the function $s \longmapsto \dfrac{g(x, s)}{s}$ is nondecreasing in $(-a, 0)$ and nonincreasing $(0, a)$.

(33) $g(x, s) \leq 0$ for $s > a$ and $g(x, s) \geq 0$ for $s \leq -a$.

Observe that the solutions u of (30) are such that $\|u\|_{L^\infty} \leq a$.

<u>Setting the operator equation</u>. Let $X = C_0^{1,\alpha}(\bar{\Omega})$ be the Banach space of functions $u:\bar{\Omega} \to R$ which are continuously differentiable and their first derivatives are Hölder continuous with exponent $0 < \alpha < 1$ satisfying the boundary condition $u = 0$ on $\partial\Omega$; the norm is the maximum of the sup of the function, of its derivatives and of theirs Hölder quocients. Let $K:X \to X$ be defined as follows: given $v \in X$, let u be the unique solution of the Dirichlet problem

$$Lu + ku = g(x, v) + kv \text{ on } \Omega \quad u = 0 \text{ on } \partial\Omega,$$

for some $k > 0$ fixed.

By the Schauder theory one knows that $u \in C^{2,\alpha}$. So K is a compact mapping. The solutions of (30) are the zeros of $I-K$. Now we proceed to compute the degree of $I-K$ in certain subsets of X.

<u>Lemma 2.10</u>. *Under the above assumptions there is an* $R > 0$ *such that* $\deg(I-K, B_R, 0) = 1$, *where* $B_R = \{u \in X: \|u\|_X < R\}$.

<u>Proof</u>. We claim that there is an $R > 0$ such that $(I-tK)u \neq 0$ for all $t \in [0, 1]$ and all $\|u\| = R$. Once this has been done the result follows from the homotopy invariance of the Leray-Schauder degree. To prove that claim observe that $(I-tK)u = 0$ is equivalent to $Lu + ku = tg(x, u) + tku$ in Ω, $u = 0$ on $\partial\Omega$. By the argument used in Lemma 2.9 we see that $\|u\|_{L^\infty} \leq a$, which implies $\|u\|_X \leq$ const through a standard combination of L^p estimates, Sobolev imbedding theorems and Schauder estimates. Take R larger than this constant.

Q.E.D.

Now recall that the positive solution u_1 is obtained by the monotone iteration method (see the proof of Theorem 2.2): let us denote by $\underline{u}$ the subsolution and by $\bar{u}$ the supersolution obtained there. We can choose them in such a way that

(34) $\quad L\bar{u} > g(x, \bar{u})$ in Ω, $L\underline{u} < g(x, \underline{u})$ in Ω, $\bar{u} = \underline{u} = 0$ on $\partial\Omega$.

Lemma 2.11. *There is an* $r > 0$ *such that* $\deg(I-K, \mathcal{O}_1, 0) = 1$ *where* $\mathcal{O}_1 = \left\{u \in X: \underline{u} < u < \bar{u} \text{ in } \Omega;\ \frac{\partial \bar{u}}{\partial \nu} < \frac{\partial u}{\partial \nu} < \frac{\partial \underline{u}}{\partial \nu} \text{ on } \partial\Omega;\ \|u\|_X < r\right\}$.

The proof of this lemma follows the general line of ideas of the proof of Lemma 12 in [23]. It will be presented below for convenience of the reader. Let us assume Lemma 2.11 and finish the proof of Theorem 2.8.

Proof of Theorem 2.8 completed. Observe that $\mathcal{O}_1$ is an open set in X containing the solution u_1. In a similar way we obtain an open set $\mathcal{O}_2$ containing u_2. It is easy to see that $\mathcal{O}_1 \cap \mathcal{O}_2 = \emptyset$ and $0 \notin \mathcal{O}_1 \cup \mathcal{O}_2$. Now by a well know result going back to Leray-Schauder (see for instance [24]) we have that the index $i(I-K, 0, 0) = (-1)^2 = 1$. Next let B_R be a ball centered at 0 in X containing both $\mathcal{O}_1$ and $\mathcal{O}_2$ and let B_ε be a small ball centered at 0 and disjoint of both $\mathcal{O}_1$ and $\mathcal{O}_2$. Then

$$\deg(I-K, B_R \backslash \overline{\mathcal{O}_1 \cup \mathcal{O}_2 \cup B_\varepsilon}, 0, 0) = -2$$

by the additivity property of the degree. So there is at least one further solution u_3 of (30) in $B_R \backslash \overline{\mathcal{O}_1 \cup \mathcal{O}_2 \cup B_\varepsilon}$. If g is odd then $-u_3$ is still another solution. Q.E.D.

Proof of Lemma 2.11. Let $k = \max\{|g_s'(x, s)| : x \in \bar{\Omega},\ \underline{u}(x) \leq s \leq \bar{u}(x)\}$ and define a function $\tilde{g}: \bar{\Omega} \times R \to R$ by

$$\tilde{g}(x, s) = \begin{cases} g(x, \underline{u}(x)) + k\underline{u}(x), & \text{if } s \leq \underline{u}(x) \\ g(x, s) + ks, & \text{if } \underline{u}(x) \leq s \leq \bar{u}(x) \\ g(x, \bar{u}(x)) + k\bar{u}(x), & \text{if } s \geq \bar{u}(x) \end{cases}$$

So $\tilde{g}$ is bounded and nondecreasing in the variable s. From the Schauder theory for the linear Dirichlet problem, we conclude that for each $v \in X$ there is a unique solution $u \in C^{2,\alpha}(\bar{\Omega})$ of

(35) $$Lu + ku = \tilde{g}(x, v) \text{ in } \Omega \quad u = 0 \text{ on } \partial\Omega.$$

So we have defined a compact mapping $\tilde{K}: X \to X$ with the property that $\tilde{K} = K$ in $\bar{\mathcal{O}}_1$. Now we claim that $\tilde{K}$ maps X into $\mathcal{O}_1$, provided r is properly chosen. First let us show how to choose r. Since $\tilde{g}$ is a bounded function it follows from the L^p estimates that the solutions of (35) are uniformly bounded in X for all $v \in X$. Take $r > \sup\{\|\tilde{K}v\|_X : v \in X\}$. Next take $v \in X$ and let us prove that $u = \tilde{K}v \in \mathcal{O}_1$. Clearly $\|u\|_X < r$. From (34) and (35) we have

$$(L+k)(\bar{u}-u) > g(x, \bar{u}) + k\bar{u} - \tilde{g}(x, v) \geq 0 \text{ in } \Omega, \quad \bar{u}-u = 0 \quad \text{on } \partial\Omega.$$

It follows from the strong maximum principle that $u < \bar{u}$ in Ω and $\frac{\partial \bar{u}}{\partial \nu} < \frac{\partial u}{\partial \nu}$ on $\partial\Omega$. A similar reasoning for the subsolution, which then completes the proof of the assertion $\tilde{K}(X) \subset \mathcal{O}_1$. Now let $\Psi \in \mathcal{O}_1$ and consider the compact homotopy $H_\theta(u) = \theta\tilde{K}(u) + (1-\theta)\Psi$, $0 \leq \theta \leq 1$. Since $u \neq H_\theta(u)$ for all $u \in \partial\mathcal{O}_1$ and all $\theta \in [0, 1]$ we conclude that $\deg(I-H_1, \mathcal{O}_1, 0) = \deg(I-H_0, \mathcal{O}_1, 0)$. But H_0 is a constant mapping and clearly $\deg(I-H_0, \mathcal{O}_1, 0) = 1$. So $\deg(I-K, \mathcal{O}_1, 0) = 1$.

Q.E.D.

Chapter 3

SUPERLINEAR ELLIPTIC PROBLEMS

3.1 THE FIXED POINT INDEX. Let C be a closed convex subset of a Banach space X, and $W \subset C$ a relatively open subset of C, that is, $W = \mathcal{O} \cap C$ for some open subset $\mathcal{O}$ of X. Let $\Phi: \bar{W} \to C$ be a compact mapping such that $\Phi(x) \neq x$ for $\bar{W}\backslash W$. Associated with each such a mapping we define an integer $i_C(\Phi, W)$, called the *fixed point index* of Φ, as follows. By a theorem of Dugundji [25] the mapping Φ has a compact extension $\tilde{\Phi}: \bar{\mathcal{O}} \to C$. Then define

$$(1) \qquad i_C(\Phi, W) = \deg(I-\tilde{\Phi}, \mathcal{O}, 0).$$

To see that this is in fact a good definition we have to settle the

the three following points: (i) $\tilde{\Phi}(x) \neq x$ for all $x \in \partial\mathcal{O}$, (ii) the degree in the right side of (1) is independent of the particular extension $\tilde{\Phi}$, (iii) it does not depend either on the particular open set $\mathcal{O}$. These facts are easily proved using the homotopy invariance of the degree and the excision property. The usual properties of the Leray-Schauder degree are transferred immediately to the fixed point index. So we have the following properties.

I) Normalization. Let $\Phi: \bar{W} \to W$ be a constant mapping, that is, $\Phi(x) = a \in W$ for all $x \in \bar{W}$ and some fixed $a \in W$. Then $i_C(\Phi, W) = 1$.

II) Additivity. Let W_1 and W_2 be two disjoint (relatively) open subsets of W, and $\Phi: \bar{W} \to C$ a compact mapping such that $\Phi(x) \neq x$ for all $x \in \bar{W}\backslash(W_1 \cup W_2)$. Then

$$i_C(\Phi, W) = i_C(\Phi, W_1) + i_C(\Phi, W_2).$$

III) Homotopy invariance. Let $I \subset \mathbb{R}$ be a compact interval and $h: I\times\bar{W} \to C$ a compact mapping such that $h(t, x) \neq x$ for all $x \in \bar{W}\backslash W$ and all $t \in I$. Then $i_C(h(t, \cdot), W) =$ constant for $t \in I$.

IV) Excision. Let $V \subset W$ be relatively open, and $\Phi: \bar{W} \to C$ be a compact mapping such that $\Phi(x) \neq x$ for $x \in \bar{W}\backslash V$. Then $i_C(\Phi, V) = i_C(\Phi, W)$.

V) Solution property. $i_C(\Phi, W) \neq 0 \Longrightarrow \exists\ x \in W$ such that $\Phi(x) = x$.

We shall apply the previous facts to the case when C is a cone. Let us recall that a *cone* C in a Banach space X is a closed subset of X such that (i) if $x,y \in C$ and $\alpha,\beta \geq 0$, then $\alpha x+\beta y \in C$, (ii) if $x \in C$ and $x \neq 0$, then $-x \in C$. A cone induces a partial order in X defined as follows: $x \leq y$ if and only if $y-x \in C$. In the present section we use the following notation: for $r > 0$,

$B_r = \{x \in C: \|x\| < r\}$. A cone C is said to be *normal* if there is a constant $k > 0$ such that $0 \leq x \leq y$ implies $\|x\| \leq k\|y\|$.

Theorem 3.1. *Let* C *be a cone in a Banach space* X, *and* $\Phi: C \to C$ *a compact mapping. Assume that there are real numbers* $r, R > 0$ *such that*

(2) $\qquad x \neq t\Phi(x)$ *for* $0 \leq t \leq 1$ *and* $\|x\| = r$, $x \in C$,

(3) *there exists a compact mapping* $F: \bar{B}_R \times [0, \infty) \to C$ *such that* $F(x, 0) = \Phi(x)$ *for* $\|x\| = R$, $F(x, t) \neq x$ *for* $\|x\| = R$ *and* $t \geq 0$, *and* $F(x, t) = x$ *has no solution* $x \in \bar{B}_R$ *for* $t \geq t_0$.

Then: i) (2) $\Longrightarrow i_C(\Phi, B_r) = 1$, *and* ii) (3) $\Longrightarrow i_C(\Phi, B_R) = 0$.

Proof. i) A direct consequence of the homotopy invariance and the normalization properties of the fixed point index.

ii) Let us denote by $F_t: \bar{B}_R \to C$ the mapping $F_t(x) = F(t, x)$. The second part of (3) implies that $i_C(F_t, B_R) = \text{const}$, and the third part implies that this constant is zero. So $i_C(F_0, B_R) = 0$. Since $F_0(x) = \Phi(x)$ for $x \in \bar{B}_R \setminus B_R$, the result follows from the homotopy invariance property of the index. Q.E.D.

A sufficient condition for (2):

(2') There exists a bounded linear operator $A: X \to X$ such that $A(C) \subset C$, where C is a normal cone, A has spectral radius $r_\sigma(A)$ stricIty less than 1 and $\Phi(x) \leq Ax$ for $x \in C$ and $\|x\| = r$.

Indeed, suppose that $x = t\Phi(x)$ for some $\|x\| = r$ and $0 < t \leq 1$. Then $x \leq tAx$, which by iteration gives $x \leq t^n A^n x$. If C is a normal cone then $\|x\| \leq k\|t^n A^n x\|$ where $k > 0$ is the normality constant which depends only on the cone C. We then have $1 \leq kt^n\|A^n\|$. Taking n^{th} root and passing to the limit we come to

$1 \le tr_\sigma(A)$, which is a contradiction.

<u>Another suficient condition for (2)</u>:

(2") $\Phi(x) \not\geq x$ for all $x \in C$ and $\|x\| = r$

<u>A sufficient condition for (3)</u>:

(3') There exists $v \in C\setminus\{0\}$ such that $x \neq \Phi(x) + tv$ for $\|x\| = R$ and $t \geq 0$.

Indeed, let $\mu = \sup\{\|\Phi(x)\| : x \in C, \ \|x\| \leq R\}$ and take $t_0 > (R+\mu)/\|v\|$. The function $F(x, t) = \Phi(x) + \psi(t)v$, where $\psi(t) = t$ if $0 \leq t \leq t_0$ and $\psi(t) = t_0$ if $t \geq t_0$, satisfies all the conditions required in (3).

<u>Another sufficient condition for (3)</u>:

(3") $\Phi(x) \not\leq x$ for all $x \in C$ and $\|x\| = R$.

The following set of corollaries are easy consequences of Theorem 3.1.

<u>Corollary 3.2</u>. *Let* C *be a cone in a Banach space* X *and* $\Phi: C \to C$ *a compact mapping. Assume that there are numbers* $0 < r < R$ *for which* (2) *and* (3) *hold. Then* $i_C(\Phi, U) = -1$, *where* $U=\{x\in C: r<\|x\|<R\}$. *In particular,* Φ *has a fixed point in* U. [*Similar considerations can be made if* $0 < R < r$].

<u>Corollary 3.3</u>. (*Compression of a cone*). *Let* $\Phi: C \to C$ *be a compact mapping in the cone* C. *Assume that there are numbers* $0 < R < r$ *such that* (2") *and* (3") *hold. Then* Φ *has a fixed point* $x \in C$, *with* $R < \|x\| < r$.

<u>Corollary 3.4</u> (*Expansion of a cone*). *Let* $\Phi: C \to C$ *be a compact*

mapping in the cone C. *Assume that there are numbers* $0 < r < R$ *such that* (2") *and* (3") *hold. Then* Φ *has a fixed point* $x \in C$, *with* $r < \|x\| < R$.

3.2 DIFFERENTIABLE MAPPINGS IN CONES. A mapping $\Phi: C \to C$ defined in a cone C of a Banach space X is said to be (right) *differentiable at a point* $x \in C$ if there is a bounded linear operator $T: \overline{C-C} \to X$ such that

$$(4) \qquad \lim_{\substack{h \to 0 \\ h \in C}} \frac{\Phi(x+h) - \Phi(x) - Th}{\|h\|} = 0$$

[Recall that $\overline{C-C}$ is the closure of the set $C-C = \{x-y: x,y \in C\}$]. This operator T is called the (right) derivative of Φ at x and it is denoted by $\Phi'_+(x)$.

Proposition 3.5. *If* $\Phi: C \to C$ *is a compact mapping in the cone which is differentiable at* $x \in C$, *then* $\Phi'_+(x)|_C$ *is a compact mapping in* C.

Remark. If the cone C is generating, that is, $X = C-C$, then $\Phi_+(x): \overline{C-C} \to X$ is a compact operator, cf. Krasnoselskii [1, p. 102].

Proof of Proposition 3.5. Suppose by contradiction that there are $h_n \in C$, $\|h_n\| = 1$, such that $\|Th_n - Th_m\| \geq \varepsilon_0 > 0$ for $n \neq m$. It follows from (4) that there is a $\delta > 0$ such that

$$\|\Phi(x+h) - \Phi(x) - Th\| \leq \frac{1}{3}\varepsilon_0\|h\|, \quad \text{for} \quad h \in C, \ \|h\| \leq \delta.$$

Using this expression with $h = h_m$ to estimate the expression $\varepsilon_0\delta \leq \|T(\delta h_n) - T(\delta h_m)\|$, we are led to

$$\|\phi(x + \delta h_n) - \phi(x + \delta h_m)\| \geq \frac{1}{3}\varepsilon_0\delta \quad \text{for all} \quad n \neq m$$

which contradicts the compactness of Φ.

Q.E.D.

A mapping $\Phi: C \to C$ defined in a cone C of a Banach space X is said to be *asymptotically linear* if there is a bounded linear operator $T: \overline{C-C} \to X$ such that

$$\lim_{\substack{\|x\|\to\infty \\ x\in C}} \frac{\Phi(x) - Tx}{\|x\|} = 0. \tag{5}$$

In this case T is called the *asymptotic derivative* of Φ and it is denoted by $\Phi'(\infty)$.

Proposition 3.6. *If* $\Phi: C \to C$ *is compact asymptotically linear, then* $\Phi'(\infty)|_C$ *is a compact mapping in* C. [If C is a generating cone, then $\Phi'(\infty): \overline{C-C} \to X$ is a compact operator, cf. Krasnosel'skii, loc. cit.].

Proof. By contradiction: $x_n \in C$, $\|x_n\| = 1$ and $\|Tx_n - Tx_n\| \geq \varepsilon_0 > 0$. It follows from (5) that there is a $\delta > 0$ such that $\|\phi(x) - Tx\| \leq \frac{\varepsilon_0}{3}\|x\|$ for $\|x\| \geq \delta$. As in the previous proposition we come to the contradictory statement: $\|\Phi(\delta x_n) - \Phi(\delta x_m)\| \geq \frac{1}{3}\varepsilon_0\delta$.

Q.E.D.

Proposition 3.7. *Let* $\Phi: C \to C$ *be a compact mapping in a normal cone* C, *which is differentiable at* 0, *and* $\Phi(0) = 0$. i) *Suppose that the eigenvalue problem*

$$\Phi'_+(0)h = \lambda h, \quad h \in C, \tag{6}$$

has no eigenvalue $\lambda \geq 1$; *then there exists a number* $r_0 > 0$ *such that* (2) *holds for all* $0 < r \leq r_0$. ii) *On the other hand, suppose that* $\lambda = 1$ *is not an eigenvalue of* (6) *and that there is an eigenvalue strictly greater than* 1; *then there exists a number* $R_0 > 0$ *such that* (3') *holds for all* $0 < R \leq R_0$.

Proof. i) By contradiction: there exist $r_n \to 0$, $x_n \in C$, $\|x_n\| = r_n$, $0 < t_n \leq 1$ such that $x_n = t_n\Phi(x_n)$. So

$$\frac{1}{t_n} x_n = \Phi(x_n) = \Phi_+'(0)x_n + o(x_n)$$

or

$$\frac{1}{t_n} y_n = \Phi_+'(0)y_n + \frac{o(x_n)}{\|x_n\|}, \quad y_n = x_n/\|x_n\|.$$

Using Proposition 3.5 we conclude that $t_n^{-1}y_n \to z \in C$ and $t_n \to t^* > 0$. So $z = t^*\Phi_+'(0)z$, which is impossible.

ii) Let $v \in C\backslash\{0\}$ be the eigenvector of (6) corresponding to the eigenvalue $\lambda > 1$ announced in the hypothesis. We claim that (3') holds with this v and some positive R. Let us prove it by contradiction: there exist $R_n \to 0$, $x_n \in C$, $\|x_n\| = R_n$, $t_n \geq 0$ such that $x_n = \Phi(x_n) + t_n v$. Then $v_n = \Phi_+'(0)x_n + o(x_n) + t_n v$. Dividing by $\|x_n\|$ we get $y_n = \Phi_+'(0)y_n + \frac{o(x_n)}{\|x_n\|} + \frac{t_n}{\|x_n\|} v$, which shows that $\{t_n/\|x_n\|\}$ is a bounded sequence. Passing to a subsequence we may assume that it converges to α^*. Using the compactness of $\phi_+'(0)$ it follows that $y_n \to y \in C$ and $y = \Phi_+'(0)y + \alpha^* v$. Iterating we get $\Phi_+'(0)^n y = y - \alpha^*(1 + \lambda + \ldots + \lambda^{n-1})v$, which is impossible (in the case $\alpha^* > 0$), since the left side is in C and the right side is not for large n. Indeed, if it were one would get $\|\alpha^*(1+\lambda+\ldots+\lambda^{n-1})v\| \leq k\|y\|$, where k is the constant of normality of the cone C. Consequently $\alpha^* = 0$ and $y = \Phi_+'(0)y$ which is also a contradiction to the hypothesis.

Q.E.D.

Proposition 3.8. *Let* $\Phi: C \to C$ *be an asymptotically linear compact mapping in a normal cone* C. (i) *Suppose that the eigenvalue problem*

$$(7) \qquad \Phi'(\infty)h = \lambda h, \quad h \in C,$$

has no eigenvalue $\lambda \geq 1$; *then there is a number* $r_0 > 0$ *such that* (2) *holds for all* $r \geq r_0$. (ii) *Suppose that* $\lambda = 1$ *is not an eigenvalue of* (7) *and that there is an eigenvalue strictly greater than* 1; *then there is a number* $R_0 > 0$ *such that* (3') *holds for all* $R \geq R_0$.

Proof. Analogous to the proof of Proposition 3.7.

Application to elliptic problems with asymptotically linear nonlinearities. We shall apply the preceding results to the Dirichlet problem

(8) $$Lu = g(x, u) \text{ in } \Omega, \quad u = 0 \text{ on } \partial\Omega,$$

where L is a second order elliptic operator as in Sections 1.2 and 2.1, and $g: \bar{\Omega}\times R^+ \to R^+$ is a C^α function. We also assume that:

(i) $g(x, 0) = 0$; (ii) $\lim\limits_{s\downarrow 0} \frac{g(x, s)}{s} = a(x)$; (iii) $\lim\limits_{s\to+\infty} \frac{g(x, s)}{s} = m(x)$,

and $a(x)$ and $m(x)$ are L^r functions with $r > N$, and the limits are assumed to be uniform in Ω. Condition (ii) is verified if we ask that the right partial derivative $g_s'(x, 0)$ exists and it is continuous. Let $C_0^0(\bar{\Omega})$ be the Banach space of the continuous real valued functions in $\bar{\Omega}$ which vanish on $\partial\Omega$, with the sup norm. Now let $\Phi: C_0^0(\bar{\Omega}) \to C_0^0(\bar{\Omega})$ be the mapping defined as follows: given $v \in C_0^0(\bar{\Omega})$, then $u = \Phi(v)$ is the solution of

(9) $$Lu = g(x, v) \text{ in } \Omega, \quad u = 0 \text{ on } \partial\Omega.$$

By a solution we mean a function $u \in W^{2,p}$. Since the right side of (9) is continuous, and consequently in L^p for all $p > 1$, we have by the Sobolev imbedding theorem that $u \in C^{1,\alpha}$. This shows that mapping Φ is compact. Also by the strong maximum principle $\Phi(C) \subset C$, where $C = \{u \in C_0^0(\bar{\Omega}): u \geq 0 \text{ in } \Omega\}$.

Lemma 3.9. *The mapping* Φ *is differentiable at* 0 *and its derivative* $\Phi_+'(0)$ *is defined as follows: given* $v \in C$, *let* $u = \phi_+'(0)v$ *be the solution of*

(10) $$Lu = a(x)v \text{ in } \Omega, \quad u = 0 \text{ on } \partial\Omega.$$

Moreover the eigenvalue problem (6) is equivalent to the following eigenvalue problem:

(11) $$Lh = \lambda^{-1}a(x)h, \quad h \geq 0 \quad \text{in} \quad \Omega, \; h = 0 \quad \text{on} \quad \partial\Omega.$$

Proof. It follows from (9) and hypothesis (i) that $\Phi(0) = 0$. From (9) and (10) we obtain

$$L[\Phi(v) - \Phi_+'(0)v] = g(x, v) - a(x)v \quad \text{in} \quad \Omega,$$

and the Dirichlet boundary condition for the function in brackets. From the L^p-estimates and Sobolev imebedding theorem we obtain

$$\|\Phi(v) - \Phi_+'(0)v\|_{L^\infty} \leq \|g(x, v)-a(x)v\|_{L^p}$$

for $p > N$ and the result follows readily from (ii). (11) is immediate.

Q.E.D.

Lemma 3.10. *The mapping* Φ *is asymptotically linear, and* $\Phi'(\infty)$ *is defined as follows: given* $v \in C$, *let* $u = \Phi'(\infty)v$ *be the solution of*

(12) $$Lu = m(x)v \quad \text{in} \quad \Omega, \quad u = 0 \quad \text{on} \quad \partial\Omega.$$

Moreover the eigenvalue problem (7) *is equivalent to the following eigenvalue problem*

(13) $$Lh = \lambda^{-1}m(x)h, \quad h \geq 0 \quad \text{in} \quad \Omega, \quad h = 0 \quad \text{on} \; \partial\Omega.$$

The proof of the above lemma is completely analogous to the proof of Lemma 3.9. Now we can use Propositions 3.7 and 3.8 and make the following assertions with respect to problem (8):

Theorem 3.11. *Suppose that* $g:\bar{\Omega}\times R^+ \to R^+$ *is a* C^α-*function satisfying hypotheses* (i), (ii) *and* (iii) *above. Then problem* (8) *has a positive solution* $u \in C^{2,\alpha}(\Omega)$ *if either one of the following assumptions hold*

(S) $\mu_1(a) > 1$ and $\mu_1(m) < 1$ (Superlinear case)

or

(s) $\mu_1(a) < 1$ and $\mu_1(m) > 1$ (sublinear case)

Remark. Compare this result with Theorem 2.2. Also as in Section 2.1 we may discuss the eigenvalue problem (14) of that section complementing the discussion we started there. We particularize a little bit: let $f\colon \bar{\Omega}\times R^{+} \to R^{+}$ be a C^{α} function, and consider the eigenvalue problem

(14) $$Lu = \lambda f(x, u) \quad \text{in} \quad \Omega, \quad u = 0 \quad \text{on } \partial\Omega.$$

Suppose that $f(x, 0) = 0$ and the limits below exist and as L^{r} functions with $r > N$

$$\lim_{s \downarrow 0} \frac{f(x, s)}{s} = f_{0}(x) \quad \text{and} \quad \lim_{s \to +\infty} \frac{f(x, s)}{s} = f_{\infty}(x).$$

Then it is easy to conclude from Theorem 3.11 that if $\inf f_{0} < \sup f_{\infty}$, then problem (14) has a positive solution for all λ such that $\mu_{1}(1)/\sup f_{\infty} < \lambda < \mu_{1}(1)/\inf f_{0}$.

Remark. Observe that the second half of hypothesis (S) implies through Propositions 3.8 and 3.10 an a priori bound on the solutions of $Lu = g(x, u) + t\phi_{1}$ in Ω, $u = 0$ on $\partial\Omega$, for all $t \geq 0$, where ϕ_{1} is the (positive) eigenfunction corresponding to the first eigenvalue $\mu_{1}(m)$ of (12). In the next two sections we develop techniques to obtain a priori bounds for positive solutions of superlinear problems, which are not necessarily asymptotically linear.

3.3 A PRIORI BOUNDS A LA BRÉZIS-TURNER. Let us consider the Dirichlet problem

(15) $$Lu = g(x, u) \quad \text{in} \quad \Omega, \quad u = 0 \quad \text{on} \quad \partial\Omega.$$

where L is a second order elliptic operator as in Sections 1.2 and 2.1. The results of the present section are due to Brézis-Turner [8] that proved them in the more general case of second order operators which are not self-adjoint. The function $g\colon \bar{\Omega}\times R^{+} \to R^{+}$ is supposed to be continuous and to satisfy the following assumptions:

(16) $$\liminf_{s\to+\infty} \frac{g(x, s)}{s} > \lambda_1, \quad \text{uniformly for} \quad x \in \bar{\Omega},$$

(17) $$\lim_{s\to+\infty} \frac{g(x, s)}{s^{\beta}} = 0, \quad \text{uniformly for} \quad x \in \bar{\Omega}, \quad \text{where} \quad \beta = \frac{N+1}{N-1}.$$

[$\lambda_1 = \mu_1(1)$, see the end of Section 1.2]. Let ϕ_1 be a positive eigenfunction corresponding to the first eigenvalue λ_1; for simplicity let us omit the index 1 and write just ϕ.

Theorem 3.12. *Under the above assumptions, there is a constant (independent of* $t \geq 0$*) such that* $\|u\|_{L^\infty} \leq k$ *for all solutions of*

(18) $$Lu = g(x, u) + t\phi \quad \textit{in} \quad \Omega, \quad u = 0 \quad \textit{on} \quad \partial\Omega.$$

In the proof of the above theorem one uses the *Hardy-Sobolev inequality:* if $u \in H_0^1(\Omega)$, then $u/\phi^\tau \in L^q$, where $q^{-1} = 2^{-1}-(1-\tau)N^{-1}$, $0 \leq \tau \leq 1$, and there is a constant $c > 0$ such that

(19) $$\left\|\frac{u}{\phi^\tau}\right\|_{L^q} \leq c\|\nabla u\|_{L^2}, \quad \forall\, u \in H_0^1(\Omega).$$

Observe that the extreme case $\tau = 0$ is the Sobolev imbedding theorem $H_0^1 \subset L^{2^*}$, where $2^* = 2N/(N-2)$. The other extreme case $\tau = 1$ is a fact already observed in Hardy-Littlewood-Polya [26], (see also Lions-Magenes [27]), that the behavior of a function $u \in H_0^1(\Omega)$ near the boundary $\partial\Omega$ is such that u/ϕ is in $L^2(\Omega)$.

Proof of Theorem 3.12. First step. It follows from (16) that there are $\lambda > \lambda_1$ and an $s_0 > 0$ such that $g(x, s) \geq \lambda s$ for $s \geq s_0$. So there is a constant $c_1 > 0$ such that

(20) $$g(x, s) \geq \lambda s - c_1 \quad \forall\, x \in \Omega, \quad \forall\, s \geq 0.$$

Multiplying (18) by ϕ and integrating by parts one obtains

(21) $$\lambda_1 \int u\phi = \int g(x, u) + t\int \phi^2.$$

Using (20) it follows that

$$\lambda_1\int u\phi \geq \lambda\int u\phi - c_1\int\phi + t\int\phi^2,$$

which implies that there are positive constants such that

$\int u\phi \leq c$ and $\int g(x, u)\phi \leq c$, for all solutions u of (18), and

$t \leq c$, if problem (18) is to have a solution $u \geq 0$.

(We use the same letter c to denote different constants).

<u>Second step</u>. Multiplying equation (18) by u, using the coerciveness of the bilinear form $a[u, u]$ associated with L, and estimating u by Poincaré's inequality we have

$$\|\nabla u\|^2_{L^2} \leq c\int g(x, u)u + c\|\nabla u\|_{L^2}. \tag{22}$$

Now to estimate the first term in the right side of (22) we write

$$\int g(x, u)u = \int g(x, u)^{\alpha}\phi^{\alpha}g(x, u)^{1-\alpha}\phi^{-\alpha}u$$

where the parameter $\alpha \in (0, 1)$ will be determined shortly. Using Hölder's inequality one obtains

$$\int g(x, u)u \leq \left[\int g(x, u)\phi\right]^{\alpha}\left[\int \frac{g(x, u)}{\phi^{\frac{\alpha}{1-\alpha}}}\, u^{\frac{1}{1-\alpha}}\right]^{1-\alpha} \tag{23}$$

The first term in the right side of (23) is bounded. To estimate the second one we use (17): for each $\varepsilon > 0$ there is $c_\varepsilon > 0$ such that $g(x, s) < \varepsilon s^{\beta} + c_\varepsilon$ for all $x \in \bar{\Omega}$ and $s \geq 0$. Then

$$\int g(x, u)u \leq c\varepsilon^{1-\alpha}\left[\int \frac{u^{\beta+\frac{1}{1-\alpha}}}{\phi^{\frac{\alpha}{1-\alpha}}}\right]^{1-\alpha} + c'_\varepsilon\left[\int \frac{u^{\frac{1}{1-\alpha}}}{\phi^{\frac{\alpha}{1-\alpha}}}\right]^{1-\alpha}$$

Taking $\alpha = 2/(N+1)$ and using Hardy-Sobolev inequality we obtain

$$\int g(x, u)u \leq c\varepsilon^{\frac{N-1}{N+1}}\|\nabla u\|^2_{L^2} + c\|\nabla u\|_{L^2}.$$

Using this estimate in (22) we get $\|\nabla u\|_{L^2} \leq c$.

Third step. By the Sobolev imbedding theorem $\|u\|_{L^{2^*}} \le c$. Using a bootstrap argument we conclude that $\|u\|_{L^\infty} \le c$.

Q.E.D.

3.4 SHARPER A PRIORI BOUNDS. In this section we describe a technique due do P.L. Lions, R.D. Nussbaum and the writer [28] for obtaining a priori bounds. It rests on two basic ideas, one is the use of Pohozaev identity [29] for that purpose and the other is the use of results by Gidas, Ni and Nirenberg [30] relative to the maximum principle. Unfortunately there are some restrictions on the class of second order operators L considered as well as on the geometry of the domain Ω. Those are due to the necessity of applying arguments based in [30]. For simplicity we restrict at the present work to the case $L = -\Delta$ and to *convex domains* Ω. More general cases are discussed in [28]. Recently, Gidas and Spruck [32] have obtained similar results by a different technique. So let us consider the Dirichlet problem

(24) $$-\Delta u = f(u) \quad \text{in} \quad \Omega, \quad u = 0 \quad \text{on} \quad \partial\Omega.$$

We assume the following hypotheses on the function f:

(25) $f\colon R^+ \to R^+$ is locally Lipschitzian and

(26) $$\liminf_{s\to+\infty} \frac{f(s)}{s} > \lambda_1,$$

where λ_1 is the first eigenvalue of the Laplacian acting in $H_0^1(\Omega)$. Assumption (26) is just one half of the superlinearity condition for problem (24); the other half, that is, the behavior of f near zero is not needed for the a priori estimates. That will be used only at the moment of proving existence of positive solutions for (24) in Section 3.5.

If Ω is convex then it follows, using arguments from [30], that the following condition holds:

(27) there are positive numbers γ, ε and η (depending only on Ω)

such that for all $x \in \{y \in \bar{\Omega}: d(y, \partial\Omega) < \varepsilon\}$ there exists a measurable set I_x with (i) meas $(I_x) \geq \gamma$, (ii) $I_x \subset \{y \in \bar{\Omega}: d(y, \partial\Omega) > \eta\}$, and (iii) $u(\xi) \geq u(x)$ for all $\xi \in I_x$ and all positive solutions of (24).

Theorem 3.13. *Assume that* f *satisfies* (25) *and* (26) *and that condition* (27) *holds. Suppose the following conditions in addition:*

(28) $\lim_{s\to+\infty} \frac{f(s)}{s^\sigma} = 0$, *where* $\sigma = \frac{N+2}{N-2}$ *if* $N \geq 3$ (*or* $\sigma < \infty$ *if* $N = 2$), *and*

(29) $$\limsup_{s\to+\infty} \frac{sf(s)-\theta F(s)}{s^2 f(s)^{2/N}} \leq 0 \text{ for some } 0 \leq \theta < \frac{2N}{N-2},$$

where $F(s) = \int_0^s f$. [*In the case of* $N = 2$ *this condition is vacuous*]. *Then there is a constant* c *such that* $\|u\|_{L^\infty} \leq c$ *for all positive solutions of* (24).

Remarks: Condition (29) is verified in the special case when

(30) $$\limsup_{s\to+\infty} \frac{f(s)}{s^\tau} = 0, \quad \text{where} \quad \tau = \frac{N}{N-2}.$$

Indeed, taking $\theta = 0$, (29) reduces to $\limsup\, [f(s)s^{-\tau}]^{1/\tau} \leq 0$, which is an obvious consequence of (30). Observe that if (17) holds then (28) and (29) are verified. So the above theorem extends Theorem 3.12 at least in the special case of the Laplacian in convex domains.

The proof of Theorem 3.13 is split in a series of lemmas in the sequel.

Lemma 3.14. *There is a constant* $c > 0$ *such that*

(31) $\int u\phi \leq c$ *and* $\int f(u)\phi \leq c$, *for all solutions* u *of* (24).

This is precisely the contents of the first step in the proof of Theorem 3.12. Observe that it depends only on hypothesis (26). [The above statement implies that both u and $f(u)$ are bounded in $L^1_{loc}(\Omega)$].

Lemma 3.15. u *and* ∇u *are bounded in the* L^∞*-norm in a neighborhood of* $\partial\Omega$*, for all solutions* u *of* (24).

Proof. Using (27) and (31) we have that for all $x \in \{y \in \bar{\Omega}: d(y, \partial\Omega) < \varepsilon\}$

$$c \geq \int_{I_x} u(\xi)\phi(\xi) \geq u(x)[\min_{\xi \in I_x} \phi(\xi)]\gamma.$$

Also from (27) we see that there is a positive constant c (independent of x) such that $\min_{I_x} \phi(\xi) \geq c$. We will be using the same letter c to denote distinct constants. Then

$$(32) \qquad u(x) \leq c \quad \text{for all} \quad x \in \{y \in \bar{\Omega}: d(y, \partial\Omega) < \varepsilon\}.$$

Next we use a theorem of the Giorgi-Nash type (see Theorem 14.1 of [31, p. 201]) to deduce that there is $0 < \alpha < 1$ such that

$$\|u\|_{C^{0,\alpha}(\Omega_1)} \leq c, \quad \text{where} \quad \Omega_1 = \left\{y \in \Omega: \frac{\varepsilon}{4} < d(y, \partial\Omega) < \frac{3\varepsilon}{4}\right\}.$$

Now using Schauder interior estimates (see Theorem 6.2 of [16, p. 85]) we have for $\frac{\varepsilon}{4} < r_1 < \frac{\varepsilon}{2} < r_2 < \frac{3\varepsilon}{4}$ that

$$\|u\|_{C^{2,\alpha}(\Omega_2)} \leq c, \quad \text{where} \quad \Omega_2 = \left\{y \in \Omega: r_1 < d(y, \partial\Omega) < r_2\right\}.$$

Finally Schauder boundary estimates gives

$$\|u\|_{C^{2,\alpha}(\Omega_3)} \leq c, \quad \text{where} \quad \Omega_3 = \left\{y \in \Omega: d(y, \partial\Omega) < r_2\right\},$$

which implies

$$(33) \qquad |\nabla u(x)| \leq c \quad \text{for all} \quad x \in \left\{y \in \bar{\Omega}: d(y, \partial\Omega) < \frac{\varepsilon}{2}\right\}.$$

Thus (32) and (33) give the result. Q.E.D.

Remark. (31) and (32) imply that $\|f(u)\|_{L^1} \leq c$ for all positive solutions of (24).

Lemma 3.16 (*Pohozaev identity* [29]). *Let* $u \in C^2(\bar{\Omega})$ *be a solution of* (24). *Then*

(34) $$2N\int_\Omega F(u) - (N-2)\int_\Omega f(u)u = \int_{\partial\Omega} (x\cdot\nu)|\nabla u|^2,$$

where $\nu = \nu(x)$ *is the outward unit normal to* $\partial\Omega$ *at* x.

Proof. The idea is to multiply equation (24) by $x.\nabla u$ and integrate, using next the divergence theorem. For that matter we first write down some terms as divergences:

$$\Delta u(x.\nabla u) = \frac{N-2}{2}|\nabla u|^2 + \sum_i\Big(\sum_j x_j u_{x_i} u_{x_j}\Big)_{x_i} - \frac{1}{2}\sum_j\Big(\sum_i x_j u_{x_i}^2\Big)_{x_j}$$

and

$$f(u)(x\cdot\nabla u) = \sum_j \Big(x_j F(u)\Big)_{x_j} - NF(u).$$

So we obtain

$$\frac{2-N}{2}\int_\Omega |\nabla u^{\;2} + \frac{1}{2}\int_{\partial\Omega}\sum_{i,j} x_j\nu_j u_{x_i}^2 - \int_{\partial\Omega}\sum_{i,j} x_j\nu_i u_{x_i}u_{x_j} = \int_{\partial\Omega}\sum_i x_j\nu_j F(u) - N\int_\Omega F(u),$$

which gives readily identity (34), by observing that

$$\int_\Omega f(u)u = -\int_\Omega \Delta u.u = \int_\Omega |\nabla u|^2. \qquad \text{Q.E.D.}$$

Lemma 3.17. *The following estimate on the positive solutions of* (24) *holds:*

(35) $$\|\nabla u\|_{L^2} \le c$$

Proof. i) Case $N \ge 3$. Using (33) and (34) we deduce that

$$\left|2N\int_\Omega F(u) - (N-2)\int_\Omega uf(u)\right| \le c$$

which is used in the estimate below:

(36) $$\int_\Omega |\nabla u|^2 = \int_\Omega uf(u) = \frac{1}{1-\lambda}\int_\Omega [uf(u)-\theta F(u)] - \frac{\lambda}{1-\lambda}\int_\Omega [uf(u) - \frac{2N}{N-2}F(u)]$$
$$\le c\int_\Omega |uf(u)-\theta F(u)| + c$$

where λ is chosen in such a way that $\theta = \lambda\frac{2N}{N-2}$. So $0 \le \lambda < 1$.

Now we use hypothesis (29) [n.b.: that is the only point where we use this hypothesis]: given $\varepsilon > 0$ there is $s_0 > 0$ such that $sf(s) - \theta F(s) \leq \varepsilon s^2 f(s)^{2/N}$ for $s \geq s_0$. Thus we obtain from (36):

$$\int_\Omega |\nabla u|^2 \leq c_\varepsilon + c\varepsilon \int_\Omega u^2 f(u)^{2/N}. \tag{37}$$

To estimate the last term in the right side of (37) we use Hölder's inequality, the Sobolev imbedding theorem and the boundedness of $f(u)$ in L^1:

$$\int_\Omega u^2 f(u)^{2/N} \leq \left|\int f(u)\right|^{2/N} \left|\int_\Omega u^{\frac{2N}{N-2}}\right|^{\frac{N-2}{N}} \leq c\|\nabla u\|^2_{L^2}.$$

This estimate together with (37) gives (35), by taking $\varepsilon > 0$ in such a way that $\varepsilon c < 1$.

ii) Case $N = 2$. Choose $0 < \gamma < 1$ and use hypothesis (28) with $\sigma = 1/(1-\gamma)$, which then implies

$$\lim_{s\to+\infty} \frac{sf(s)}{s^2 f(s)^\gamma} = 0$$

Consequently for $\varepsilon > 0$ there is an $s_0 > 0$ such that $sf(s) \leq \varepsilon s^2 f(s)^\gamma$ for $s \geq s_0$. This is then used in the estimate below

$$\int_\Omega |\nabla u|^2 = \int_\Omega uf(u) \leq c_\varepsilon + \varepsilon \int_\Omega u^2 f(u)^\gamma,$$

and we proceed as above. Q.E.D.

Lemma 3.18. *The following estimate holds for all positive solutions of* (24):

$$\|u\|_{L^\infty} \leq c \tag{38}$$

Proof. i) Case $N = 2$. It follows from (35) using the Sobolev imbedding theorem that $\|u\|_{L^p} < c$ for all $1 \leq p < \infty$. Then in virtue of (28) we see that $\|f(u)\|_{L^p} \leq c$ for all p. By the L^p theory of the Dirichlet problem and again Sobolev imbedding theorem we conclude that $\|u\|_{L^\infty} \le c$.

ii) Case $N \ge 3$. Multiplying equation (24) by u^p, $p > 1$, and integrating we have

$$p\int_\Omega |\nabla u|^2 u^{p-1} = \int_\Omega f(u)u^p$$

or

$$\frac{4p}{(p+1)^2}\int_\Omega \left|\nabla u^{\frac{p+1}{2}}\right|^2 = \int_\Omega f(u)u^p. \tag{39}$$

Now using (28) we estimate the right side of (39) to get

$$\int_\Omega \left|\nabla u^{\frac{p+1}{2}}\right|^2 \le c_\varepsilon + c\varepsilon\int_\Omega u^{\sigma-1}u^{p+1}. \tag{40}$$

Using the Sobolev imbedding theorem to estimate the left side of (40) and Hölder's inequality to the right side we have

$$\left(\int_\Omega u^q\right)^{\frac{N-2}{N}} \le c_\varepsilon + c\varepsilon\left(\int_\Omega u^{\frac{2N}{N-2}}\right)^{2/N}\left(\int_\Omega u^q\right)^{\frac{N-2}{N}}, \quad q = \frac{N(p+1)}{N-2}.$$

Using the Sobolev imbedding theorem once more and (35) we conclude that

$$\|u\|_{L^q} \le c$$

by taking $\varepsilon c < 1$. We may choose $p > 1$ in such a way that $q > \frac{N}{2}\sigma$. Then $\|f(u)\|_{L^{\frac{q}{\sigma}}} \le c$ with $\frac{q}{\sigma} > \frac{N}{2}$. By the regularity theory and Sobolev again we get

$$\|u\|_{L^\infty} \le c.$$

Q.E.D.

3.5 EXISTENCE OF POSITIVE SOLUTIONS FOR SUPERLINEAR PROBLEMS.

The following result is due to P.L. Lions, R.D. Nussbaum and the writer [28].

Theorem 3.19. *Let* $f: R^+ \to R^+$ *be a function satisfying* (25), (26), (28), *and* (29), *with* $f(0) = 0$. *Suppose that* (27) *holds and*

$$\limsup_{s \downarrow 0} \frac{f(s)}{s} < \lambda_1. \tag{41}$$

Then problem (24) *has at least one positive solution.*

Proof. Let $X = \{u \in C^0(\bar{\Omega}) : u = 0 \text{ on } \partial\Omega\}$ and $C = \{u \in X : u \geq 0\}$. Let us denote by $\Phi: X \to X$ the mapping defined as follows: for each $v \in X$ let $u = \Phi(v)$ be the solution of $-\Delta u = f(v)$ in Ω and $u = 0$ on $\partial\Omega$. It is well known that $u \in X$ and that the mapping Φ is compact. By the strong maximum principle $\Phi(C) \subset C$. Now we proceed in order to apply Theorem 3.1. Using (41) let $0 < \alpha < \lambda_1$ be such that $f(s) \leq \alpha s$ for $0 \leq s \leq s_0$, some $s_0 > 0$. Define the linear operator $A: X \to X$ as follows: for each $v \in X$ let $u = Av$ be the solution of $-\Delta u = \alpha v$ in Ω and $u = 0$ on $\partial\Omega$. It follows that A is bounded and its spectral radius $r_\sigma(A) < 1$. Also by the maximum principle $A(C) \subset C$. Now observe that for $\|u\|_X \leq s_0$ one has $\Phi(u) \leq Au$. So condition (2'), which implies condition (2) of Theorem 3.1, holds. Next we verify condition (3) of that theorem. Let $F: C \times [0, \infty) \to C$ be defined as follows: for each $v \in C$ and $t \geq 0$ let $u = F(v, t)$ be the solution of $-\Delta u = f(v+t)$ in Ω and $u = 0$ on $\partial\Omega$. Now an argument similar to the one in lemma 3.14 implies that $u = F(u, t)$ has no solution $u \in C$ for $t \geq t_0$, some convenient $t_0 > 0$. By the a priori estimates of the previous section there is a constant $R > 0$ such that $u \neq F(u, t)$ for all $\|u\| > R$ with $u \in C$ and all $t \geq 0$. This concludes the proof of the theorem.

Q.E.D.

Now we show how to obtain existence of a positive solution for (24) without assuming (29). In this case we have to proceed differently since we do not know if there are a priori bounds in this case. The idea here is to use the techniques developed in the previous section plus some variational methods.

Theorem 3.20. [28] *Let* $f: R^+ \to R^+$ *be a function satisfying* (25),

(26) *and* (28), *with* $f(0) = 0$. *Assume that* (27) *and* (41) *hold. Then* (24) *has a positive solution.*

Proof. First step: Construction of a sequence of approximate problems. Let us choose, using (26), a sequence $s_n \to +\infty$ such that $s^{-1}f(s) \geq \lambda_1+\beta$ for all $s \geq s_n$ and some fixed $\beta > 0$. Let $1 < \gamma < (N+2)/(N-2)$ be some fixed constant, and define

$$f_n(s) = \begin{cases} 0 & \text{for } s \leq 0 \\ f(s) & \text{for } 0 \leq s \leq s_n \\ f(s_n) + \dfrac{f(s_n)}{s_n^\gamma}(s-s_n)^\gamma & \text{for } s \geq s_n. \end{cases}$$

It easy to see that f_n satisfies conditions (25), (26), (28), (29) and also $\lim_{s\to+\infty} s^{-1}f_n(s) = +\infty$. Moreover there exist $0 < \theta < \frac{1}{2}$ and $t_n > 0$ such that

(42) $$F_n(s) \leq \theta s f_n(s) \quad \text{for} \quad s \geq t_n$$

where $F_n(s) = \int_0^s f_n$. The approximate problems considered are

(43) $$-\Delta u_n = f_n(u_n) \quad \text{in} \quad \Omega, \quad u_n = 0 \quad \text{on} \quad \partial\Omega.$$

We shall need positive solutions of (43) given by a variational method, namely the mountain pass theorem of Ambrosetti-Rabinowitz [7, Thm 2.1].

Second step. Let us consider the functional $J_n: H_0^1(\Omega) \to R$ defined by

$$J_n(v) = \int_\Omega \left[\frac{1}{2} |\nabla v|^2 - F_n(v)\right].$$

In view of the conditions on f_n it follows that J_n is a well defined C^1 functional. Now we observe that J_n satisfies the hypotheses of Theorem 2.1 in [7], see Lemmas 3.3, 3.4 and 3.6 of that paper. In particular there exists an $R > 0$ (independent of n) such that $J_n(R\phi) \leq 0$, where ϕ is the first eigenfunction of $-\Delta$. So

$$b_n = \inf_{g\in\Gamma} \sup_{t\in[0,1]} J_n[g(t)]$$

is a critical value of J_n, where $\Gamma = \{g\in C([0,1]; H_0^1(\Omega)) : g(0) = 0, g(1) = R\phi\}$. It follows readily that $0 \leq b_n \leq c$ for some constant c independently of n; indeed, use the path $g(t) = tR\phi$. Observe that the critical point u_n corresponding to the critical value b_n which is a solution of (43), belongs in fact to $C^{2,\alpha}(\bar{\Omega})$.

<u>Third step</u>. There is a constant $c > 0$ (independent of n) such that $\|\nabla u_n\|_{L^2} \leq c$. Indeed, using arguments similar to the ones employed in the proofs of Lemmas 3.14 and 3.15 we see that $|\nabla u_n(x)| \leq c$ for all $x \in \partial\Omega$, uniformly in n. Using this information we obtain from Pohozaev identity

$$\frac{1}{2}\int_\Omega |\nabla u_n|^2 - \frac{N}{N-2}\int_\Omega F_n(u_n) \geq -c.$$

This together with

$$0 \leq b_n = \frac{1}{2}\int_\Omega |\nabla u_n|^2 - \int_\Omega F_n(u_n) \leq c$$

gives the result. Since (28) is satisfied uniformly on n, we obtain as in Lemma 3.8 that $\|u_n\|_{L^\infty} \leq c$.

<u>Forth step</u>. Passage to the limit in (43). Since (41) holds uniformly in n we see that there exists an $\delta > 0$ such that $\|u_n\|_{L^\infty} \geq \delta$. So, for large n, u_n is a positive solution of (24).

Q.E.D.

REFERENCES

[1] M. A. Krasnosels'kii, "Positive solutions of Operator equations". P. Noordhoff, Groningen (1964).

[2] Herbert Amann, "Fixed point equation and nonlinear eigenvalue problems in ordered Banach spaces", S.I.A.M. Review 18 (1976), p. 620-709.

[3] Antonio Ambrosetti, "On exact number of positive solutions of convex nonlinear problems". Bolletino U.M.I. vol 15-A (1978) p. 610-615.

[4] Antonio Ambrosetti - Giovanni Mancini, "Sharp nonuniqueness results for some nonlinear problems". Nonlinear Analysis, Theory, Methods & Applications, vol. 3 (1979), p. 635-645.

[5] Henri Berestycki, "Contributions à l'étude des problèmes aux limites elliptiques non-linéaires". Thèse de Doctorat d'État ès Sciences Mathématiques présentée à l'Université Pierre et Marie Curie, Paris (Février 1980).

[6] Peter Hess - Tosio Kato, "On some linear and nonlinear eigenvalue problems with an indefinite weight function". Comm. P.D.E. 5 (1980) p. 999-1030.

[7] Antonio Ambrosetti - Paul H. Rabinowitz, "Dual variational methods in critical point theory and applications", J. Functional Analysis 14 (1973), p. 369-381.

[8] Haim Brezis - Robert E. L. Turner, "On a class of superlinear elliptic problems", Comm. in P.D.E., 2 (1977), p. 601-614.

[9] Michael G. Crandall - Paul H. Rabinowitz, "Some continuation and variational methods for positive solutions of nonlinear elliptic eigenvalue problems", Arch. Rat. Mech. Anal. 58 (1975), p. 207-218.

[10] Roger D. Nussbaum, "Positive Solutions of nonlinear elliptic boundary value problems", J. Math. Anal. Appl. 51 (1975), p. 461-482.

[11] F. Mignot - Jean Pierre Puel, "Sur une classe de problèmes non linéaires avec nonlinearité positive, croissante, convexe". To appear.

[12] Robert E. L. Turner, "A priori bounds for positive solutions of nonlinear elliptic equations in two variables", Duke Math. J. 41 (1974), p. 759-774.

[13] Adele Manes - Ana Maria Micheletti, "Un'estensione della teoria variazionale classica degli autovalori per operatori ellittici del secondo ordine". Bolletino U.M.I. vol. 7 (1973) p. 285-301.

[14] Guido Stampacchia, "Equations Elliptiques du Second Ordre à Coefficients Discontinues". Les Presses de l'Université de Montréal (1966).

[15] Guido Stampacchia, "Le problème de Dirichlet pour les équations elliptiques du second ordre à coefficients discontinus". Ann. Inst. Fourier 15 (1965), p. 189-258.

[16] David Gilbarg - Neil S. Trudinger, "Elliptic Partial Differential Equations of Second Order". Springer-Verlag (1977).

[17] Herbert Amann, "On the existence of positive solutions of nonlinear elliptic boundary value problems". Indiana Univ. Math. J. vol. 21 (1971), p. 125-146.

[18] David H. Sattinger, "Monotone methods in nonlinear elliptic and parabolic boundary value problems". Indiana Univ. Math. J. vol. 21 (1972), pp. 979-1000.

[19] David H. Sattinger, "Topics in Stability and Bifurcation Theory". Lecture Notes vol. 309, Springer Verlag (1973).

[20] Antonio Ambrosetti - Peter Hess, "Positive solutions of Asymptotically Linear Elliptic Eigenvalue Problems". Journal of Math. Anal. Appl. vol 73 (1980), pp. 411-422.

[21] Melvyn S. Berger, "Nonlinear Problems with exactly three solutions". Indiana University Mathematics Journal. Vol 28 (1979), p. 689-698.

[22] Alan C. Lazer - P. J. McKenna, "On multiple solutions of a nonlinear Dirichlet problem". Differential Equations. Academic Press (1980) p. 199-214.

[23] Djairo Guedes de Figueiredo, "Lectures on Boundary Value Problems of the Ambrosetti-Prodi type". Atas do 12º Seminário Brasileiro de Análise. São José dos Campos S.P. (Oct. 1980).

[24] Henri Berestycki, "Méthodes Topologiques et Problèmes aux Limites non-linéaires". Université de Paris (1975).

[25] James Dugundji, "An extension of Tietze's theorem" Pacific J. Math. vol. 1 (1951), pp. 353-367.

[26] G. H. Hardy - J. Littlewood - G. Polya, "Inequalities". Cambridge University Press.

[27] J.-L. Lions - G. Magenes, "Nonhomogeneous Boundary Value Problems and Applications", Vol. 1. Springer, Berlin (1970).

[28] Djairo Guedes de Figueiredo - Pierre-Louis Lions - Roger D. Nussbaum, "A priori estimates for positive solutions of semilinear elliptic equations". To appear.

[29] S. I. Pohozaev, "Eigenfunctions of $\Delta u+\lambda f(u) = 0$", Soviet Math. Dokl. 6 (1965), p. 1408-1411.

[30] Basilis Gidas - Wei ming Ni - Louis Nirenberg, "Symmetry and related properties via the maximum principle", Comm. Math. Phys. 68 (1979), p. 209-243.

[31] O. Ladyzhenskaya - N. Uraltseva, "Linear and Quasilinear Elliptic Equations", translated by Scripta Technica, Academic Press, New York, 1968.

[32] Basilis Gidas - Joel Spruck, "A priori bounds for positive solutions of nonlinear elliptic equations". Comm. P.D.E. vol 6 (1981) p. 883-901.

A REGULARITY THEOREM FOR INVERSE BOUNDED AND ACCRETIVE OPERATORS IN ABSTRACT HILBERT SPACE

Saul Hahn-Goldberg[1]

Centro de Investigación del IPN
Departamento de Matemáticas
Apartado Postal 14-740
México 14, D.F.

Accretive linear differential equations or symmetric positive systems were originally introduced by Friedrichs [2] in order to handle problems of mixed type. Nonlinear accretive equations have been studied by the author [6], [7] in the case the domain is the full space $\mathbb{R}^n$ or the torus, i.e. the problem with periodic coefficients.

An important step for the construction of the solutions uses a special regularity Theorem which involves the inverse boundedness of a certain pseudodifferential operator associated with the original equations. This argument obtains a weak solution which already has built-in the required regularity properties.

In this note we shall state a result which gives an abstract version of the above regularity Theorem. This is a synthesis, in abstract space, of diverse regularity arguments employed by several

1 The research in this paper was partially supported by the "Consejo Nacional de Ciencia y Tecnología" (CONACyT) under project PCCBNAL 790015.

authors, [4], [7], [8]. By specializing the Hilbert spaces involved to adequate functions spaces one obtains, in particular, either Lax's a priori inequality with respect to a norm of negative order or the inverse boundedness of the pseudodifferential operator as stated in [4], [6], [7].

We proceed now to describe the regularity Theorem in abstract space.

Let L be a closed linear map from the Hilbert space K into itself, with domain $\mathcal{D}_L$ dense in K, $L:\mathcal{D}_L \subset K \to K$. We denote by $L^{\dagger}$ a formal adjoint map to L, $L^{\dagger}:\mathcal{D}_{L^{\dagger}} \subset K \to K$, which satisfies the relation

$$<L^{\dagger}u,v> = <u,Lv> \quad \text{for all} \quad u,v \in \mathcal{D}_L, \mathcal{D}_{L^{\dagger}}.$$

The bracket denotes the inner product in K. The operator $L^{\dagger}$ is also assumed to be closed and the domain $\mathcal{D}_L \cap \mathcal{D}_{L^{\dagger}}$ dense.

If the relation

(I)... $<(L+L^{\dagger})u,u> \geqq 2c<u,u>, \quad \forall\, u \in \mathcal{D}_L \cap \mathcal{D}_{L^{\dagger}},$ holds for some

positive constant c then L is said to be accretive, cf. [2], [4]. The operator L is said to be inverse bounded if the inequality $\|Lu\| \geqq c\,\|u\|$ holds for all $u \in \mathcal{D}_L$, with c a positive constant.

Applying Schwarz's inequality to the left side of (I) the following two relations are obtained:

(a) $\|Lu\| \geqq c\,\|u\|$

and

(b) $\|L^{\dagger}u\| \geqq c\,\|u\|$

for all $u \in \mathcal{D}_L \cap \mathcal{D}_{L^{\dagger}}$.

The first relation, (a), implies that the operator L is one to one, i.e, the equation $Lu = f$, $f \in K$, has at most one solution.

The second relation, inequality (b), implies the existence of a weak solution $\tilde{u}$, i.e. $\tilde{u} \in \mathcal{D}_{L^{\dagger *}}$ or $\langle L^{\dagger}u,\tilde{u}\rangle = \langle u,f\rangle$, $\forall\, u \in \mathcal{D}_{L^{\dagger}}$. (The symbol $L^{\dagger *}$ denotes the strict adjoint to the formal adjoint $L^{\dagger}$). We shall reinterpret the equation $Lu = f$ is some appropriate subspace Z of K which is 'smoother' than K, and obtain a weak solution in Z. This is done by obtaining the inverse boundedness of some suitable operator associated with L. Let Z be a dense subset of K equipped with an inner product $\langle\ ,\ \rangle_Z$ so that Z is a Hilbert space with respect to this form. Furthermore, we assume that the norm $\|\cdot\|_Z$, induced by this inner product, dominates the original norm in K, i.e, $\|\cdot\|_Z \geqq \|\cdot\|_K$.

Let $\mathcal{D}_\infty \subset Z \cap \mathcal{D}_L \cap \mathcal{D}_{L^{\dagger}}$ be a densely defined subset, and $\tilde{L}_Z$ the restriction of the orginal map L to $\mathcal{D}_\infty$ so that

$$\tilde{L}_Z : \mathcal{D}_\infty \subset Z \to Z; \quad \tilde{L}_Z = L|\mathcal{D}_\infty .$$

If u_r is a sequence in Z such that $u_r \to 0$ and $\tilde{L}_Z u_r \to u$, with convergence with respect to Z, then the convergence of both sequences also holds with respect to the norm in K. Since L is closed in K it follows that $u = 0$. This trivial remark shows that $\tilde{L}_Z$ is preclosed with respect to Z. We denote by L_Z its closure with domain $\mathcal{D}_{L_Z}$,

$$L_Z : \mathcal{D}_{L_Z} \subset Z \to Z.$$

Because of the assumed properties of the subspace Z the following basic facts of functional analysis follow, cf. [3].

There exists a subspace K_π dense in Z, and an operator Π mapping K_π onto K, with bounded inverse B, $\|B\| \leqq 1$ from K onto K_π such that

(II)... $\langle \phi_Z, \phi_\pi \rangle_Z = \langle \phi_Z, \Pi\phi_\pi \rangle_K$, $\forall\ \phi_Z, \phi_\pi \in Z, K_\pi$.

Furthermore, Π and B are self adjoint maps. For each $\phi \in K$, identity (II) implies

$$\text{(III)}\ldots \quad \langle\phi_Z,\phi\rangle_K = \langle\phi_Z,B\phi\rangle_Z, \quad \forall\ \phi,\phi_Z \in K,\ K_Z.$$

We claim now that the map $\tilde{L}_Z^{\dagger} \equiv B\,L^{\dagger}\Pi$ predefined in the domain $\mathcal{D}_{\tilde{L}_Z^{\dagger}} = \{B\ v:\ v \in \mathcal{D}_{L^{\dagger}}\}$ is a formal adjoint to the operator L_Z with respect to the space Z.

Indeed, let $u,v \in \mathcal{D}_L,\ \mathcal{D}_{\tilde{L}_Z^{\dagger}}$, and $V = Bv$.

Then

$$\langle\tilde{L}_Z^{\dagger}\ V,u\rangle_Z = \langle BL^{\dagger}\Pi Bv,u\rangle_Z = \langle BL^{\dagger}v,u\rangle_Z$$

$$= \langle L^{\dagger}v,u\rangle_K = \langle\Pi V,L_Z u\rangle_K$$

$$= \langle V,L_Z u\rangle_Z.$$

With the spaces Z, K and the maps B, L as above we can now easily prove the following

<u>Theorem</u>. Let L be an accretive operator.

Then the condition

$$(*)\ldots \quad \langle BL^{\dagger}u,L^{\dagger}u\rangle_K \geqq c\ \langle Bu,u\rangle_K, \quad \forall\ u \in \mathcal{D}_{L^{\dagger}} \text{ with } c$$

some positive constant implies that the equation $Lu = f$, for each $f \in Z$, possesses a weak solution $\tilde{u} \in Z$.

<u>Remark</u>. By a weak solution here it is implied a weak solution with respect to the space K where the equation $Lu = f$ is originally defined.

<u>Proof</u>. Let $v = Bu$, with $u \in \mathcal{D}_{L^{\dagger}}$. Then substituting in relation (*),

(IV)... $\langle BL^{\dagger}\Pi v, L^{\dagger}\Pi v\rangle_K \geqq c\ \langle v, \Pi v\rangle_K$.

We now apply identity (II) to the right side of (IV), and identity (III) to the left hand side of inequality (IV) to obtain,

$$\|L_Z^{\dagger}v\|_Z \geqq c^{1/2}\|v\|_Z.$$

Thus, assumption (*) implies the inverse boundedness, with respect to Z, of the map $L_Z^{\dagger}$ which as shown before is a formal adjoint to L_Z with respect to Z. Hence the equation $L_Z u = f$, with $f \in Z$, possesses a weak solution $u^\wedge \in Z$. It is easy to see that $u^\wedge$ is also a weak solution of the original equation $Lu = f$, i.e. $u^\wedge$ is a weak solution with respect to the original space K. Indeed, with f and $u^\wedge$ in Z,

$$\langle f, V\rangle_Z = \langle u^\wedge, L_Z^{\dagger}V\rangle_Z, \quad \forall\ V \in \mathcal{D}_{L_Z^{\dagger}} \supset B\mathcal{D}_{L^{\dagger}}.$$

Let $V = Bv$, with $v \in \mathcal{D}_{L^{\dagger}}$. Then

$$\langle f, BV\rangle_Z = \langle u^\wedge, BL^{\dagger}v\rangle_Z \quad \text{or} \quad \langle f, v\rangle_K = \langle u^\wedge, L^{\dagger}v\rangle_K.$$

Thus $u^\wedge$ is a weak solution with respect to K.

Remark. The fact that the function $u^\wedge \in Z$ is a weak solution with respect to K is quite important in applications where it is common to take K as the L^2-space. Friedrichs [1] has shown that for a large class of partial differential equations of the first order, weak solutions with respect to the L^2-space, are also strong solutions.

Examples.

Let $s \geqq 0$, $K = L^2$ and $Z = K_s$, the usual Sobolev space. The underlying space is $\mathbb{R}^n$. Then in this case $Z \subset K$, $\|\cdot\|_Z \geqq \|\cdot\|_K$ and the map Π is simply given by

$$\Pi = (I-\Delta)^s : K_{2s} \to L^2$$

with $K_\pi = K_{2s}$. The condition (*) is given by

$$<(I-\Delta)^{-s}L^{+}u, L^{+}u>_{L^2} \geqq c <(I-\Delta)^{-s}u,u>_{L^2}.$$

This condition is equivalent to Lax's a priori inequality of negative order, $\|L^{+}u\|_{-s} \geqq c\,\|u\|_{-s}$, $\forall\ u \in \mathcal{D}_{L^{+}}$. This follows immediately from the identity $\|u\|_{-s} = \|(I-\Delta)^{-s}\ u\|_s$. In this example the adjoint map L_Z^{+} is given by the pseudo-differential operator $L_Z^{+} = BL^{+}\Pi = (I-\Delta)^{-s}L^{+}(I-\Delta)^{s}$.

<u>Example</u>. cf [4].

Let $A:\mathcal{D}_A \subset K \to K$ be a closed densely defined operator in a Hilbert space K. In $\mathcal{D}_A$ we can define a bilinear form $<u,v>_A = <Au,Av> + <u,v>$, with u,v in $\mathcal{D}_A$. Then $\mathcal{D}_A$ is a Hilbert space with respect to $< , >_A$.

Here $\mathcal{D}_\pi = \mathcal{D}_{A*A}$, with A^* the adjoint (strict) map to A and $\Pi = (A^*A+I)$.

Then with $B = (A^*A+I)^{-1}$, the condition

$<BL^{+}u,u>_K \geqq <Bu,u>_K$ implies the existence of a weak solution $u\hat{}\ \in \mathcal{D}_A$ of the equation $Lu = \delta$. As a special case take A = gradient, i.e $u \to \text{grad}\ u = (\partial_{x_1}u,\dots\partial_{x_m}u)$ (say in $C_0^\infty(\mathbb{R}^m)$). A formal adjoint to A is the divergence map, $A^{+}v = -\sum_{\mu=1}^{m} \partial_{x_\mu} v^{(\mu)}$ where $v = (v^{(1)},\dots,v^{(m)})$, i.e

$-<\text{grad}\phi,\varphi>_{L^2} = <\phi,\text{div}\,\varphi>_{L^2}$. Taking the corresponding closed extensions of these maps, the operator Π is given by $\Pi = (A^*A+I) = (I-\Delta)$ with Δ = Laplacian, which corresponds to the previous example for $\delta = 1$.

In applications to differential equations one has to give local conditions on the coefficients which will assure the validity of the global a priori inequality (*) in the Theorem above. In [6], [7] we state these local conditions for the nonlinear equations

(V)... $\sum_{\mu=1}^{m} A^{\mu}(x,u)\partial_{x_\mu} u+B(x,u) = 0$, where A^{μ} are $k\times k$ symmetric matrices, $\mu = 1,\dots,m$. The vector x lies in the full space $\mathbb{R}^m$, and $B = B(x,y)$ is a $k\times 1$ vector with $y = \{y^{\kappa}:\kappa = 1,\dots,k\}$ a vector belonging to some neighbourhood V of the origin in $\mathbb{R}^{\kappa}$. The k-vector $u = u(x)$ stands for the unknown function.

We briefly restate these conditions here.

With systems (V) we associate a generalized linear system with coefficients parametrized by certain given functions w (These functions w belong to a small neighbourhood of zero in K_s, with $s > \frac{m}{2} + 1$),

(VI) $\sum_{\mu=1}^{m} \hat{A}^{\mu}(x,w)\partial_{x_\mu} u+B_y(x,w)u = B_y(x,w)w-B(x,w) \equiv f(x,w)$

where B_y is the Jacobian matriz

$$B_y(x,y) = \left\{\partial_{y_{\kappa'}} B^{(\kappa)}(x,y);\ \kappa,\kappa' = 1,2\dots k\right\}.$$

Under suitable conditions system (VI) defines a linear transformation mapping the function w into the solution u which as shown in [7] has a fixed point $\tilde{u} \in K_s$, which is in fact a strong solution of (V). To obtain the corresponding local accretiveness conditions, we formally differentiate equations (VI) to obtain

$$\sum_{\mu=1}^{m} \hat{A}^{\mu}(x,w)\partial_{x_\mu} U^s + P_0(x,w)U^s + P_1(x,W^s)U^s = f^{(s)}(x,W^s).$$

The vector U^s, W^s and $f^{(s)}$ denote the vectors formed by taking u and all its distinct derivatives up to order s respectively. The matrix $\hat{A}^\mu$ can be expressed in the form $\hat{A}^\mu = \mathrm{diag}\{A^\mu\}$ where the original matrices A^μ appear along the diagonal and the remaining elements are zero. The matrix P_0 is a function of only x and w while the derivatives of w are grouped into the matrix P_1.

The original equations (V) are said to be accretive in the generalized sense if the matrix

$$Q(x,y) \equiv P_0(x,y) + P_0^*(x,y) - \sum_{\mu=1}^{m} \partial_{x_\mu} \hat{A}^\mu(x,y)$$

is uniformly positive definite for all $x \in \mathbb{R}^m$ and for all vectors $y \in \mathbb{R}^k$ sufficiently small.

The positivity of the matrix Q is the local condition which allows to obtain the inequalities as described in (*) in the Theorem above, cf [7].

REFERENCES

[1] K.O. Friedrichs, The identity of weak and strong extensions of differential equations, Trans. Amer. Math. Soc., 55, 1944, 132-155.

[2] , Positive symmetric differential equations, Comm. Pure Appl. Math, *11*, 1958, 333-418.

[3] , Spectral Theory of Operators in Hilbert Space, Applied Mathematical Series, Springer Verlag, New York, 9, 1973.

[4] , On the differentiability of accretive linear differential equations, Contributions to Analysis, (A collection of Papers Dedicated to Lipman Bers.), Academic Press, ed. L.V. Ahlfors et. al. 1974, 147-150.

[5] , Von Neumann's Hilbert space theory and partial differential equations, SIAM Review, No. *4*, Oct. 1980.

[6] S. Hahn-Goldberg, Quasilinear Accretive Systems of partial differential, III, Mexican - U.S. Symposium on Differential Equations, Mexican Mathematical Society, 1976, 25-86.

[7] , Generalized Linear and Quasilinear Accretive Systems of Partial Differential Equations, Comm. in Partial Differential Equations, 2(2), 109-164 (1977).

[8] P.D. Lax, On Cauchy's Problem for hyperbolic equations and the differentiability of solutions of elliptic equations, Comm. Pure Appl. Math, 8, 1955, 615-633.

HOW TO REMEMBER THE SOBOLEV INEQUALITIES

Daniel B. Henry[§]

Instituto de Matemática e Estatística
Universidade de São Paulo
São Paulo, SP - Brazil

The Sobolev spaces $W^{m,p}(\mathbb{R}^n)$ (integer $m \geq 0$, $1 \leq p \leq \infty$) consist of functions in $L_p(\mathbb{R}^n)$ with distributional derivatives of order $\leq m$ in $L_p(\mathbb{R}^n)$. But one should resist the temptation to consider "m" as the measure of smoothness; the important measure of smoothness is $m-n/p$, which we call the "net smoothness":

$m-n/p$ = the *net smoothness* of functions in $W^{m,p}(\mathbb{R}^n)$.

The net smoothness describes the local behavior, and the only remaining role for p is to measure the rate of approach to zero as $|x| \to \infty$. The condition of "smallhess at infinity" becomes more stringent when p decreases, or $1/p$ increases.

Described in terms of net smoothness and rate of decrease at ∞, the relations among the Sobolev spaces become understandable and therefore easy to remember. The net smoothness is not merely an aid to memory, but also a guide in proving new theorems, such as the interpolation theorem below. But the main point is that *the net smoothness is the natural way to describe Sobolev spaces*. We list some

§ [Partially supported by FAPESP].

standard results to illustrate this point. For proofs, see references [1, 3, 6].

(1) We have the continuous inclusion $W^{m,p}(\mathbb{R}^n) \subset W^{k,q}(\mathbb{R}^n)$ provided the functions in the smaller space are at least as smooth as those in the larger space, and tend to zero at ∞ at least as rapidly: $m-n/p \geq k-n/q$ and $1/p \geq 1/q$. The only exception is when the net smoothness $m-n/p$ is a non-negative integer and $1 < p < \infty$, in which case we do not allow $q = \infty$, $m-n/p = k$. Note $p = 1$ and $p = \infty$ are exceptions to the exception: for example $W^{n,1}(\mathbb{R}^n) \subset C^0(\mathbb{R}^n)$ since $u(x) = \int_{-\infty}^{x_1} \cdots \int_{-\infty}^{x_n} \partial^n u/\partial x_1 \ldots \partial x_n$.

For integer-order Sobolev spaces these exceptional cases occur only when the net smoothness is a non-negative integer or when we restrict to a subspace (see (5) below). Fractional order spaces may also have exceptional cases for $p = 1$ or $p = \infty$.

(2) If $K \subset \mathbb{R}^n$ is a compact set and $B < \infty$, then

$$\left\{u \in W^{m,p}(\mathbb{R}^n) \;\middle|\; \|u\|_{W^{m,p}(\mathbb{R}^n)} \leq B \text{ and } \operatorname{supp} u \subset K\right\}.$$

Is contained in a compact set in $W^{k,q}(\mathbb{R}^n)$ provided

$$m-n/p > k-n/q \quad \text{and} \quad 1/p \geq 1/q.$$

(The support of u is abbreviated supp u; thus $u(x) = 0$ a.e. in $\mathbb{R}^n \backslash K$).

(3) $W^{m,p}(\mathbb{R}^n) \subset C^{\mu}(\mathbb{R}^n)$ is $m-n/p \geq \mu$, but we require strict inequality when these are non-negative integers. In fact, if $p < \infty$ we have $W^{m,p}(\mathbb{R}^n) \subset C_0^{\mu+}(\mathbb{R}^n)$, where $u \in C_0^{\mu+}(\mathbb{R}^n)$ when u has bounded continuous derivatives to order k ($\mu = k+\alpha$, k = integer ≥ 0, $0 \leq \alpha < 1$) and

$$|D^k u(x) - D^k u(y)|/|x-y|^{\alpha} \to 0 \quad \text{as} \quad x-y \to 0 \qquad (x,y \in \mathbb{R}^n)$$

($D^k u$ is any $k^{\underline{th}}$ order derivative), and $D^j u(x) \to 0$ as $|x| \to \infty$ for $0 \leq j \leq k$. If $\alpha = 0$, the above condition says $D^k u$ is uniformly continuous. $C_0^{\mu+}(\mathbb{R}^n)$ is the closure of $C_c^\infty(\mathbb{R}^n)$ in the C^μ-norm,

$$\|u\|_{C^\mu} = \max\left\{\sup_x |D^j u(x)|, \quad 0 \leq j \leq k; \quad \sup_{x \neq y} \frac{|D^k u(x) - D^k u(y)|}{|x-y|^\alpha}\right. .$$

Note $C_0^{\mu+}(\mathbb{R}^n)$ is separable, while $C_0^\mu(\mathbb{R}^n)$ is not separable if $\mu \neq$ integer.

(4) If $K \subset \mathbb{R}^n$ is compact, $\left\{u \;\middle|\; \|u\|_{W^{m,p}} \leq B, \ \operatorname{supp} u \subset K\right\}$ is in a compact set in $C^\mu(\mathbb{R}^n)$ provided $m-n/p > \mu$.

(5) Let $0 < d < n$ and consider $R^d = R^d \times 0$ as a subspace of $\mathbb{R}^n$; the restriction map $u \longmapsto u|\mathbb{R}^d \times 0$: $W^{m,p}(\mathbb{R}^n) \to W^{k,q}(\mathbb{R}^d)$ is continuous if $m-n/p > k-d/q$ and $1/p \geq 1/q$. Here we require strict decrease of net smoothness, to avoid complications. If we used Besov spaces instead [1], we could allow equality of net smoothness.

(6) If $0 \leq a \leq 1$, $1 \leq p,q,r \leq \infty$ and m,k are non negative integers with

$$k-n/q = a(m-n/p) + (1-a)(-n/r)$$

and $1/q \leq a/p + (1-a)/r$

except that we require $a \neq 1$ when $m-n/p = k$, $1 < p < \infty$, then there is a constant C such that

$$\|D^k u\|_{L_q(\mathbb{R}^n)} \leq C \, \|D^m u\|^a_{L_p(\mathbb{R}^n)} \, \|u\|^{1-a}_{L_r(\mathbb{R}^n)}$$

for all u in $C_c^\infty(\mathbb{R}^n)$.

Because of the homogeneity of the seminorms, the net

smoothness must balance exactly: if $u_\lambda(x) = u(\lambda x)$ for $\lambda > 0$, $x \in \mathbb{R}^n$, then

$$\|D^k u_\lambda\|_{L_q} = \lambda^{k-n/q} \|D^k u\|_{L^q} \quad \text{for all} \quad \lambda > 0.$$

If $0 < \alpha < 1$, $k+\alpha = a(m-n/p) + (1-a)(-n/r)$, then $H_\alpha(D^k u) \leq C \|D^m u\|^a_{L^p} \|u\|^{1-a}_{L^r}$ for all $u \in C_c^\infty(\mathbb{R}^n)$. Here $H_\alpha(v) = \sup_{x \neq y} |v(x)-v(y)|/|x-y|^\alpha$, and the net smoothness must again balance exactly: $H_\alpha(D^k u_\lambda) = \lambda^{k+\alpha} H_\alpha(D^k u)$ for all $\lambda > 0$.

(7) The extension theorem of Calderon and Stein [6] says that, if $\Omega \subset \mathbb{R}^n$ is an open set lying on one side of its boundary $\partial\Omega$ and $\partial\Omega$ is a Lipschitz surface (locally, a Lipschitz-continuous graph over a portion of a hyperplane), with certain uniformty requirements when $\partial\Omega$ is not compact (ee [6], sec. 6.3) there is a linear extension map E_Ω with the following properties. When u is the restriction to Ω of a function in $C_c^\infty(\mathbb{R}^n)$, $E_\Omega(u) \in C_c^\infty(\mathbb{R}^n)$, $E_\Omega(u) = u$ on Ω, and

$$\|u\|_{W^{m,p}(\Omega)} \leq \|E_\Omega(u)\|_{W^{m,p}(\mathbb{R}^n)} \leq K \|u\|_{W^{m,p}(\Omega)}$$

$$\|u\|_{C^\mu(\Omega)} \leq \|E_\Omega(u)\|_{C^\mu(\mathbb{R}^n)} \leq K' \|u\|_{C^\mu(\Omega)}$$

for certain constants K, K' depending on Ω, m, p, μ ($m \geq 0$, $1 \leq p \leq \infty$, $0 \leq \mu < \infty$). Using this map, the above results transfer without difficulty to $W^{m,p}(\Omega)$. For example

$$W^{m,p}(\Omega) \subset \begin{cases} W^{k,q}(\Omega) \text{ if } m-n/p \geq k-n/q,\ 1/p \geq 1/q & \text{(unless } m-n/p = k,\ q = \infty) \\ \\ C^{\mu+}(\Omega) \text{ if } m-n/p \geq \mu & \text{(unless } m-n/p = \mu = \text{integer)} \end{cases}$$

and if Ω is bounded, the inclusion is compact when $m-n/p > k-n/q$ or $m-n/p > \mu$, respectively. Also

$$\|u\|_{W^{k,q}(\Omega)} \leq C \, \|u\|^{a}_{W^{m,p}(\Omega)} \, \|u\|^{1-a}_{L_r(\Omega)}$$

if $0 \leq a \leq 1$, $k-n/q \leq a(m-n/p) + (1-a)(-n/r)$ and $1/q \leq a/p + (1-a)/r$ (except for the case $a = 1$, $m-n/p = k$, $q = \infty$).

If Ω is bounded, then trivially $L_q(\Omega) \subset L_s(\Omega)$ when $q \geq s$, so other inclusion relations are possible. These are easily obtained from those valid for unbounded regions, but such arguments should only occur at the last step. *The restrictions arising from "smallness at infinity" are needed even when Ω is bounded.* It is simplest to write everything initially for $\mathbb{R}^n$.

(8) Let $-n < s < \infty$ and fix $\phi \in C_c^{\infty}(\mathbb{R}^n)$ with $\phi(x) = 1$ for x near 0; let $\psi_s(x) = |x|^s \phi(x)$. We may consider ψ_s (or $x \longmapsto |x|^s$ near 0) as representative of the local behavior of a function with net smoothness s, in the sense that for all $m \geq 0$, $1 \leq p \leq \infty$, $\mu \geq 0$,

$\psi_s \in W^{m,p}(\mathbb{R}^n)$ if $s > m-n/p$, $\quad \psi_s \in W^{m,p}(\mathbb{R}^n)$ if $s < m-n/p$,

$\psi_s \in C^{\mu}(\mathbb{R}^n)$ if $s > \mu$, $\quad \psi_s \in C^{\mu}(\mathbb{R}^n)$ if $s < \mu$.

This is true even when m is not an integer.

Note for fixed $s_1,\dots,s_N$ and various choices of $y_1,\dots,y_N$ in $\mathbb{R}^n$, $x \longmapsto \prod_{j=1}^{N} |x-y_j|^{s_j}$ may behave locally like

$x \longmapsto |x-y|^{\delta}$ where δ can take any of the values $\sum_{j\in\sigma} s_j$, σ any nonempty subset of $\{1,\dots,N\}$.

(9) Multiplication $\{u_1,\ldots,u_N\} \longmapsto u_1 \ldots u_N : \prod_{j=1}^{N} W^{m_j,p_j}(\Omega) \to W^{k,q}(\Omega)$ is continuous if $\Omega \subset \mathbb{R}^n$ satisfies the conditions of (7) above - for example, $\Omega = \mathbb{R}^n$ - the m_j and k are non-negative integers, $1 \leq p_j, q \leq \infty$,

$$k \leq \min\{m_j \mid 1 \leq j \leq N\}, \quad 1/q \leq \sum_{j=1}^{N} 1/p_j, \quad \text{and}$$

$$k-n/q \leq \min\left\{\sum_{j\in\sigma} m_j-n/p_j \mid \sigma \text{ any nonempty subset of } \{1,\ldots,N\}\right\},$$

except that we require strict inequality in the last condition if some m_j-n/p_j is an integer ≥ 0 with $1 < p_j < \infty$. When $\Omega = \mathbb{R}^n$, these restrictions on k and q are all necessary for continuity, except possibly the "strict inequality" requirement.

We sketch the proof, supposing $s_j = m_j-n/p_j$ is never an integer ≥ 0, for simplicity. We must estimate the L_q-norm of $D^{k_1}u_1 \ldots D^{k_N}u_N$, whenever the $k_j \geq 0$, $\sum_1^N k_j \leq k$, in terms of $\prod_{j=1}^{N} \|u_j\|_{W^{m_j,p_j}}$. Now $D^{k_j}u_j \in W^{m_j-k_j,p_j} \subset L_{q_j}(\mathbb{R}^n)$ if $1 < q_j \leq \infty$ and

$$s_j-k_j \geq -n/q_j \geq -n/p_j, \quad s_j = m_j-n/p_j, \quad \text{so}$$

$$1/q = \sum 1/q_j \quad \text{should have} \quad \sum_j (s_j-k_j)_- \geq -n/q \geq -n\sum_j 1/p_j,$$

where $x_- = \min\{x,0\}$. We will have continuity provided

$$1/q \leq \sum_{j=1}^{N} 1/p_j \quad \text{and}$$

$$k-n/q \leq \min_{\{k_j\}} \left\{k + \sum_j (s_j-k_j)_- \;\middle|\; \sum k_j \leq k\right\}$$

$$= \min_{\{k_j\}} \left\{\sum_j (s_j-k_j)_- + k_j \;\middle|\; \sum k_j = k\right\}$$

$$= \min_{\{k_j\},\sigma} \left\{ \sum_{j\in\sigma} S_j + \sum_{j\in\sigma} k_j \;\middle|\; \sum k_j = k,\ \sigma \subset \{1,\dots,N\} \right\}$$

$$= \min_{\sigma} \left\{ k \ (\text{if } \sigma = \phi);\ \sum_{j\in\sigma} s_j \ (\text{if } \sigma \neq \phi),\ \sigma \subset \{1,\dots,N\} \right\}.$$

The essentials of this argument are in Palais [5, p. 27-30].

Now we present a general interpolation inequality, which includes the Nirenberg-Gagliardo inequality described in (6) above.

Nirenber [3] observed a connection between the L_p-norms and the Hölder seminorms $H_\alpha(\cdot)$, which we describe in a different form. Define

$$\{u\}_\alpha = \begin{cases} H_\alpha(u) = \sup\limits_{x\neq y} |u(x)-y(y)|/|x-y| & \text{when } 0 < \alpha < 1, \\ \|u\|_{L_p(\mathbb{R}^n)} & \text{when } \alpha = -n/p \leq 0 \end{cases}$$

for any $u \in C_c^\infty(\mathbb{R}^n)$, $-n \leq \alpha < 1$. Then if $u_\lambda(x) = u(\lambda x)$ $(\lambda > 0,\ x \in \mathbb{R}^n)$ we have $\{D^j u_\lambda\}_\alpha = \lambda^{j+\alpha}\{D^j u\}_\alpha$ for any $\lambda > 0$ $j = 0,1,2,\dots, -n \leq \alpha < 1$ which suggests the following.

<u>Theorem</u>. Suppose $-n \leq \alpha, \beta, \gamma < 1$ and j,k are nonnegative integers, $0 \leq \theta \leq 1$, and

$$j+\beta = \theta\,(k+\alpha) + (1-\theta)\gamma$$

$$1/p_\beta \leq \theta/p_\alpha + (1-\theta)/p_\gamma$$

where $p_\delta = \infty$ if $\delta \geq 0$, $= -n/p_\delta$ if $\delta < 0$. When $k+\alpha$ is an integer $\geq j$, and $-n < \alpha < 0$ (i.e. $1 < p_\alpha < \infty$) we require $\theta \neq 1$. Then there is a constant C such that

$$\{D^j u\}_\beta \leq C\, \{D^k u\}_\alpha^\theta\, \{u\}_\gamma^{1-\theta} \quad \text{for all } u \in C_c^\infty(\mathbb{R}^n).$$

Remarks. If $\theta = 1$, such an inequality cannot hold with $k+\alpha$ = integer $\geq j$, $-n < \alpha < 0$, as is well known. The net smoothness must balance, as noted above, and when this is true the condition $1/p_\beta \leq \theta/p_\alpha + (1-\theta)/p_\gamma$ is equivalent to requiring $j+\beta_+ \leq \theta(k+\alpha_+) + (1-\theta)\gamma_+$, where $x_+ = \max\{x,0\}$. The necessity of this condition is proved by considering $u(x) = \phi(x)\cos(\xi x_n)$ as $\xi \to \infty$ for fixed $\phi \in C_c^\infty(\mathbb{R}^n)$: we have $\{D^j u\}_\beta = \xi^{j+\beta_+}(C + o(1))$ for some $C = C(\phi,j,\beta) > 0$, if $\phi \not\equiv 0$. Thus all the stated conditions are necessary.

Note $j \leq j+\beta_+ < k+1$, so $0 \leq j \leq k$ in every case.

If $\alpha \geq 0$, $\gamma \geq 0$ then $1/p_\beta \leq 0$ so $\beta \geq 0$ also, and we have the familiar interpolation of Hölder seminorms.

If $\alpha \leq 0$, $\gamma \leq 0$ then $j \leq j+\beta_+ \leq \theta k$ $(\alpha_+ = \gamma_+ = 0)$ so $j/k \leq \theta \leq 1$ and we have the inequality of Nirenberg [3] and Gagliardo [2].

Nirenberg [4] also proved the result when $\alpha \leq 0 < \gamma < 1$ and $1 \leq j < k$, and claimed (without proof) we need only require $\theta \neq 1$ in the exceptional case ($m+\alpha$ = integer $\geq j$, $-n < \alpha < 0$). I am indebted to Paul Rabinowitz for this reference. We will outline the proof for the cases $\alpha \leq 0 < \gamma < 1$ and $\gamma \leq 0 < \alpha < 1$, as well as treating the exceptional cases. For this we use the "simple lemma" of Nirenberg [3, p. 126], which is the case $k = j = 0$ above together with $k \geq j \geq 0$ when $\alpha \geq 0$ and $\gamma \geq 0$. A proof of this lemma (occupying only three pages) is in the appendix to my course notes "Evolution Equations", IME-USP (1981).

Outline of proof. We must prove

(A): $\{D^j u\}_\beta \le C \, \|D^k u\|_{L_p}^{\theta} \, H_\gamma(u)^{1-\theta}$, if $0 \le \theta \le 1$, $0 \le j \le k$,

$0 < \gamma < 1$, $-n \le \beta < 1$,

$J+\beta = \theta(k-n/p) + (1-\theta)\gamma$, $1/p_\beta \le \theta/p$

(except $\theta \neq 1$ if $1 < p < \infty$ and $k-n/p$ = integer $\ge j$)

and

(B): $\{D^j u\}_\beta \le C \, H_\alpha(D^k u)^\theta \, \|u\|_{L_r}^{1-\theta}$ if $0 \le \theta \le 1$, $0 \le j \le k$,

$0 < \alpha < 1$,

$j+\beta = \theta(k+\alpha) + (1-\theta)(-n/r)$, $1/p_\beta \le (1-\theta)/r$.

If $k = 0$, these are included in Nirenberg's lemma, so assume $k \ge 1$. We prove the result by interpolation in θ (using Nirenberg's lemma) between the minimal and maximal values of θ. If $j = 0$, we may take $\theta = 0$ as the minimal value. If $j=k > 0$, we have $\theta = 1$ in case (A), but in case (B), $k/(k+\alpha) \le \theta \le 1$ are possible values. In case (B) and in case (A), unless $k-n/p$ = = integer $\ge j$ and $1 < p < \infty$, $\theta = 1$ is the maximal value; for the exceptional cases of (A) we show (A) holds for arbitrarily close to 1. We also prove (A), (B) for minimal θ, specifically:

(A_{min}): $\|D^j u\|_{L_q} \le C \, \|D^k u\|_{L_p}^{p/q} \, H_\gamma(u)^{1-p/q}$

when $1 \le j < k$ and $j = \frac{p}{q}k + (1-p/q)\gamma$, i.e. $p/q = (j-\gamma)/(k-\gamma)$;

(B_{min}): $\|D^j u\|_{L_q} \le C \, H_\alpha(D^k u)^{1-r/q} \, \|u\|_{L_r}^{r/q}$

when $1 \le j \le k$ and $j = (1-r/q)(k+\alpha)$.

Thus for (A), $\theta_{min} = (j-\gamma)/(k-\gamma)$, and for (B), $\theta_{min} = j/(k+\alpha)$.

Suppose $0 < j < k$; for any $x,h \in \mathbb{R}^n$ and $u \in C_c^\infty(\mathbb{R}^n)$

let $\Delta_h u(x) = u(x+h)-u(x)$, so $\Delta_h^j u(x) = \sum_{\ell=0}^{j} \binom{j}{\ell} u(x+\ell h)(-1)^{j-\ell} =$

$$= \int_{[0,1]^j} dt\, D^j u(x+|t|h)h^j \quad \text{where} \quad |t| = t_1+\ldots+t_j.$$ Then

$$(*) \qquad D^j u(x)h^j = \sum_{m=1}^{k-j} \frac{(-1)^m}{m^j} \binom{k-j}{m} \Delta_{mh}^j u(x)$$

$$+ (-1)^{k-j} \int_{[0,1]^j} dt \int_{[0,1]^{k-j}} ds |t|^{k-j} D^k u(s+|s||t|h)h^k.$$

It is sufficient to prove $(A)_{min}$ $(B)_{min}$ for $n = 1$ if $1 \leq j < k$; we may then simply integrate to get the results for $n > 1$ and pure derivatives $(\frac{\partial}{\partial x_m})^j u(x)$, and by rotating coordinates obtain estimates of mixed derivatives.

Suppose therefore $n = 1$, $u \in C_c^\infty(\mathbb{R})$, and for $(A)_{min}$ estimate the integral on the right-side of the identity (*) as $C|h|^{k-1/p} (\int_x^{x+Nh} |D^k u|^p)^{1/p}$, where $N = j(k-j)$ and the integral is $\int_{x+Nh}^x \ldots$ when $h < 0$. Fix any $\varepsilon > 0$; let $h = \varepsilon$ when $a-N\varepsilon \leq x \leq a$ and $h = -\varepsilon$ when $a \leq x \leq a+N\varepsilon$, so

$$\varepsilon^{j-1/q}\left(\int_{a-N\varepsilon}^{a+N\varepsilon} |D^j u|^q\right)^{1/q} \leq C\left[\varepsilon^\gamma H_\gamma(u) + \varepsilon^{k-1/p}\left(\int_{a-N\varepsilon}^{a+N\varepsilon} |D^k u|^p\right)^{1/p}\right].$$

This holds for every real a and every $\varepsilon > 0$, and we replace a-N by a, $b = 2N\varepsilon$, to obtain

$$b^{j-1/q} \left(\int_a^{a+b} |D^j u|^q\right)^{1/q} \leq C'\left[b^\gamma H_\gamma(u) + b^{k-1/p}\left(\int_a^{a+b} |D^k u|^p\right)^{1/p}\right].$$

Unless $u \equiv 0$ on (a,∞), we may choose $b = b(a) > 0$ to make the terms on the right equal, i.e.

$$\int_a^{a+b} |D^j u|^q \leq (2C')^q \int_a^{a+b} |D^k u|^p . H_\gamma(u)^{q-p}.$$

Summing over adjacent intervals $[a, a+b(a)]$ covering the support of u proves (A_{min}).

For (B_{min}), first note the identity (*) has the form $D^j u(x)h^j = cu(x) + \sum(h)$, where c is constant, $\sum(h)$ is a sum of terms $u(x+sh)$ for $1 \le s \le N$ and an integral of $D^k u(x+\sigma h)h^k$ over $0 \le \sigma \le N$. We then have $(2^j-1)D^j u(x)h^j =$ $= \sum(2h)-\sum(h)$ and $(2^j-2^k)(2^j-1)D^j u(x)h^j = \sum(4h)-2^k\sum(2h)-\sum(2h)+2^k\sum(h)$ and since $1 \le j < k$

$$|D^j u(x)h^j| \le C\left\{\sum_{s=1}^{4N} |u(x+sh)| + H_\alpha(D^k u)|h|^{k+\alpha}\right\}.$$

Taking the seminorm $(\int|\cdot|^r dh)^{1/r}$ over $(-\varepsilon,0)$ or $(0,\varepsilon)$, then the seminorm $\left\{\int_{a-4N\varepsilon}^{a+4N\varepsilon} |\cdot|^q\, dx\right\}^{1/q}$ we find

$$\varepsilon^{j-1/q}\, \|D^j u\|_{L_q(a-4N\varepsilon,a+4N\varepsilon)} \le C'\, (\varepsilon^{-1/r}\, \|u\|_{L_r(a-4N\varepsilon,a+4N\varepsilon)} + \varepsilon^{k+\alpha} H_\alpha(d^k u)).$$

The proof is completed as for $(A)_{min}$ above.

If $j = k$, the simpler identity

$$D^k u(x)h^k = \Delta_h^k u(x) - \int_{[0,1]^k} dt(D^k u(x+|t|h)-D^k u(x))h^k \quad \text{yields}$$

$$b^{k-1/q}\, \|D^k u\|_{L_q(a,a+b)} \le C'\, (b^{-1/r}\, \|u\|_{L_r(a,a+b)} + b^{k+\alpha} H_\alpha(D^k u))$$

for all real a and $b > 0$, which completes the proof of (B_{min}).

It remains only to examine the "exceptional" cases of (A), when $1 < p < \infty$, $k-n/p = q = \text{integer} \ge j$. We have

$\|D^{q+1}u\|_{L_n} \le C\, \|D^k u\|_{L_p}$, so it is enough to consider the case

$p = n$. First suppose $j = 0$, $k = 1$, $0 < \gamma < 1$; if $\phi \in C_c^\infty(\mathbb{R}^n)$,

$\phi \geq 0$, $\int_{\mathbb{R}^n} \phi = 1$, $\phi_\varepsilon(x) = \varepsilon^{-n}\,\phi(x/\varepsilon)$ for $\varepsilon > 0$, then $u_\varepsilon = u*\phi_\varepsilon$ satisfies $|u_\varepsilon(x) - u(x)| \leq \{u\}_\gamma \varepsilon^\gamma$ and

$|u_\varepsilon(x+h)-u_\varepsilon(x)| \leq |h| \int_0^1 \phi_\varepsilon * |Du|(x+th)dt \leq |h|\,\|\phi_\varepsilon\|_{L_{n'}}\,\|Du\|_{L_n} = C|h|\,{}^{-1}\|Du\|_{L_n}$

so $|u(x+h)-u(x)| \leq C|h|\varepsilon^{-1}\,\|Du\|_{L_n} + 2\varepsilon^\gamma\{u\}_\gamma$ for every h and $\varepsilon > 0$. Minimizing over $\varepsilon > 0$ we find

$$\{u\}_{\gamma/(1+\gamma)} \leq C_1\,\|Du\|_{L_n}^{\frac{\gamma}{1+\gamma}}\,\{u\}_\gamma^{\frac{1}{1+\gamma}}.$$

Iterating this inequality,

$$\{u\}_{\gamma/(1+m\gamma)} \leq C_m\,\|Du\|_{L_n}^{m\gamma/(1+m\gamma)}\,\{u\}_\gamma^{1/(1+m\gamma)} \qquad (m = 1,2,\ldots)$$

so (A) holds with $j=0$, $k=1$ and $\theta = m\gamma/(1+m\gamma)$ arbitrarily close to 1.

Next suppose $j=1$, $k=2$. From (A_{min}) we have

$$\|Du\|_{L_q} \leq C\,\|D^2u\|_{L_n}^{n/q}\,\{u\}_\gamma^{1-n/q}$$

if $n/q = (1-\gamma)/(2-\gamma)$, hence

$$\{u\}_{\frac{1}{2-\gamma}} \leq C'\,\|Du\|_{L_q} \leq C''\,\|D^2u\|_{L_n}^{1-\frac{1}{2-\gamma}}\,\{u\}_\gamma^{\frac{1}{2-\gamma}}.$$

Since $\gamma < \frac{1}{2-\gamma} < 1$ whenever $0 < \gamma < 1$, iteration yields

$$\{u\}_{\gamma_m} \leq C_m\,\|D^2u\|_{L_n}^{(\gamma_m-\gamma)/(1-\gamma)}\,\{u\}_\gamma^{(1-\gamma_m)/(1-\gamma)} \qquad (m=1,2,\ldots)$$

and $\gamma_m \to 1$ as $m \to \infty$ $(\gamma_0 = \gamma,\ \gamma_m = 1/(2-\gamma_{m-1})$ for $m \geq 1)$

$$\|Du\|_{L_{q_m}} \leq C'_m\,\|D^2u\|_{L_n}^{\theta_m}\,\{u\}_\gamma^{1-\theta_m}$$

$\theta_m \to 1$ and $n/q_m = (1-\gamma_m)/(2-\gamma_m) \to 0$ as $m \to \infty$, so (A) holds with $j=1$, $k=2$ and θ arbitrarily close to 1. In general, if $D^k u \in L_n$ and $\{u\}_\gamma < \infty$, $0 < \gamma < 1$, we estimate $\|D^{k-1}u\|_{L_r}$ in terms of $\|D^k u\|_{L_n}$ and $\{D^{k-2}u\}_\alpha$ (from the case $j=1$, $k=2$) with r arbitrarily large, and $\{D^{k-2}u\}_\alpha \le C \|D^{k-1}u\|_{L_q}$, which may be estimated initially from (A_{min}) if $\alpha = 1-n/q = 1/(k-\gamma)$.

REFERENCES

[1] R. Adams, *Sobolev Spaces*, Academic Press (1975).

[2] E. Gagliardo, Proprietā di alcune classi di funzioni in piu variabili, *Richerche di Mat., Napoli* 7 (1958), 102-137.

[3] L. Nirenberg, On elliptic partial differential equations (lecture II), *Annali della Scuola Norm. Sup. - Pisa*, Ser. 3, V. 13 (1959).

[4] L. Nirenberg, An extended interpolation inequality, *Annali della Scuola Norm. Sup. - Pisa*, Serv. 3, V. 20 (1966), 733-737.

[5] R. Palais, Foundations of Global Nonlinear Analysis, W. A. Benjamin, New York (1968).

[6] E. Stein, *Singular Integrals and Differentiability Properties of Functions*, Princeton Univ. Press (1970).

THE ADJOINT EQUATION OF A LINEAR VOLTERRA STIELTJES-INTEGRAL EQUATION WITH A LINEAR CONSTRAINT

Chaim Samuel Hönig

Instituto de Matemática e Estatística
Universidade de São Paulo
São Paulo, SP - Brazil

Abstract - Let X and Y be Banach spaces. $G([a, b], X)$ denotes the space of all functions $f: [a, b] \to X$ that are regulated, i.e., that have only discontinuities of the first kind. Let $K: [a, b]\times[a, b] \to L(X)$ be a kernel such that for every $x(t_o) \in X$ and $f \in G([a, b], X)$ the linear Volterra Stieltjes-integral equation

$$(K) \qquad x(t) - x(t_o) + \int_{t_o}^{t} \cdot\, d_s K(t, s)\cdot x(s) = f(t) - f(t_o), \quad a \leq t \leq b,$$

has one and only one solution $x \in G([a, b], X)$, i.e., K has a unique resolvent (see §1). Let $\alpha: [a, b] \to L(X, Y)$ be a function of bounded variation or, more generally, of bounded semivariation (see §1) such that $\alpha(t_o) = 0$. We consider the linear constraint

$$(F_\alpha) \qquad F_\alpha[x] \equiv \int_a^b \cdot\, d\,(t)\cdot x(t) = c.$$

We prove that the adjoint (or transposed) equation of the system $(K)+(F_\alpha)$ (see §2) is given by

$$(K^*) \qquad \begin{aligned} &y(s) + \int_s^b \cdot K(t, s)^*\cdot dy(t) - \int_{t_o}^b \cdot\, K(t, t_o)^*\cdot dy(t) - \\ &-Y(t_o-s)\int_a^b \cdot\,[K(t, s)-K(t, t_o)]^*\cdot dy(t)+\alpha(s)^* d = h(s), \quad a\leq s\leq b \end{aligned}$$

where $d \in Y'$ and $h, y \in BV^{a,b}_{t_o}([a, b], X')$; we denote by $BV^{a,b}_{t_o}([a, b], X')$ the space of all functions $g: [a, b] \to X'$ that are of bounded variation and such that $g(t_o) = 0$ and $g(b) = g(a)$. This space appears naturally as a dual (see example 2 of §2), hence the condition $g(b) = g(a)$ is of functional-analytic nature and is not a constraint or a boundary condition.

Our main result is the Theorem 4.1: the following properties are equivalent

A - For $(f, c) \in G([a, b], X)\times Y$ the system $(K)+(F_\alpha)$ ha a solution x iff for every $(y, d) \in BV^{a,b}_{t_o}([a, b], X')\times Y'$ that is a solution of (K^*_α) with $h = 0$, we have

$$(*) \qquad \int_a^b \cdot <dy(s), f(s)>+<d, c> = 0$$

B - $J(t_o)X$ is a closed vector subspace of Y, where $J(t_o) = \int_a^b \cdot d\alpha(t)\circ R(t, t_o) \in L(X, Y)$ (R is the resolvent of K).

C - $J(t_o)^*Y'$ is a closed vector subspace of X'.

D - For $h \in BV^{a,b}_{t_o}([a, b], X')$ the equation (K^*_α) has a solution $(y, d) \in BV^{a,b}_{t_o}([a, b]), X')\times Y'$ iff for every $z\in G([a, b], X)$ that is a solution of the system $(K)+(F_\alpha)$ with $f = 0$ and $c = 0$ we have

$$(**) \qquad \int_a^b \cdot <dh(s), z(s)> = 0.$$

(A and D say that the system $(K)+(F_\alpha)$ and the equation (K^*_α) are "normally solvable").

§1. Introduction

1. We keep the notations of [6].

For the notion of regulated function and the corresponding notations, see [6], p. 175 and 176.

For the definitions of function of bounded variation, of function of bounded semivariation, of the interior integral $\int \cdot$ and their properties see [6]: p. 176, 177, Theorems 1.9, 1.10 and 1.12.

For the operators defined by kernels K see [6], p. 179 and 181. For the definition of the kernel $K \in G^{\sigma}_{\Delta} \cdot SV^{u}(\Gamma_{t_0}, L(X))$ and its properties see [6], p. 182, Theorems 2.6 and 2.10 (we may always suppose that $K(t, t) = 0$ and then we write $K \in G^{\sigma}_{0} \cdot SV^{u}(\Gamma_{t_0}, L(X))$).

For the definition of the resolvent of the kernel K and its main properties see [6]: p. 196 and 1), 7) and 8) of Theorem 3.4.

2. In [4] we gave a particular case of Theorem 4.1: when $t_0 = a$ and the kernel K satisfies some restrictions, for instance, continuity in the first variable. These results extend those known for systems $(L)+(F_\alpha)$ where

$$(L) \qquad x(t) - x(a) + \int_a^t \cdot dA(s) \cdot x(s) = f(t) - f(a), \qquad a \leq t \leq b.$$

For these systems see [8]. There are many articles on these systems with particular instances of the constraint (F_α); see the references given in [8].

Our results encompass those of [4] and [8] and are obtained in a functional-analytic way and thus prove the real nature of the condition $y(b) = y(a)$. See also [9].

Elsewhere we will apply the results of this paper to particular equations (for instance, linear delay differential equations, see [4], p. 91, [1] and [2]) and/or particular constraints.

§2. Auxiliary Results

1. Let E and F be complex vector spaces. We say that a

bilinear form $B: F\times E \to \mathbb{C}$ is *separating at the left*, if only $y = 0$ has the property that $B(y, x) = 0$ for all $x \in E$.

Example 1 - The bilinear form

$$(y, x)\in BV_{t_o}([a, b], X')\times G([a, b], X)\longmapsto\int_a^b \cdot<dy(t), x(t)> \in \mathbb{C}$$

is separating at the left (cf. [6], Theorem 1.12 with $Y = \mathbb{C}$).

We define

$$BV_{t_o}^{a,b}([a, b], X') = \left\{y\in BV_{t_o}([a, b]), X') \mid y(a) = y(b)\right\}.$$

Example 2 - The bilinear form

$$((y, d), (x, c))\in(BV_{t_o}^{a,b}([a, b], X')\times Y')\times(G_{t_o}([a, b], X)\times Y)\longmapsto$$

$$\longmapsto\int_a^b \cdot<dy(t), x(t)> + <d, c> \in \mathbb{C}$$

is separating at the left.

Let $B_1: F_1\times E_1 \to \mathbb{C}$ and $B_2: F_2\times E_2 \to \mathbb{C}$ be bilinear forms. Given a linear mapping $u: E_1 \to E_2$ we say that a linear mapping $u^*: F_2 \to F_1$ is a *adjoint* (or *transposed*) of u (with respect to B_1 and B_2) if we have

$$B_2(y_2, ux_1) = B_1(u^*y_2, x_1) \quad \text{for all} \quad x_1 \in E_1 \quad \text{and} \quad y_2 \in F_2.$$

It is immediate the

Theorem 2.1 - With the notations above if B_1 is separating at the left, the adjoint u^* is unique whenever it exists.

2. Given $K \in G_o^\sigma\cdot SV^u(\Gamma_{t_o}, L(X))$ and $\alpha \in SV_{t_o}([a, b], L(X, Y))$ we consider the system

$$\text{(K)} \qquad x(t)-x(t_o) + \int_{t_o}^t \cdot\, d_\sigma K(t, \sigma)\cdot x(\sigma) = f(t)-f(t_o), \qquad a \le t \le b$$

$$(F_\alpha) \qquad F_\alpha[x] \equiv \int_a^b \cdot\, d\alpha(t)\cdot x(t) = c$$

where $x, f \in G([a, b], X)$. This system defines a linear continuous mapping

$$K_\alpha : x \in G([a, b], X) \longmapsto (x - x(t_0) + k_{t_0} x,\ F_\alpha[x]) \in G_{t_0}([a, b], X) \times Y.$$

With the notations of the preceding examples 1 and 2 we have the

<u>Theorem 2.2</u> - K_α has one and only one adjoint

$$K_\alpha^* : (y, d) \in BV_{t_0}^{a,b}([a, b], X') \times Y' \longmapsto K_\alpha^*(y, d) \in BV_{t_0}([a, b], X')$$

where

$$K_\alpha^*(y, d)(s) = y(s) + \int_s^b \cdot K(t, s)^* \cdot dy(t) - \int_{t_0}^b \cdot\, K(t, t_0)^* \cdot dy(t) -$$

$$(1) \qquad - Y(t_0 - s) \int_a^b \cdot [K(t, s) - K(t_0, s]^* \cdot dy(t) + \alpha(s)^* d, \quad a \leq s \leq b,$$

($Y(\sigma) = 1$ for $\sigma \geq 0$ and $Y(\sigma) = 0$ for $\sigma < 0$).

<u>Proof</u> - The unicity of the adjoint follows immediately from Theorem 2.1. In order to prove that K_α^* defined by (1) is the adjoint of K_α we have to show that for every $x \in G([a, b], X)$, $d \in Y'$ and $y \in BV_{t_0}^{a,b}([a, b], X')$ we have

$$(2) \qquad \int_a^b \cdot <d_s K_\alpha^*(y, d)(s), x(s)> = \int_a^b \cdot <dy(t), x(t) - x(t_0) + (k_{t_0} x)(t)> + <d, F_\alpha[x]>.$$

If we replace (1) in the first integral of (2) we obtain

$$\int_a^b \cdot <dy(s), x(s)> + \int_a^b \cdot <d_s \int_s^b \cdot K(t, s)^* \cdot dy(t), x(s)> -$$

$$- \int_a^{t_0} \cdot <d_s \int_a^b \cdot K(t, s)^* \cdot dy(t), x(s)> + \int_a^b \cdot <d\alpha(s)^* d, x(s)>.$$

Since $y(b) = y(a)$ implies $\int_a^b \cdot <dy(t), x(t_0)> = 0$ and since

$\int_{t_o}^{t} \cdot = \int_a^t \cdot - \int_a^{t_o} \cdot$ we have to prove that

$$\int_a^b \cdot <d_s \int_s^b \cdot K(t, s)^* \cdot dy(t), x(s)> - \int_a^{t_o} \cdot <d_s \int_a^b \cdot <K(t, s)^* \cdot dy(t), x(s)> =$$

$$= \int_a^b \cdot <dy(t), \int_a^t \cdot d_s K(t, s) \cdot x(s)> - \int_a^b \cdot <dy(t), \int_a^{t_o} \cdot d_s K(t, s) \cdot x(s)>$$

and this is immediate since the first and second integrals of the left hand are equal, respectively, to the first and second integrals of the right hand, respectively by the Dirichlet formula ([6], Theorem 2.13) and by the Theorem of Bray ([6], Theorem 2.12).

<u>Corollary 2.3</u> - If $(f, c) \in G([a, b], X) \times Y$ is such that the system $(K)+(F_\alpha)$ has a solution then for every $(y, d) \in BV_{t_o}^{a,b}([a, b], X') \times Y'$ that is a solution of

$$(K_\alpha^*)_o \quad \begin{aligned} & y(s) + \int_s^b \cdot K(t, s)^* \cdot dy(t) - \int_{t_o}^b \cdot K(t, t_o)^* \cdot dy(t) - \\ & -Y(t_o - s) \int_a^b \cdot [K(t, s) - K(t, t_o)]^* \cdot dy(t) + \alpha(s)^* d = 0, \quad a \leq s \leq b \end{aligned}$$

we have

$$(*) \qquad \int_a^b \cdot <dy(t), f(t)> + <d, c> = 0.$$

<u>Proof</u> - Given (f, c) let x be a solution of $(K)+(F_\alpha)$. The result follows immediately from (1) and (2) since (y, d) is a solution of $(K_\alpha^*)_o$ iff we have $K_\alpha^*(y, d) = 0$.

<u>Corollary 2.4</u> - Let $y, g \in BV_{t_o}^{a,b}([a, b], X')$ be such that we have

$$(3) \quad \begin{aligned} & y(s) + \int_s^b \cdot K(t, s)^* \cdot dy(t) - \int_{t_o}^b \cdot K(t, t_o)^* \cdot dy(t) - \\ & -Y(t_o - s) \int_a^b \cdot [K(t, s) - K(t, t_o)]^* \cdot dy(t) = g(s), \quad a \leq s \leq b. \end{aligned}$$

Then for every $z \in G([a, b], X)$ that satisfies (K) with $f = 0$ we have

$$\int_a^b \cdot <dg(t), z(t)> = 0.$$

Proof - Follows immediately from (2) since $d = 0$, $K_\alpha^*(y, 0) = g$ and $z - z(t_o) + k_{t_o} z = 0$.

§3. Kernels with resolvents

In what follows we suppose that the kernel $K \in G_o^\sigma \cdot SV^u(\Gamma_{t_o}, L(X))$ has a unique resolvent. We recall that in this case for every $t \in [t_o, b[$ ($t \in]a, t_o]$) the operator $I_X - K(t\dot{+}, t)$ ($I_X - K(t\dot{-}, t)$) is invertible in $L(X)$: see [6], the Remark d) that follows the Theorem 3.4. This condition is also sufficient for the existence of the resolvent if the operator k_{t_o} defined by K is compact (see [6], III of Theorem 3.4) or even under more general conditions (see [7]).

Theorem 3.1 - Let $K \in G_o^\sigma \cdot SV^u(\Gamma_{t_o}, L(X))$ have a resolvent R. Then for $g \in BV_{t_o}([a, b], X')$ the equation (3) has one and only one solution $y \in BV_{t_o}([a, b], X')$; this solution is given by

$$\text{(4)}\quad y(s) = \int_{t_o}^t \cdot R(\sigma, t_o)^* \cdot dg(\sigma) - \int_s^b \cdot R(\sigma, s)^* \cdot dg(\sigma) + Y(t_o - s)\int_a^b \cdot [R(\sigma, s) - R(\sigma, t_o)]^* \cdot dg(\sigma),$$

$a \le s \le b$, and we have

$$\text{(5)}\qquad y(b) - y(a) = -\int_a^b \cdot R(\sigma, t_o)^* \cdot dg(\sigma).$$

Proof - Unicity of the solution: if there where two solutions y_1 and y_2, their difference y would satisfy (3) with $g = 0$. For any $f \in G([a, b], X)$ let $x \in G([a, b], X)$ satisfy (K); since we have $d = 0$ from (*) it follows that

$$\int_a^b \cdot\langle dy(t), f(t)\rangle = 0 \quad \text{for every} \quad f \in G([a, b], X)$$

hence y is constant but $y(t_o) = 0$, hence $y = 0$.

In order to prove that y given by (4) satisfies (3) let us take (3) and (4) in the intervals $[a, t_o]$ and $[t_o, b]$:

(3a) $$y(s) - \int_a^s \cdot K(t, s)^* \cdot dy(t) + \int_a^{t_o} \cdot K(t, t_o)^* \cdot dy(t) = g(s), \quad a \leq s \leq t_o$$

(3b) $$y(s) + \int_s^b \cdot K(t, s)^* \cdot dy(t) - \int_{t_o}^b \cdot K(t, t_o)^* \cdot dy(t) = g(s), \quad t_o \leq s \leq b$$

(4a) $$y(s) = \int_a^s \cdot R(\sigma, s)^* \cdot dg(\sigma) - \int_a^{t_o} \cdot R(\sigma, t_o)^* \cdot dg(\sigma), \quad a \leq s \leq t_o$$

(4b) $$y(s) = \int_{t_o}^b \cdot R(\sigma, t_o)^* \cdot dg(\sigma) - \int_s^b \cdot R(\sigma, s)^* \cdot dg(\sigma), \quad t_o \leq s \leq b.$$

Let us prove that y given by (4b) satisfies (3b), i.e., that we have

(6) $$\int_{t_o}^b \cdot R(\sigma, t_o)^* \cdot dg(\sigma) - \int_s^b \cdot R(\sigma, s)^* \cdot dg(\sigma) - \int_s^b \cdot K(t, s)^* \cdot d_t \int_t^b \cdot R(\sigma, t)^* \cdot dg(\sigma) + {} \\ + \int_{t_o}^b \cdot K(t, t_o)^* \cdot d_t \int_t^b \cdot R(\sigma, t)^* \cdot dg(\sigma) = g(s), \quad t_o \leq s \leq b.$$

We recall that for $u, v \in L(X)$ and $x' \in X'$ we have $u^* \cdot x' = x' \circ u$ since $\langle u^* \cdot x', x\rangle = \langle x', ux\rangle$. Hence $v^* \cdot (u^* \cdot x') =$ $= (u^* \cdot x') \circ v = x' \circ u \circ v$. From this remark and from the Dirichlet formula ([6], Theorem 2.13) it follows that

$$\int_s^b \cdot K(t, s)^* \cdot d_t \int_t^b \cdot R(\sigma, t)^* \cdot dg(\sigma) = \int_s^b \cdot d_t \left[\int_t^b \cdot dg(\sigma) \circ R(\sigma, t)\right] \circ K(t, s) =$$

$$= \int_s^b \cdot dg(\sigma) \circ \left[\int_s^\sigma \cdot d_t R(\sigma, t) \circ K(t, s) - K(\sigma, s)\right] =$$

$$= \int_s^b \cdot \left[\int_s^\sigma \cdot d_t R(\sigma, t) \circ K(t, s) - K(\sigma, s)\right]^* \cdot dg(\sigma) = \int_s^b \cdot [I_X - R(\sigma, s)]^* \cdot dg(\sigma)$$

where for the last equality we used the relation (R_*) of Theorem 3.4

of [6]. Analogously we have

$$\int_{t_0}^{b} \cdot K(t, t_0)^* \cdot d_t \int_{t}^{b} \cdot R(\sigma, t)^* \cdot dg(\sigma) = \int_{t_0}^{b} \cdot [I_X - R(\sigma, t_0)]^* \cdot dg(\sigma)$$

and these relations imply immediately (6). In an analogous way we prove that y given by (4a) satisfies (3a).

If $\alpha \in SV([a, b], L(X, Y))$ and $d \in Y'$ it follows from 3.6 of [3] that the function

$$s \in [a, b] \longmapsto \alpha(s)^* d = d \circ \alpha(s) \in X'$$

belongs to $BV([a, b], X')$.

Theorem 3.2 - Under the hypothesis of Theorem 3.1 we give $\alpha \in SV_{t_0}([a, b], L(X, Y))$ and $d \in Y'$. Equation (3) with $g(s) = -\alpha(s)^* d$ has a solution $y \in BV_{t_0}^{a,b}([a, b], X')$ iff $J(t_0)^* d = 0$, where

$$J(t_0) = \int_a^b \cdot d\alpha(\sigma) \circ R(\sigma, t_0) \in L(X, Y).$$

In this case we have

$$\begin{aligned} y(s) = \Big\{ & \int_s^b \cdot d\alpha(\sigma) \circ R(\sigma, s) - \int_{t_0}^b \cdot d\alpha(\sigma) \circ R(\sigma, t_0) - \\ & - Y(t_0 - s) \int_a^b \cdot d\alpha(\sigma) \circ [R(\sigma, s) - R(\sigma, t_0)] \Big\}^* d, \quad a \le s \le b. \end{aligned} \tag{7}$$

Proof - Follows immediately from Theorem 3.1 if we recall that by (5) we have $y(b) - y(a) = J(t_0)^* d$.

§4. The main theorem

Theorem 4.1 - Given $K \in G_0^{\sigma} \cdot SV^u(\Gamma_{t_0}, L(X))$ that has a unique resolvent R and $\alpha \in SV_{t_0}([a, b], L(X, Y))$ the following properties are equivalent:

A - For $(f, c) \in G([a, b], X)\times Y$ the system $(K)+(F_\alpha)$ has a solution $x \in G([a, b], X)$ iff for every $(y, d)\in BV_{t_o}^{a,b}([a, b], X')\times Y'$ that satisfies $(K_\alpha^*)_o$ we have (*).

B - $J(t_o)X$ is a closed vector subspace of Y.

C - $J(t_o)^*Y'$ is a closed vector subspace of X'.

D - For $h \in BV_{t_o}^{a,b}([a, b], X')$ the equation

$$(K_\alpha^*) \qquad \begin{aligned} &y(s) + \int_s^b \cdot K(t, s)^* \cdot dy(t) - \int_{t_o}^b \cdot K(t, t_o)^* \cdot dy(t) - \\ &-Y(t_o-s)\int_a^b \cdot [K(t, s)-K(t, t_o)]^* \cdot dy(t)+\alpha(s)^*\cdot d = h(s), \quad a\leq s\leq b \end{aligned}$$

has a solution $(y, d) \in BV_{t_o}^{a,b}([a, b], X')\times Y'$ iff for every $z \in G([a, b], X)$ that is a solution of the system $(K)+(F_\alpha)$ with $f = 0$ and $c = 0$ we have

$$(**) \qquad \int_a^b \cdot \langle dh(s), z(s)\rangle = 0.$$

<u>Proof</u> - We recall that in Corollary 2.3 we proved the necessity of the condition given in A. In order to prove A $\Longleftrightarrow$ B we need 3 Lemmas

<u>Lemma 4.2</u> - If for $(f, c) \in G([a, b], X)\times Y$ the system $(K)+(F_\alpha)$ has a solution $x \in G([a, b], X)$ then we have $c = J(t_o)x(t_o) + c_f$ where

$$c_f = -J(t_o)f(t_o)-\int_a^b \cdot d_\sigma\left[\int_\sigma^b \cdot d\alpha(\tau)\, R(\tau, \sigma)\right]\cdot f(\sigma) + \int_a^{t_o} \cdot d_\sigma\left[\int_a^b \cdot d\alpha(\tau)\, R(\tau, \sigma)\right]\cdot f(\sigma)$$

<u>Proof</u> - By (8) of Theorem 3.4 of [6] we have

$$(\rho) \quad x(t) = f(t)+R(t, t_o)[x(t_o)-f(t_o)] - \int_{t_o}^t \cdot d_\sigma R(t, \sigma)\cdot f(\sigma), \qquad a \leq t \leq b.$$

By the formula of Dirichlet (see Theorem 2.13 of [6]) and by the Theorem of Bray (see Theorem 2.12 of [6]) we have

$$\int_a^b \cdot d\alpha(t)\cdot\int_{t_0}^{t} \cdot\ d_\sigma R(t,\sigma)\cdot f(\sigma) =$$

$$= \int_a^b \cdot d\alpha(t)\cdot\int_a^{t} \cdot\ d_\sigma R(t,\sigma)\cdot f(\sigma) - \int_a^b \cdot d\alpha(t)\cdot\int_a^{t_0} \cdot\ d_\sigma R(t,\sigma)\cdot f(\sigma)$$

$$= \int_a^b \cdot d_\sigma\left[\int_\sigma^b \cdot d\alpha(t)\circ R(t,\sigma)\right]\cdot f(\sigma) + \int_a^b \cdot d\alpha(t)\cdot f(t) -$$

$$- \int_a^{t_0} \cdot\ d_\sigma\left[\int_a^b \cdot d\alpha(t)\circ R(t,\sigma)\right]\cdot f(\sigma).$$

If we apply F_α to (ρ) and take into account the equality we just proved we obtain the result.

From Lemma 4.2 follows immediately the

<u>Lemma 4.3</u> - For $(f, c) \in G([a, b], X)\times Y$ the system $(K)+(F_\alpha)$ has a solution iff $c-c_f \in J(t_0)X$. If $c-c_f = J(t_0)x_0$ then a solution $x \in G([a, b], X)$ of the system $(K) + (F_\alpha)$ is given by

$$x(t) = f(t)+R(t, t_0)[x_0-f(t_0)] - \int_{t_0}^{t} \cdot\ d_\sigma R(t,\sigma)\cdot f(\sigma), \quad a \leq t \leq b.$$

Let Y be a Banach space and Y' its dual. We recall that for $A \subset Y$ and $B \subset Y'$ their polarsets A° and $^\circ B$ are defined by

$$A^\circ = \left\{y' \in Y' \mid \langle y', a\rangle = 0 \quad \text{for all} \quad a \in A\right\}$$

$$^\circ B = \left\{y \in Y \mid \langle b, y\rangle = 0 \quad \text{for all} \quad b \in B\right\}.$$

A° and $^\circ B$ are closed vector subspaces of Y' and Y respectively. We have $A^\circ = [^\circ(A^\circ)]^\circ$ and $^\circ B = {}^\circ[(^\circ B)^\circ]$ and from the Theorem of Hahn-Banach it follows that $A = {}^\circ(A^\circ)$ iff A is a closed vector subspace of Y and analogously $B = (^\circ B)^\circ$ iff B is a weak $*$ closed vector subspace of Y' (see [10]).

<u>Lemma 4.4</u> - $(f, c) \in G([a, b], X)\times Y$ satisfies $(*)$ for every $(y, d) \in BV_{t_0}^{a,b}([a, b], X')\times Y'$ that is a solution of $(K_\alpha^*)_0$ iff we have $c-c_f \in (\ker J(t_0)^*)$.

Proof - From Theorem 3.2 it follows that $(K^*_\alpha)_o$ has a solution $(y, d) \in BV^{a,b}_{t_o}([a, b], X')\times Y'$ iff $J(t_o)^*d = 0$ and in this case y is given by (7). If we replace this expression of y in (*) we obtain

$$\int_a^b \cdot <dy(s), f(s)>+<d, c> = \int_a^{t_o} \cdot <dy(s), f(s)> + \int_{t_o}^b \cdot <dy(s), f(s)> + <d, c> =$$

$$= -\int_a^{t_o} \cdot <d_s\left[\int_a^s \cdot d\alpha(\sigma)\circ R(\sigma, s)\right]^* d, f(s)> + \int_{t_o}^b \cdot <d_s\left[\int_s^b \cdot d\alpha(\sigma)\circ R(\sigma, s)\right]^* d, f(s)+<d, c> =$$

$$= <d, c - \int_a^{t_o} \cdot d_s\left[\int_a^s \cdot d\alpha(\sigma)\circ R(\sigma, s)\right]\cdot f(s) + \int_{t_o}^b \cdot d_s\left[\int_s^b \cdot d\alpha(\sigma)\circ R(\sigma, s)\right]\cdot f(s)> =$$

$$= <d, c + \int_a^b d_s\left[\int_s^b \cdot d\alpha(\sigma)\circ R(\sigma, s)\right]\cdot f(s) - \int_a^{t_o} \cdot d_s\left[\int_a^b \cdot d\alpha(\sigma)\ R(\sigma, s)\right]\cdot f(s)> =$$

$$= <d, c-c_f - J(t_o)f(t_o)>;$$

since by Theorem 3.2 we have

$$<d, J(t_o)f(t_o)> = <J(t_o)^*d, f(t_o)> = 0$$

it follows that we have (*) i.e. that we have

$$<d, c-c_f-J(t_o)f(t_o)> = 0 \quad \text{for every} \quad d \in \ker J(t_o)^*$$

iff we have

$$<d, c-c_f> = 0 \quad \text{for} \quad d \in \ker J(t_o)^*$$

i.e., iff we have $c-c_f \in$ $(\ker J(t_o)^*)$.

Proof of A <==> B - From the Lemmas 4.4 and 4.3 it follows that we have A iff $J(t_o)X = {}^\circ(\ker J(t_o)^*)$ and since ${}^\circ B = {}^\circ[({}^\circ B)^\circ]$ for $B \subset Y'$ (see the remarks that precede Lemma 4.4), we have $J(t_o)X = {}^\circ(\ker J(t_o)^*)$ iff $J(t_o)X = {}^\circ[(J(t_o)X)^\circ]$ i.e., iff $J(t_o)X$ is a closed vector subspace of Y.

In order to prove C <==> D we need the 2 following Lemmas:

<u>Lemma 4.5</u> - For $h \in BV_{t_o}^{a,b}([a, b], X')$ we have

$$h \in K_\alpha^*[BV_{t_o}^{a,b}([a, b], X')\times Y'] \text{ iff } \int_a^b \cdot R(\sigma, t_o)^* \cdot dh(\sigma) \in J(t_o)^* Y'.$$

<u>Proof</u>. We have $h \in K_\alpha^*[BV_{t_o}^{a,b}([a, b], X')\times Y']$ iff there exists $(y, d) \in BV_{t_o}^{a,b}([a, b], X')\times Y'$ such that $h = K_\alpha^*(y, d)$. By Theorem 3.1 this equation has always a solution $y \in BV_{t_o}([a, b], X')$ and by (5) we have

$$y(b) - y(a) = -\int_a^b \cdot R(\sigma, t_o)^* \cdot dh(\sigma) + \int_a^b \cdot [d\alpha(\sigma)\circ R(\sigma, t_o)]^* d;$$

hence we have $y(b) = y(a)$ iff $\int_a^b \cdot R(\sigma, t_o)^* \cdot dh(\sigma) = J(t_o)^* d$ for some $d \in Y'$, i.e., iff $\int_a^b \cdot R(\sigma, t_o)^* \cdot dh(\sigma) \in J(t_o)^* Y'$.

<u>Lemma 4.6</u> - For $h \in BV_{t_o}^{a,b}([a, b], X')$ we have

$$h \in (\ker K_\alpha)^\circ \text{ iff } \int_a^b \cdot R(\sigma, t_o)^* \cdot dh(\sigma) \in (\ker J(t_o))^\circ$$

<u>Proof</u> - We have $h \in (\ker K_\alpha)^\circ$ iff $\langle h, z\rangle = 0$ for every $z \in G([a, b], X)$ that is a solution of the system $(K) + (F_\alpha)$ with $f = 0$ and $c = 0$ i.e. (by (ρ) of Lemma 4.2) iff $\langle h, z\rangle = 0$ for every $z \in G([a, b], X)$ that is of the form $z(t)=R(t, t_o)z(t_o)$ with $F_\alpha[z] = 0$ (i.e. with $J(t_o)z(t_o) = \int_a^b \cdot d\alpha(t)\cdot R(t, t_o)z(t_o) = 0$ i.e. with $z(t_o) \in \ker J(t_o)$) i.e., iff

$$\int_a^b \cdot \langle dh(\sigma), R(\sigma, t_o)z(t_o)\rangle = 0 \text{ for every } z(t_o) \in \ker J(t_o)$$

i.e., iff $\int_a^b \cdot R(\sigma, t_o)^* \cdot dh(\sigma) \in (\ker J(t_o))^\circ$.

By the Lemmas 4.5 and 4.6 we have

$$K_\alpha^*[BV_{t_o}^{a,b}([a, b], X')\times Y'] = (\ker K_\alpha)^\circ \text{ iff } J(t_o)^* Y' = (\ker J(t_o))^\circ$$

hence we have C iff $J(t_o)^* Y' = (\ker J(t_o))^\circ$. Since $A^\circ = [{}^\circ(A^\circ)]^\circ$

for $A \subset X$ it follows that we have $J(t_0)^*Y' = (\ker J(t_0))^\circ$ iff $J(t_0)^*Y' = [{}^\circ(J(t_0)^*Y')]^\circ$, i.e., iff $J(t_0)^*Y'$ is a weak * closed vector subspace of X' (see the remarks that precede Lemma 4.4). But by Theorem 11.3.3 of [10] the following properties are equivalent

a) $J(t_0)X$ is a closed in Y

b) $J(t_0)^*Y'$ is closed in X'

c) $J(t_0)^*Y'$ is weak * closed in X'

and this completes the proof of $C \Longleftrightarrow D$ and proves $B \Longleftrightarrow C$.

Remark - When the equivalent properties of Theorem 4.1 are satisfied the solution x of the system $(K) + (F_\alpha)$ may not be unique: if we take $\alpha = 0$ it follows from Theorem 3.2 that for any $(f, 0)$ the hypothesis of A are satisfied but any solution x of (K) is a solution of the system $(K) + (F_\alpha)$.

Examples where B is satisfied:

1. $\dim Y < \infty$ or, more generally, $\dim J(t_0)X < \infty$.

2. $J(t_0) = j+k$, where j is bicontinuous and k is compact.

Remark - If (K) has no resolvent then in general (even in the numerical case, $X = Y = \mathbb{R}$) it is not true that the system $(K) + (F_\alpha)$ has a solution if (f, c) satisfies (*) for every solution (y, d) of $(K_\alpha^*)_0$. Example: we take the kernel K of Example 1, p. 189 of [6] (where $a = t_0 = 0$ and $b = 1$). Then (K) becomes

$$x(t)-x(0)+x(0+)-x(t-) - \int_0^{t/2} x(s)ds + \int_{t/2}^{t} x(s)ds = f(t)-f(0), \quad 0<t\leq 1,$$

and $(K_\alpha^*)_0$ is

$$y(s) + \int_s^1 \cdot K(t, s)\cdot dy(t) - \int_0^1 \cdot K(t, 0)\cdot dy(t) + \alpha(s)d = 0, \quad 0 \leq s \leq 1,$$

i.e.,

$$y(s) + \int_s^{2s} (1-t+s)dy(t) + \int_{2s}^{1} (1-s)dy(t)+\alpha(s)d = 0 \quad \text{for} \quad 0 \le s \le \frac{1}{2},$$

$$y(s) + \int_s^{1} (1-t+s)dy(t)+\alpha(s)d = 0 \quad \text{for} \quad \frac{1}{2} \le s \le 1,$$

that a short calculation reduces to

$$\int_s^{2s} y(t)dt + \alpha(s)d = 0 \quad \text{for} \quad 0 \le s \le \frac{1}{2},$$

$$\int_s^{1} y(t)dt + \alpha(s)d - 0 \quad \text{for} \quad \frac{1}{2} \le s \le 1.$$

If we take $\alpha = 0$ then the solutions of $(K_\alpha^*)_0$ are $(0, d)$, $d \in Y' = \mathbb{R}$, hence for any $(f, 0)$, $f \in G([0, 1])$, we have (*). However if f is continuous (K) becomes

$$-\int_0^{t/2} x(s)ds + \int_{t/2}^{t} x(s)ds = f(t) - f(0),$$

hence (K) has no solution x if f is not the primitive of a regulated function.

REFERENCES

[1] J. Hale, *Theory of Functional Differential Equations*, Springer-Verlag, 1977.

[2] D. Henry, "The adjoint of a linear functional differential equation and boundary-value problems", *J. Diff. Eq.* 9 (1971), 55-66.

[3] C. S. Hönig, *The abstract Riemann-Stieltjes integral and its applications to linear differential equations with generalized boundary conditions*, Notas do Instituto de Matemática e Estatística da Universidade de São Paulo, Série Matemática, nº 1, 1973.

[4] C. S. Hönig, "The adjoint system of a Volterra-Stieltjes integral equation with a linear constraint", *Atas do 4º Seminário Brasileiro de Análise* (1976), 152-162. Sociedade Brasileira de Matemática.

[5] C. S. Hönig, *Volterra Stieltjes-Integral Equations*, Mathematics Studies 16, North-Holland Publ. Comp., 1975.

[6] C. S. Hönig, "Volterra Stieltjes-integral equations", in *Functional Differential Equations and Bifurcation*, Proceedings of the São Carlos Conference, 1979, Springer Lecture Notes in Mathematics, 799, p. 173-216.

[7] C. S. Hönig, "The resolvent of linear Volterra Stieltjes-integral equations", *Atas do 11º Seminário Brasileiro de Análise* (1980), 361-381. Sociedade Brasileira de Matemática.

[8] St. Schwabik & M. Tvrdý, "Boundary value problems for generalized linear differential equations", Czech. Math. J. 29 (104) 1979, 451-477.

[9] M. Tvrdý, "Linear functional-differential operators; normal solvability and adjoints", in *Colloquia Mathematica Soc. János Bolyai, 15. Differential Equations*, Keszthely (Hungary), 1975, 379-389.

[10] A. Wilansky, *Modern Methods in Topological Vector Spaces*, MacGraw-Hill, 1978.

ON A FIXED POINT INDEX METHOD FOR THE ANALYSIS OF THE ASYMPTOTIC BEHAVIOR AND BOUNDARY VALUE PROBLEMS OF PROCESS AND SEMIDYNAMICAL SYSTEMS

A. F. Izé

Universidade de São Paulo
Instituto de Ciências Matemáticas de São Carlos
Departamento de Matemática
São Paulo, SP - Brazil

1. Introduction

Ważewski principle [23] plays an important role in the study of ordinary differential equations. Its applicability is largely due to the fact that in a finite dimensional euclidean space, the unity sphere is not a retract of the closed unit ball. Since this is no longer true in infinite-dimensional Banach space the direct extension of Ważewski's principle to processes or semidynamical systems on infinite dimensional Banach spaces has a very limited applicability.

Since in finite dimensional spaces the fact that the unity sphere is not a retract of the closed unit ball is equivalent to the fact that every continuous mapping of the unity closed convex ball has a fixed point, the main idea of this work is to develop a method based on fixed point index properties instead of retraction properties.

Our fixed point formulation, Theorem 2, is essentially equivalent, in finite dimension, to Ważewski theorem. Although in infinite dimension, Ważewski theorem is no longer applicable, Theorem 2 and also Theorem 3 are applicable and give deeper results since fixed point methods have proved to be very useful in the solution of differential equations either in finite or infinite dimensional spaces. After that we go further and generalize theorems 2 and 3 using Leray-Schauder degree theory or the fixed point index theory for compact or condensing maps. These generalizations, theorems 4, 5 and 6 are stronger even in finite dimension than Ważewski theorem.

After Ważewski paper several papers arose applying Ważewski principle to the asymptotic behavior of ordinary differential equations, C. Olech [17], V. Pliss [20], Mikolajska [14], N. Onuchic [18], A. F. Izē [11] and others. Kaplan, Lasota and Yorke [12] applied Ważewski method to boundary value problem and C. Conley [3] also applied Ważewski method to a boundary value problem for a difusion equations in biology. Since our approach uses Ważewski basic ideas in connection with degree theory it should give, even in finite dimensions much better results and can be applied also to boundary problems in Hilbert spaces.

2. Processes

Definition 1: [2]. Suppose X is a Banach space $R^+ = [0,\infty)$, $u: R \times X \times R^+ \to X$ is a given mapping and define $U(\sigma,t): X \to X$ for $\sigma \in R$, $t \in R^+$ by

$$U(\sigma,t)x = u(\sigma,x,t)$$

A process on X is a mapping $u: R \times X \times R^+ \to X$ satisfying the following property

i) u is continuous

ii) $U(\sigma,0) = I$ (identidy)

iii) $U(\sigma+s,t)U(\sigma,s) = U(\sigma,s+t)$

A process is said to be an autonomous process or a semidynamical system if $U(\sigma,t)$ is independent of σ, that is, $T(t) = U(0,t)$, $t \geq 0$. Then $T(t)x$ is continuous for $(t,x) \in R^+ \times X$.

The results presented here are also true for local processes or local semidynamical system, that is, when the interval of definition is $[0,\omega_u)$, $\omega_u \leq \infty$.

Processes arise from ordinary differential equations evolution equations, retarded and neutral functional differential equations and partial differential equations.

<u>Definition 2</u>: Suppose u is a process on X. The trajectory $\tau^+(\sigma,x)$ through $(\sigma,x) \in R \times X$ is the set in $R \times X$ defined by

$$\tau^+(\sigma,x) = \{(\sigma+t,\ U(\sigma,t)x \mid t \in R^+\}.$$

The orbit $\gamma^+(\sigma,x)$ through (σ,x) is the set in X defined by

$$\gamma^+(\sigma,x) = \{U(\sigma,t)x,\quad t \in R^+\}$$

<u>Definition 3</u>: If u is a process on X then an integral of the process on R is a continuous function $y:R \to X$ such that for any $\sigma \in R$, $\tau^+(\sigma,\ y(\sigma)) = \{(\sigma+t,\ y(\sigma+t)) \mid t \geq 0\}$. An integral y is an integral through $(\sigma,x) \in R \times X$ if $y(\sigma) = x$.

We assume in the following that the integral through each $(\sigma,x) \in R \times X$ is unique.

We define

$$\tau^{-1}(x) = \{(\sigma,y) \in R \times X \mid \exists\, t > 0 \text{ such that } U(\sigma,t)y = x\}.$$

If $P_0 = (\sigma,x) \in R \times X$ and $z \in \gamma^+(\sigma,x)$, we define

$$t_z = \inf\{t \geq 0 \mid U(\sigma,t)x = z\}$$

$$Q_z = (\sigma+t_z,\ U(\sigma,t_z)x)$$

$$[P_0,\ Q_z] = \{(\sigma+t,\ U(\sigma,t)x \mid 0 \leq t < t_z\}$$

$$[P_o, Q_z) = \{(\sigma+t, U(\sigma,t)x) | 0 \leq t < t_z\}$$

$$(P_o, Q_z] = \{(\sigma+t, U(\sigma,t)x) | 0 < t \leq t_z\}$$

$$(P_o, Q_z) = \{(\sigma+t, U(\sigma,t)x) | 0 < t < t_z\}.$$

Let Ω be an open set of $R \times X$, ω an open set of Ω, $\omega \subset \Omega$, $\omega \neq \phi$ and $\partial\omega = \bar{\omega} \cap (\overline{\Omega-\omega})$ the boundary of ω with respect to Ω. We put:

$$S^o = \Big\{P_o = (\sigma,x) \in \partial\Omega \mid \exists t > 0 \text{ and } z \in \gamma^+(\sigma,x), \text{ with }$$

$$(P_o, Q_z) \neq \emptyset \text{ and } (P_o, Q_z) \cap \bar{\omega} = \emptyset\Big\}$$

$$S = \Big\{Q \in \partial\Omega \mid \exists P_o = (\sigma,x) \in \omega, \text{ with } Q \in \tau^+(\sigma,x) \text{ and }$$

$$[P, Q) \subset \omega\Big\}$$

$$S^* = S^o \cap S.$$

The points of S are called egress points, the points of S^* are called strict egress points.

Given a point $P_o = (\sigma,x) \in \omega$, if the trajectory $\tau^+(\sigma,x)$ of the process is contained in ω for every $t > 0$, we say that the trajectory is asymptotic with respect to ω; if the trajectory is not asymptotic with respect to ω then there is a $t > 0$ such that $(\sigma+t, U(\sigma,t)x) \in \partial\omega$. Taking

$$t_{P_o} = \Big\{\min t > 0 \mid (\sigma+t, U(\sigma,t)x) \in \partial\omega\Big\}$$

$$Q = (\sigma+t_{P_o}, U(\sigma,t_{P_o})x = C(P_o)$$

we have

$$[P_o, Q) \subset \omega.$$

The point $C(P_o)$ is called the consequent of P_o.

Define G to be the set of all $P_o = (\sigma,x) \in \omega$ such that there is $C(P_o)$ and $C(P_o) \in S^*$.

Consider the mapping

$$K:G\cup S^* \to S^*.$$

$$K(P_o) = C(P_o), \text{ if } P_o \in \omega \text{ and } K(P_o) = P_o, \text{ if } P_o \in S^*.$$

The proof of the following is standard, see for example [2], [19].

<u>Lemma 1</u>. The mapping $K:G\cup S^* \to S^*$ is continuous.

<u>Theorem 1</u>. (First form) Assume that there exists sets ω open in Ω, $S \subset \partial\omega$ and $Z \subset \omega\cup S$, $Z \neq \emptyset$, satisfying the conditions

i) $S = S^*$ (that is, the points of S are strict egress points)

ii) Z is compact and convex

iii) There is a retraction $r:S \to Z \cap S$

iv) There exists a constinuous mapping $\Phi:Z \cap S \to Z \cap S$ such that $\Phi(P) \neq P$ for every $P \in Z \cap S$.

Then there is at least one point $P_o = (\sigma,x) \in Z \cap \omega$ such that the trajectory $\tau^+(\sigma,x)$ is contained in ω, that is, the trajectory $\tau^+(\sigma,x)$ is asymptotic with respect to ω.

<u>Proof</u>. Assume the theorem is not true. Hence, for every $P_o \in Z \cap \omega$ the trajectory through P_o is not asymptotic with respect to ω, that is there exists $C(P_o)$. Since $S = S^*$, $C(P_o) \in S^*$. Then

$$Z \cap \omega \subset G.$$

From Lemma 1 the mapping K is continuous. From condition (iii) there is a retraction:

$$r:S \to Z \cap S.$$

Then the mapping $R = r.K:Z \cap \omega \to Z \cap S$ is continuous and takes $P_o = (\sigma,x)$ into $C(P_o) = (\sigma+t_Q, U(\sigma,t_Q)x) \in Z \cap S$.

From condition (iv) the mapping Φ takes $C(P_o)$ into $\Phi(C(P_o)) = C'(P_o) \neq C(P_o)$ and then the composite mapping $\Phi\cdot r\cdot K:Z \to Z$ is continuous and never has a fixed point. This is in contradiction with Schauder fixed point theorem. Then the trajectory $\tau^+(P_o)$ is asymptotic with respect to ω and the theorem is proved.

Remark 1: One simple situation in which condition (iv) is satisfied is when $Z \cap S$ is symmetric, that is, $-Z \cap S \subset Z \cap S$. The mapping $\Phi:Z \cap S \to Z \cap S$ is defined by $\Phi(P) = -P$.

Theorem 2: (Second form) Assume that there exists sets ω open in Ω,

$S \subset \partial\omega$ and $Z \subset \omega \cup S$, $Z \neq \emptyset$, satisfying the conditions:

i) $S = S^*$

ii) Z is closed bounded convex

iii) There is a retraction $r:S \to Z \cap S$

iv) There is a continuous mapping $\Phi:Z \cap S \to Z \cap S$ such that $\Phi(P) \neq P$ for every $P \in Z \cap S$

v) U is compact

Then there is at least one point $P_o = (\sigma,x) \in Z \cap \omega$ such that the trajectory $\tau^+(\sigma,x)$ is contained in ω, that is, the trajectory through P_o is asymptotic with respect to ω.

The proof follows as in Theorem 1. Since U is compact the mapping K that takes P_o into $C(P_o)$ is compact. r is a retraction and then $R = r.K:Z \cap \omega \to Z \cap S$ is compact. The mapping $\Phi\cdot r.K:P_o \longrightarrow C'(P_o)$ is also compact and never has a fixed point. This is in contradiction with Schauder fixed point theorem since $\Phi.r.K:Z \to Z$ and Z is closed convex, bounded.

Then the trajectory through P_0 is asymptotic with respect to ω and the theorem is proved.

In some applications, v.g. functional differential equations of the neutral type, evolution equations and some integral equations, the process described by these equations is not compact, then we should extend Theorem 2 to a more general class of mapping called α-set contractions and condensing mappings, introduced by Kuratowski [13].

If X is a bounded metrix space we define the measure of non compactness of A to be $\inf\{d > 0 \mid A$ can be covered by a finite number of sets of diameter less than or equal to $d\}$. If X is a Banach space and A a bounded subset of X, A inherites a metric from X and we can give the same definition of the measure of non compactness of A.

Let X_1 and X_2 be metrix spaces and suppose $f:X_1 \to X_2$ is a continuous map. We say that f is a k-set-contraction if given any bounded set A in X_1, $f(A)$ is bounded and $\gamma_2(f(A)) \leq k\gamma_1(A)$. Of course γ_i denotes the measure of non compactness in X_i, $i = 1, 2$. We assume that $0 \leq k < 1$. If f is a k-set contraction we define $\gamma(f) = \inf\{k \geq 0 \mid f$ is a k-set contraction$\}$. We say that $f:X \to X$ is a local strict set-contraction if for every $x \in X$ there is a neighbourhood $N(x)$ such that $f/N(x)$ is a k_x-set-contraction.

M. Furi and A. Vignoli [8] and B. N. Sadovskii gave a slight generalization of k-set-contraction. Given a continuous mapping $f:X_1 \to X_2$ we say that f is a condensing map if for every bounded set $A \subset X_1$ such that $\gamma_1(A) \neq 0$, $\gamma_2(f(A)) < \gamma_1(A)$. We say that f is a local condensing map if every $x \in X$ has a neighbourhood $N(x)$ such that $f/N(x)$ is a condensing map. If f is an k-set-contraction f is condensing but the converse is not true in general: see Nussbaum [16]. If f is linear the two concepts are equivalent.

Theorem 2 can be reformulated for k-set contraction or condensing mapping. The proof is similar to the proof given in Theorem 2.

<u>Theorem 3</u>: Assume that there exists sets ω open in Ω, $S \subset \partial\omega$ and $Z \subset \omega \cup S$, $Z \neq 0$ satisfying the conditions:

i) $S = S^*$

ii) Z is closed convex bounded

iii) $Z \cap S$ is a retract of S, that is, there exists a retraction $r: S \to Z \cap S$

iv) U is condensing

v) There is a continuous map $\Phi: Z \cap S \to Z \cap S$ such that $\Phi(P) \neq P$ for every $P \in Z \cap S$ and $\Phi.r.U$ is condensing.

Then there exists at least one point $P_o = (\sigma, x) \in Z \cap \omega$ such that the trajectory $\tau^+(\sigma, x)$ is contained in ω.

The proof follows as in Theorem 2, since U is condensing, the map K that takes P_o into $C(P_o)$ is condensing. r is a retraction and then $R = r.K: Z \cap \omega \cap Z \to S$ is condensing. The map $\Phi.r.K: P_o \to C'(P_o)$ is condensing from (v) and never has a fixed point. This is in contradiction with the fixed point theorem for condensing maps then $\tau^+(\sigma, x) \subset \omega$ and the theorem is proved.

Theorems 2 and 3 are general enough to be applied in most of the problems on the asymptotic behavior and boundary value problems on Banach or Hilbert spaces. One criticism made about the application of Ważewski theorem to boundary problems made by Kaplan, Lasota, Yorke [12] is that it should be possible to get similar results applying the Leray-Schauder degree. As was explained in the introduction, Ważewski method is essentially equivalent, in finite dimension, to theorems 1 and 2 that uses Brouwer fixed point theorem. In the following we will generalize theorems 1, 2 and 3 in the more general context of the fixed point index theory. Even for finite dimension this more general formulation is more general than Ważewski method and should give deeper results when applied to boundary value problems either in finite or infinite dimensional Hilbert spaces.

First of all let us recall the basic properties of the classical fixed point index [1] as well as the extensions made by Nussbaum [16] for k-set contraction and condensing maps. We shall say that a topological space X is an absolute neighbourhood retract (ANR) if given any metric space M, a closed sub space $A \subset M$ and a continuous map $f:A \to X$ there exists an open neighbourhood U of A and a continuous map $F:U \to X$ such that $F(a) = f(a)$ for $a \in A$. X is called an absolute retract (AR) if F as above can be defined on all of M. A theorem of Dugundji [6] asserts that any convex subset of a locally convex topological space is an AR. Let A be the category of compact metric absolute neighbourhood retract (ANR_s). Let $A \in A$, G be an open subset of A and $f:\bar{G} \to A$ be a continuous function which has no fixed points on ∂G. Then there is a unique integer valued function $i_A(f,G)$ which satisfies the following four properties:

1 - If $f:\bar{G} \to A$ has no fixed point on ∂G and the fixed points of f lie in $G_1 \cup G_2$ where G_1 and G_2 are two disjoint open sets included in G, then $i_A(f,G) = i_A(f,G_1) + i_A(f,G_2)$. In particular if f has no fixed points in G, this is meant to say that $i_A(f,G) = 0$. (the additivity property).

2 - Let I denote the closed unit interval $[0, 1]$. If $F:\bar{G} \times I \to A$

(A belongs to $\mathcal{A}$ of course) is a continuous map, and $F_t(x) = F(x,t)$ has no fixed points on ∂G for $0 \leq t \leq 1$ then $i_A(F_0, G)=i_A(F_1, G)$. (the homotopy property).

3 - If $\bar{G} = A$ then $i_A(f,G) = \Lambda(f)$, the Lefchetz number of f, equals $\sum(-1)^K$ trace (f_{*K}), where $f_{*K}:H_K(A) \to H_K(A)$ is the vector space homomorphism of $H_K(A)$ to $H_K(A)$ and $H_K(A)$ is the Cech homology of A with rational coefficients. (the normalization property).

4 - Let A and B be two spaces which belongs to $\mathcal{A}$. Let $f:A \to B$ be a continuous map. Let V be an open sub set of B and $g:\bar{V} \to A$ a continuous map. Assume fg has no fixed points on ∂V. Let $U = f^{-1}(V)$. Then gf has no fixed points on ∂U and $i_B(fg,V) = i_A(gf,U)$. (the commutativity property).

Let G be an open sub set of a Banach space X and $g:\bar{G} \to X$ a continuous map such that $g(x) \neq x$ for $x \in \partial G$. Assume that g is compact, that is, $g(G)$ has compact closure. Leray and Schauder [1] defined a fixed point index for g and consequently a degree for I - g, I the identity function. We shall denote this degree by $\deg(I-g, G, 0)$. It turns out that the Leray-Schauder degree satisfies all the four properties of the fixed point index listed above. So we can define the fixed point index by

$$i_X(g/X \cap G, X \cap G) = \deg(I-g, G, 0)$$

where $X = \overline{co}\, g(G)$.

Now we can generalize Theorem 2, using the fixed point index. A further generalization will be given in Theorem 5, extending these results to fixed point index for k-set contractions and condensing maps. We say that a set A is admissible if $A \in \mathcal{A}$. In most the applications we can use the Leray-Schauder degree and in this case A is a closed convex set of a Banach space.

Theorem 4: Assume that there exists sets ω open in Ω, $S_1 \subset S \subset \partial\Omega$ and $Z \subset \omega \cup S_1$, $Z \neq \emptyset$, $Z \cup S_1$ closed convex satisfying the conditions:

i) $S = S^*$

ii) There exists a continuous mapping $\Phi: S_1 \to S_1$ such that $\Phi(P) \neq P$ for every $P \in S_1$

iii) U is compact

iv) $i(\Phi U, Z \cap \omega) \neq 0$.

Then there exists at least one point $P_0 = (\sigma, x) \in Z \cap \omega$ such that either, $C(P_0) \in S - S_1$ or $C(P_0)$ does not exist, that is, the trajectory $\tau^+(\sigma, P_0)$ is asymptotic with respect to ω.

Proof. Assume that the theorem is not true. Then $C(P_0) \in S_1$ for every $P_0 \in Z \cap \omega$ and then $Z \cap \omega \subset G$. Then $Z = (Z \cap S_1) \cup (Z \cap \omega) \subset S \cup G$. From Lemma 1 the map K is continuous and the restriction of K to $Z \cup S_1$ is also continuous. Since U is compact the map K that takes P_0 into $C(P_0)$ is compact. The transformation ΦU is also compact and $\Phi U(P) \neq P$ for every $P \in Z \cup S_1$. Hence $i\ (\Phi U, Z \cap \omega) = 0$, what is a contradiction. Then there exists a least one point $P_0 \in Z \cap \omega$ such that the trajectory through P_0 is asymptotic with respect to ω.

Theorem 5. Assume that there exists sets ω open in Ω, $S_1 \subset S$ and $Z \subset \omega \cup S_1$, $Z \neq \emptyset$, Z closed convex, satisfying the conditions:

i) $S = S^*$

ii) There exists a continuous map $\Phi: Z \cap S_1 \to Z \cap S_1$ such that $\Phi(P) \neq P$ for every $P \in Z \cap S_1$

iii) U is compact

iv) $i\ (\Phi r U, Z \cap \omega) \neq 0$

v) There is a retraction $r:S \to Z \cap S_1$.

Then there exists at least one point $P_0 = (\sigma,x) \in Z \cap \omega$ such that either $C(P_0) \in S-S_1$ or $C(P_0)$ does not exist.

The proof is similar to Theorem 2 and 4.

Let X be a closed subset of a Banach space B. We shall say that $X \in F$ if we can write $X = \cup_{i=1}^{n} C_i$, where C_i are closed convex sets in X. The metric on X will always be that which it inherites from B.

Let $G \subset B$ and $f:G \to B$ a continuous map. Assume that the set $S = \{x \in G \mid g(x) = x\}$ is closed and bounded. Let us write $K_1(f,G) = \overline{Co}f(G)$, $K_n(f,G) = \overline{Co}(G \cap K_{n-1}(f,G))$ and

$K_\infty(f,G) = \cap_{n\geq 1} K_n(f,G)$ where $\overline{Co}$ denotes convex closure. It is easy to see that $f:G \cap K_\infty(f,G) \to K_\infty(f,G)$ and $K_\infty(f,G)$ is closed and convex. If G is bounded and $f:G \to X$ is a k-set contraction, $k < 1$, Darbo's results also implies that $K_\infty(f,G)$ is compact. Finally, assume that f is a local strict set contraction. By this we mean that for every point $x \in G$ there is a neighbourhood $N(x)$ and a real number $0 \leq k_x < 1$ such that $f/N(x)$ is k_x-set-contraction. Using these assumptions we can find a bounded open neighbourhood G_1 of S such that $f:G_1 \to X$ is a k-set contraction, $k < 1$. Let us write $K_\infty^* = K_\infty(f,G_1) \cap X$, K_∞^* is a compact metric ANR, $G_1 \cap K_\infty^*$ is an open subset of K_∞^* and $f:G_1 \cap K_\infty^* \to K_\infty^*$ is a continuous function satisfying the necessary condition, so $i_{K_\infty^*}(f, G_1 \cap K_\infty^*)$ is defined, R. Nussbaum [16]. We define $i_X(f,G) = i_{K_\infty^*}(f, G_1 \cap K_\infty^*)$. All the usual index property carry through to this setting. A similar definition of the index is given in [16] for condensing maps.

Let X be a closed convex subset of a Banach space B, G is an open sub set of X and $f:\bar{G} \to X$ is a continuous condensing map such that $f(x) \neq x$ for $x \in \partial G$. Then the fixed point index can be described in terms of the Leray-Schauder degree. First it is not hard to show that there exists $\delta > 0$ such that $\|x - f(x)\| \geq \delta$ for $x \in \partial G$. Select any fixed $x_0 \in X$ and define $f_t(x) = tf(x) + (1-t)x_0$ for $0 < t < 1$ and take t so close to 1 that $\|f(x) - f_t(x)\| < \delta$ for $x \in \partial G$. Define $K_1 = \overline{\text{Co}} f_t(G)$, $K_n = \overline{\text{Co}} f_t(G \cap K_{n-1})$ for $n > 1$ and $K_\infty = \bigcap_{n \geq 1} K_n$. One can prove that K is compact (possibly empty) and convex and that $f_t(G \cap K_\infty) \subset K_\infty$. If K_∞ is empty define $i_X(f,G) = 0$. If K_∞ is not empty let K be any compact convex set such that $K \supset K_\infty$ and $f_t(G \cap K) \subset K$. K_∞ is itself such a set so the collection of such K is non empty. Let ρ be any retraction of B onto K. (a result of Dugundji [6] guarantees the existence of such a retraction), and let H be any bounded open neighbourhood of the (compact) fixed point set of f_t in $\bar{G}$ such that $\bar{H} \subset \rho^{-1}(G \cap K)$. Then one can prove that $i_X(f,G) = \deg(I - f_t \cdot \rho, H, 0)$. In particular the integer on the right hand side is independent of the particular K chosen, the retraction ρ, H, t and x_0. We say that a set A is admissible if $A \subset F$. For example A is closed convex.

<u>Theorem 6</u>. Assume that there exists sets ω open in Ω, $S \subset \partial\omega$, $S_1 \subset S$ and $Z \subset \omega \cup S_1$, $Z \neq \emptyset$, $Z \cup S_1$ closed convex satisfying the conditions:

i) $S = S^*$

ii) There exists a continuous map $\Phi: S_1 \to S_1$ such that $\Phi(P) \neq P$ for every $P \in S_1$

iii) ΦU is condensing

iv) $i(\Phi U, Z \cap \omega) \neq 0$.

Then either $P_0 \in S - S_1$ or the trajectory $\tau^+(\sigma, x)$ through (σ, x) is contained in ω. The proof follows as in Theorem 4.

Theorem 7. Assume that there exists sets ω open in Ω, $S_1 \subset S \subset \partial\omega$ and $Z \subset \omega \cup S_1$, $Z \neq \emptyset$, Z closed convex satisfying the conditions:

i) $S = S^*$

ii) $Z \cap S_1$ is a retract of S_1, that is, there exists a retraction $r:S_1 \to Z \cap S_1$

iii) There exists a continuous map $\Phi:Z \cap S_1 \to Z \cap S_1$ such that $\Phi(P) \neq P$ for every $P \in Z \cap S_1$

iv) $\Phi.r.U$ is condensing

v) $i\,(\Phi rU,\ Z \cap \omega) \neq 0$.

Then there exists at least one point $P \in Z \cap \omega$ such that either $C(P_0) \in S-S_1$ or the trajectory $\tau^+(\sigma,x)$ through (σ,x) is contained in ω. The proof follows as in Theorem 5.

Since we can define the index and consequently a degree for condensing maps, Nussbaum [16], theorems 6 and 7 are true if we assume that $\Phi.r.U$ is condensing. In the applications of the theorems above we must give criteria to verify the condition $i_X(f,G) \neq 0$. In most applications G is a closed convex subset of a Banach space X and the index can be described in terms of the Leray-Schauder degree. If D is a closed convex subset of a locally convex topological vector space X we say that D is a wedge if $x \in D$ implies $tx \in D$ for $t \geq 0$. We call D a cone (with vertex at 0) if D is a wedge and $x \in D$, $x \neq 0$ implies that $-x \in D$. The following result is given by Nussbaum [15].

Assume that D is a wedge in a Banach space X, r and R are unequal positive numbers, $G_1 = \{x \in D \mid \|x\| < \rho_1 = \max(r,R)\}$ and $f:\bar{G}_1 \to D$ is a condensing map. Assume that there exists $h \neq 0$ such that $x - f(x) \neq th$ for all $x \in D$ with $\|x\| = R$ and all $t \geq 0$ and suppose that $x - tf(x) \neq 0$ for $x \in D$, $\|x\| = r$ and $0 \leq t \leq 1$. Then a) If $r < R$, one has $i_D(f,U) = -1$ and b) If $r > R$, $i_D(f,U) = +1$.

A more general formulation is given in [16] for local condensing maps and some other conditions are also given to verify $i_D(f,U) \neq 0$.

Some other criteria are known. For example, suppose that $h:R^2 \to R^2$ is of class C^1, $h(0) = 0$, and $x = 0$ is an isolated zero of h. Then $i\ (h,0) = 1 + \frac{1}{2}(E-H)$ where E and H are integers associated with the flow $\dot{x} = h(x)$, that is, E is equal to the number of elliptic regions and H is equal to the number of hyperbolic regions. This formula for the index is due to Poincaré.

There are several examples of processes described by functional differential equations and partial differential equations of the evolution type that are compact or α-set contractions.

Example 1 - Let $r > 0$, $C = C([-r,0], R^n)$ the space of continuous functions defined in $[-r,0]$. If $x \in C([\sigma-r, \sigma+A), R^n)$, $A > 0$, $\sigma \in R$ define $x_t(\theta) = x(t + \theta)$. Let $\Omega \subset R \times C$, Ω open, and let D, $f:\Omega \to R^n$ be a continuous functions, D is linear and $D(\phi) = \phi(0) - \int_{-r}^{0} d\mu(t,\theta)\phi(\theta)$ where μ is a matrix function of bounded variation for $\theta \in [-r,0]$. A functional differential equation of the neutral type is a relation of the form

$$\frac{d}{dt} D(t,x_t) = f(t,x_t), \quad x_\sigma = \phi. \tag{1.1}$$

We say that D is an uniformly stable operator if there are constants $K > 1$, $\alpha > 0$ such that $|D(t,\phi)| \leq Ke^{-\alpha(t-t_o)}$, $t \geq \sigma$. The solutions of this equations describes a process $U(f,\phi)=x_t(\sigma,\phi)$. If D is an uniformly stable operator and $t > r$, U is a weak α-set contraction, that is, for every bounded set $A \subset R \times C$ for which $U(A)$ is bounded $\gamma(U(A)) \leq k\gamma(A)$. When $D(t,\phi) = \phi(0)$ equation (1.1) is the equation $\dot{x} = f(t,x_t)$ and if $t > r$, the process U is compact [9].

When we try to check the condition $S = S^*$ we need to calculate the derivative of the solutions along the solutions of equations and the derivative $\dot{x}$ should appear, explicitly, what does not always happens for equation (1.1). Another general form of a neutral equation for which there is a reasonable existence and continuation of solutions theory, [5], is the equation

$$\dot{x} = f(t, x_t, \dot{x}_t)$$

where $x_0(\theta) = \psi(\theta) \in C([-r, 0], R^n)$, $\dot{x}_0(\theta) = \psi(\theta) \in L^p([-r,0], R^n)$ and f satisfies a uniform Lipchitz condition with respect to ψ in $L^p([-r,0], R^n)$, $1 \leq p \leq \infty$. The process described by these equations is also an k-set contraction.

The following example is given in [22].

Example 2 - Let X be a Banach space and $A:D(A) \to X$ be a closed, densely defined linear operator in X. A is called sectorial if there are constants ϕ, M, a, $0 < \phi < \pi/2$, $M \geq 1$, $a \in R$ such that the sector $S_{\phi,a} = \{\lambda \in C | \lambda \neq a, \phi < \arg|\lambda - a| \leq r\}$ is contained in $\rho(A)$ the resolvent set of A, and $\|(\lambda-a)^{-1}\| \leq M/(\lambda-a)$ for all $\lambda \in S_{\phi,a}$. If A is sectorial, then there is a $k \geq 0$ such that $\mathrm{Re}\, \sigma(A + kI) > 0$. Let $A_1 = A + KI$. For $0 < \alpha < 1$ define

$$A_1^{-\alpha} = \frac{\sin \pi\alpha}{\pi} \int_0^\infty \lambda^{-\alpha}(\lambda-A_1)^{-1} d\lambda$$

$A_1^{-\alpha}$ is bounded and injective. Let X^α be the range of $A_1^{-\alpha}$, $X^0 = X$, $X' = D(A)$. Let $A_1^\alpha : X^\alpha \to X$ be the inverse of $A_1^{-\alpha}$, $A^0 = Id_X$, $A' = A$. X^α is dense in X. Define the norm $\| \, \|_\alpha$ on X^α by $\|u\|_\alpha = \|A_1^\alpha u\|$ where $\| \, \|$ is the norm for X. X^α does not depend on the choice of k, and different choices of k yields equivalent norm on X^α. X^α is a Banach space under $\|\cdot\|_\alpha$.

Suppose $0 \leq \alpha < 1$, V is open in X^α and $f:V \to X$ is a locally Lipchitz continuous mapping. Consider the equation

$$\frac{du}{dt} + Au = f(u). \tag{1.2}$$

Let $u_0 \in V$. By a solution of (1.2) on $(0,A)$ through u_0 we mean a continuous mapping $u:[0,A) \to V$ such that $u(0) = u_0$. u is differentiable on $(0,A)$, $u(t) \in D(A)$ for $t \in (0,A)$, $t \to f(u(t))$ is locally Hölder continuous, $\int_0^a \|f(u(t))\|\,dt < \infty$ for some $a > 0$ and (1.2) holds for $t \in (0,A)$. In this definition "$t \to g(t)$" is locally Hölder-continuous" means that for every t_0 there exists a neighbourhood W of t_0 and L, $\theta > 0$ such that $\|g(t_1) - g(t_2)\| \le L|t_1 - t_2|^{\theta}$ for $t_1, t_2 \in W$.

It follows from [7] that under the above assumptions, for every $u_0 \in U$ there exists a unique solution $u(u_0)$ of (1.2) through u_0, defined on a maximal interval $[0,W_{u_0})$. Defining $U(t)u_0 = u(u_0, t)$ for $t < W_{u_0}$ we obtain a local autonomous process or a local semi dynamical system.

The most important example of a sectorial operator arises in the following way: Let Ω be an open, bounded set in R^n whose boundary is of class C^{2m} (m an integer). Let $X = L^2(\Omega)$, $D(A) = H^{2m}(\Omega) \cap H_0^m(\Omega)$.

$(Au)(x) = \sum_{|\alpha| \le m} a_\alpha(x) \cdot D^\alpha u(x)$ where the $a_\alpha : \bar{\Omega} \to C$ are continuous mappings and $D^\alpha u$ is understood in the distributional sense. Suppose that A is uniformly strongly elliptic on Ω, i.e., there is a $C_0 > 0$ such that $(-1)^m \mathrm{Re}\left\{ \sum_{|\alpha|=m} a_\alpha(x) \cdot \xi^\alpha \right\} \ge C_0 |\xi|^{2m}$ for all $\xi = (\xi_\alpha)_{|\alpha| \le m}$, $\xi_\alpha \in R$, and all $x \in \Omega$ then equation (1.3) is called a semi linear parabolic P.D.E. Results in [10], imply that A is sectorial and $R(\lambda, A)$ is compact for every $\lambda \in \rho(A)$.

REFERENCES

[1] Browder, F. E. - On the fixed point index for continuous mapping of locally connected spaces - Summa Brasiliensis Mathematica, vol. 4, fasc. 7, (1960).

[2] Conley, C. C. - "Isolated Invariant Sets and the Morse Index", CBMS 38, Amdr. Math. Soc., Providence, R.I.

[3] Conley, C. C. - An application of Ważewski method to a non linear boundary value problem which arises in population genetics. J. Math. Biol. 2, (1975), 241-249.

[4] Dafermos, C. M. - An invariance principle for compact processes, J. Diff. Eqs. 9, 239-252, (1971).

[5] Driver, R. - Existence and continuous dependence of a neutral functional differential equation. Arch. Rat. Mech. Anal., 19, 149-166 (1965).

[6] Dugundji, J. - An extension of Tietze's theorem, Pacific J. Math. 1, (1951), 353-367.

[7] Friedman, A. - Partial differential equations - Holt, Rinehart and Winston Inc., (1969).

[8] Furi, M. & Vignoli - A fixed point theorem in complete metric spaces. Boll. Unione Matem. Ital. Sẽrie IV (N.4-S)(1969), 505-506.

[9] Hale, J. K. - Theory of functional differential equations, Springer Verlag, (1974).

[10] Henry, D. - Geometric theory of semilinear parabolic equations, Lecture notes, University of Kentucky, Springer Verlag.

[11] Izẽ, A. F. - Asymptotic behavior of ratio between the components of solutions of a linear system of ordinary differential equations and its applications to determination of Lyapunov characteristic numbers. Annali di Matematica Pura ed Applicata, Serie IV, tomo LXXXV, p. 1, (1970).

[12] Kaplan, J. L.; Lasota, A.; Yorke, J. A. - An application of the Ważewski retract method to boundary value problems. ZESZYTY NAUK. Uniw. Jagiello. Prace Mat. Zeszyt 16 (1974), 7-14.

[13] Kuratowski, C. - Sur les espaces completes. Fund. Math. 15, (1930), 301-309.

[14] Mikolajska, Z. - Sur les mouvements asymptotiques d'un point matẽriel mobile dans le champ des forces repousants. Bull. Acad. Polon. Sci. vol. I, nº 1, 2 (1953).

[15] Nussbaum, R. - Periodic solutions of some nonlinear autonomous functional differential equations. II. J. Differential Equations 14, 360-394 (1973).

[16] Nussbaum, R. D. - "The fixed point index for local condensing maps", Ann. Mat. Pura Appl. 89, 217-258, (1971).

[17] Olech, C. - On the asymptotic behavior of the solutions of a system of ordinary differential equations. Bull. Acad. Pol. Sc. vol. IV, nº 9, p. 555, (1956).

[18] Onuchic, N. - On the Asymptotic Behavior of the Solutions of Functional Differential Equations, in "Differential Equations and Dynamical Systems" (J. K. Hale and J. P. LaSalle, Eds.), Academic Press, New York, (1967).

[19] Pelczar, A. - Some generalizations of the retract theorem of T. Ważewski, with applications to ordinary and partial differential equations of the first order. Annales Polonici Mathematici, XXIX, (1974).

[20] Pliss, A. - On a topological method for studying the behavior of the integrals of ordinary differential equations. Bull. Acad. Polon. Sci. c-1, III, p. 2, (1954).

[21] Rybakowski, K. - Ważewski principle for retarded functional differential equations. J. Differential Equations 36, 117-138, (1980).

[22] Rybakowski, K. P. - On the homotopy index for infinite dimensional semiflows - Trans. Amer. Math. Soc. to appear.

[23] Ważewski, T. - Sur un principle topologique de l'examen de l'allure asymptotique des intēgrales des ēquations differentielles ordinaires. Ann. Soc. Polon. Math. 20, (1947).

INTRODUCTION TO BIFURCATION THEORY

Jorge Ize

IIMAS - Universidad Nacional Autonoma de Mexico
Apartado Postal 20-726
Mexico DF - Mexico
and
IMSP - Nice - France

Introduction

Bifurcation theory as the rest of nonlinear functional analysis has known a very quick growth in the last decade in theory as well as in applications.

A great number of phenomena studied in Sciences lead to mathematical models of the general form

$$F(x, \lambda) = 0 \tag{1}$$

where x represents the sought solution (belonging in general to an infinite dimensional space) and λ a set of parameters (usually a finite number of them). The word bifurcation is pictorial enough to get a first idea of the problem: at some point (x_0, λ_0) one has more than one solution and one has the following "bifurcation picture"

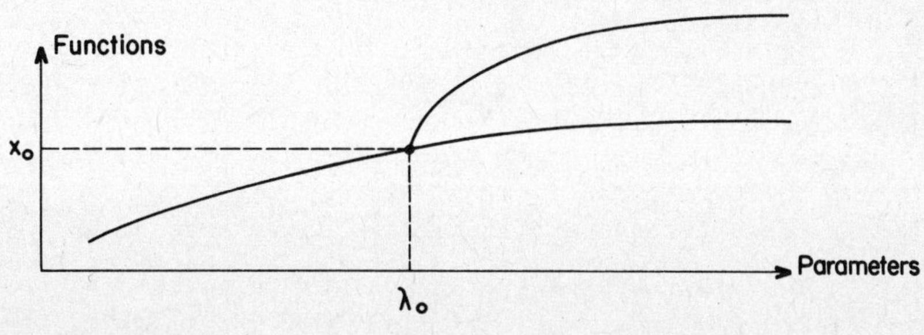

A natural idea is to linearize the equation near (x_0, λ_0) and to consider the problem as a nonlinear perturbation of a singular linear problem.

There are a great number of techniques for this problem: to the traditional perturbation theory, which is quite useful in concrete problems, several other points of view have been added:

1) singularity theory which by reducing to generic cases gives an idea of the phenomena but has the disadvantage that no one knows if the real problem is generic or not and if the stability arguments (Nature chooses generic situations) are valid or not (see [G.G.]);

2) dynamical systems theory which, by reducing to an ordinary differential system, enables one to use very powerful techniques like the center manifold theorem or attractors but with strong doubts about the validity, mainly for global behavior, of this reduction ([Sj], [M.M]);

3) variational analysis ([Be], course of A. Castro in these proceedings) which gives good results with the obvious limitations that not all problems can be cast in gradient forms. An extension of these ideas to variational inequalities and monotone operators has given interesting results ([E.T.], [Br]). For second order elliptic equations and positive solutions a combination of monotonicity and degree theory has given good advances ([Am], [Kr 2], course of D.G. de Figueiredo in these proceedings);

4) group theory gives a well defined structure to some problems, [Sa];

5) Topology theory gives quick information but, as a logical consequence, not too precise.

Clearly these techniques are not exclusive and, given a problem, it is wise to use every one of them which may bring some information.

However it is understandable that it is not possible in an introductory course to give a global view of the field and that one has to choose only one point of view, in this case the topological one. Even so this paper is a reduced version of the set of the lecture notes handed at the LatinAmerican School of differential equations (spanish written and 175 pages long) so that a certain number of assumptions on the reader's background are made: some familiarity with examples of bifurcation problems and of functional analysis applied to differential equations so that the results given in an abstract form can be translated to concrete problems. No previous knowledge of topology since the principal aim of this paper is to go from schracth to the statement of some open problems in a language accessible to an analyst.

Most of the material presented here is scattered in the litterature and the main references are [Be], [Iz 1], [Ni], [Ra], [Sa] and [Sj].

The plan of the paper is as follows: the first chapter is concerned with a brief description of the mathematical problem and with the topological tools needed. The second chapter deals with local bifurcation and the last with the global theory and some open problems are given in conclusion.

The author wishes to thank the Instituto de Matemãtica e Estatistica de la Universidade de São Paulo for its kind invitation, to the CNPq and IIMAS for their support and to the Department of Mathematics de l'Universitẽ de Nice where this version was written up.

INDEX

Chapter 1

PRELIMINARIES

In this chapter we shall use the Liapunov-Schmidt procedure to reduce the problem to a finite dimensional one and then define the generalized degree which we will be using in the rest of the paper.

§I MATHEMATICAL FORMULATION OF THE BIFURCATION PROBLEM

We want to study the zeroes of $F(x, \lambda)$ near (x_o, λ_o) (which we will take without loss of generality as $(0, 0)$) where F is defined for x in some Banach space B and for λ is some open subset Λ of $\mathbb{R}^k$ or $\mathbb{C}^k$ (noted $\mathbb{K}^k$) and with values in another

Banach space E.

We shall assume that there is a family T of solutions (x, λ), called trivial solutions and we shall say that $(0, 0)$ is a *bifurcation point* if every neighborhood of $(0, 0)$ contains solutions of (1) which do not belong to T.

The basic idea in the study of small norm solutions is to expand $F(x, \lambda)$ in Taylor series and write

(2) $$F(x, \lambda) = Ax - T(\lambda)x - g(x, \lambda) = 0$$

where A and $T(\lambda)$ are continuous linear operators from B to E, $T(0) = 0$, $T(\lambda)$ is analytic in λ and $g(x, \lambda)$ satisfies

$$\|g(x, \lambda)\|_E = o(\|x\|_B + |\lambda|^{2n+1})$$

($n \geq 0$ will be specified later on),

$$\|g(x, \lambda) - g(x', \lambda')\|_E \leq C[M(r)\|x-x'\|_B + |\lambda-\lambda'|]$$

with $M(r)$ continuous on $\mathbb{R}^+$, $M(0) = 0$, for $\|x\|_B, \|x'\|_B \leq r$, λ, λ' small.

Example: If T is given by $(x(\lambda), \lambda)$ and after linearization by $(0, \lambda)$ then $F(0, \lambda) = 0$, $A-T(\lambda) = G_x(0, \lambda) = G_x(0, 0)+G_x(0, 0)\lambda+ \ldots$ (in which case $g(x, \lambda) = o(\|x\|)$) or $A-T(\lambda)$ is the Taylor series of $G_x(0, \lambda)$ truncated after $2n$ λ derivatives.

Case 1: If A is invertible, then (2) can be written as

$$x = A^{-1}T(\lambda)x + A^{-1}g(x, \lambda).$$

Since $\|T(\lambda)\|$ tends to 0 when λ goes to 0, we leave as an exercise in applications of the contraction principle to show that there is, for each λ, an unique solution $x(\lambda)$, Lipschitz continuous in λ, and so $(0, 0)$ is *not* a bifurcation point.

Case 2: A is almost invertible, that is A is a Fredholm operator.

Definition. A is a *Fredholm* operator if

1) A is continuous from B into E

2) Ker A is of finite dimension d

3) Range A is of finite codimension d^*

($d-d^*$ is called the index of A).

Examples

a) If A is a closed operator defined in E with properties 2) and 3) then taking B as the domain of A with the graph norm, then A is Fredholm.

b) Elliptic operators in the sense of Agmon-Douglis-Nirenberg are Fredholm.

c) Integral operators lead to Fredholm operators.

d) Operators of the form I-K with K compact are Fredholm (see [Go], [Fr]).

Case 3: A is almost onto or semi-Fredholm with properties 1 and 3 only. In this case there is no general theory and studies in concrete problems are rare [Ra 1].

In these notes we shall study only the second case.

Let then P and Q be two projections, P from B onto ker A, Q from E onto R(A), then

$$B = \ker A \oplus B_2$$

$$E = E_2 \oplus R(A)$$

with B_2 a closed subspace of B and E_2 of dimension d^*.

Any x in E can be written as

$$x = x_1 + x_2$$

with $Px = x_1$ in ker A.

Since A is continuous, one to one from B_2 onto $R(A)$, there is a continuous inverse K from $R(A)$ onto B_2, that is

$$AKQ = Q, \qquad KA(I-P) = I - P.$$

One may write equation (2) as

$$Ax_2 - QT(\lambda)(x_1+x_2) = Qg(x_1 + x_2, \lambda)$$
$$-(I-Q)T(\lambda)(x_1+x_2) = (I-Q)g(x_1+x_2, \lambda).$$

Applying K to the first equation

$$(I-KQT(\lambda))x_2 = KQT(\lambda)x_1 + KQg(x, \lambda).$$

Since $T(0) = 0$ and $T(\lambda)$ is continuous in λ, $I-KQT(\lambda)$ is invertible for small λ and

(3) $$x_2 = (I-KQT(\lambda))^{-1}[KQT(\lambda)x_1 + KQg(x, \lambda)].$$

Since $T(0) = 0$ and g is Lipschitz with $M(r)$ small, we can use Banach contraction principle to get, for each (x_1, λ) in a neighborhood of $(0, 0)$ in $\mathbb{K}^d \times \mathbb{K}^k$, a unique solution $x_2(x_1, \lambda)$ with the following properties

$$\|x_2(x_1, \lambda) - x_2(x_1', \lambda')\|_B \le C[M(r)\|x_1 - x_1'\| + |\lambda-\lambda'| + |\lambda|\|x_1 - x_1'\|]$$
$$\|x_2(x_1, \lambda)\| \le C[\|x_1\|^2 + |\lambda|^{2n+1} + |\lambda|\|x_1\|]$$

for λ, λ' small and $\|x_1\|$, $\|x_1'\| \le r$. (See [Iz. 1] for details).

We have thus reduced the study of equation (1) to a finite number d^* of equations, with d unknowns and k parameters. In fact using the relations

$$I + (I-KQT(\lambda))^{-1} KQT(\lambda) = (I-KQT(\lambda))^{-1}$$
$$I + T(\lambda)(I-KQT(\lambda))^{-1}KQ = (I-T(\lambda)KQ)^{-1}$$

and the expression for $x_2(x_1, \lambda)$, the second equation gives

$$(I-Q)T(\lambda)(I-KQT(\lambda))^{-1}x_1 = -(I-Q)(I-T(\lambda)KQ)^{-1} g(x_1 + x_2(x_1, \lambda), \lambda)$$

that is (4) $$B(\lambda)x_1 = g(x_1, \lambda)$$

where $B(\lambda)$ is a $d\times d^*$ matrix, analytic in λ, $B(0) = 0$ and

$$\|\tilde{g}(x_1, \lambda)-\tilde{g}(x_1', \lambda')\| \le C[M(r)\|x_1-x_1'\|+|\lambda-\lambda'|+|\lambda|\|x_1-x_1'\|]$$

$$\|\tilde{g}(x_1, \lambda)\| \le C[\|x_1\|^2 + |\lambda|^{2n+1}]$$

for λ, λ' small, $\|x_1\|$, $\|x_1'\| \le r$. See [Iz.1] for details.

This reduction is called the Liapunov-Schmidt method and equation (4) is known as the bifurcation equation. Recall that the procedure is valid only in a neighborhood of (0, 0) possible candidate for a bifurcation point.

§II ELEMENTS OF DIFFERENTIAL TOPOLOGY

In next chapter we shall give conditions on $B(\lambda)$ so that (4) together with a certain number of side conditions, which will have the effect of specifying the sort of solutions of (4) we are looking for (i.e. bifurcating or trivial solutions), will define a mapping of the form

$$F(x) = 0$$

where F is continuous from some bounded set Ω in $\mathbb{R}^N$ into $\mathbb{R}^M$ (N and M are related to d, d^*, k and the number of side conditions) such that it is known that $F(x) \neq 0$ on $\partial\Omega$:

$$F:\partial\Omega \longrightarrow \mathbb{R}^M-\{0\}.$$

The main idea of the topological method is based on the following very simple fact:

<u>Definitions</u>. Let X be a closed bounded set in $\mathbb{R}^N$ and $F:X \to \mathbb{R}^M-\{0\}$ be continuous; let Y be bounded, closed and containing X

1) F is said to be *inessential with respect to* Y if and only if

F extends to $\tilde{F}: Y \to \mathbb{R}^M - \{0\}$.

2) F is said to be *inessential* if F is inessential with respect to any Y.

3) If F_0 and F_1 are as above, F_0 is said to be *homotopic* to F_1 if there is a homotopy

$$\tilde{F}: X \times [0, 1] \longrightarrow \mathbb{R}^M - \{0\} \text{ with } \tilde{F}(\cdot, 0) = F_0(\cdot), \tilde{F}(\cdot, 1) = F_1(\cdot).$$

Theorem. 1) Let $F_0, F_1: X \longrightarrow \mathbb{R}^M - \{0\}$ be homotopic and F_0 be inessential with respect to Y, then so is F_1.

2) F is inessential if and only if F is homotopic to $G: X \to$ point in $\mathbb{R}^M - \{0\}$, that is F is homotopic to a constant map.

Proof. 1) Let $\tilde{F}_0$ be an extension of F_0 to Y without zeroes and $F(x, t)$ be the homotopy from F_0 to F_1; set $Z \equiv X \times [0, 1] \cup Y \times \{0\}$. Define

$$\tilde{F}: Z \longrightarrow \mathbb{R}^M - \{0\} \text{ by}$$

$$\tilde{F}(x, t) = F(x, t) \quad \text{if } x \in X$$

$$\tilde{F}(y, 0) = \tilde{F}_0(y) \quad \text{if } y \in Y$$

$\tilde{F}$ is continuous on the closed set Z, then from Tietze extension theorem $\tilde{F}$ extends to

$$G: Y \times [0, 1] \longrightarrow R^M.$$

Let $A \equiv \{y \in y / G(y, t) = 0$ for some $t\}$ then $A \cap X = \phi$, A and X are closed, hence there is a separating function $\lambda: Y \longrightarrow [0, 1]$ with $\lambda(A) = 0$, $\lambda(X) = 1$. Define $\tilde{G}(y, t) \equiv G(y, \lambda(y)t)$, then $\tilde{G}: Y \times [0, 1] \longrightarrow \mathbb{R}^M - \{0\}$ since if $\tilde{G}(y, t) = 0$, then y is in A and hence $\lambda(y) = 0$, but $G(y, 0) = \tilde{F}(y, 0) = F_0(y) \neq 0$. Also $\tilde{G}(y, 0) = \tilde{F}_0(y)$ and, if y is in X, $\tilde{G}(y, 1) = G(y, 1) = F_1(y)$.

2) Since constant maps are clearly inessential it remains to prove that if F is inessential then F is deformable to a constant

map: let B be a large ball centered at the origin and containing X; then F extends to $\tilde{F}:\bar{B} \longrightarrow \mathbb{R}^M-\{0\}$. Define $H(x, t) = \tilde{F}((1-t)x)$: H is a homotopy from F to $\tilde{F}(0)$, that is we have a linear retraction of X to the origin.

Notes

1) If $X = \partial\Omega$ with Ω regular enough, we can prove the converse of (2) by choosing a tubular neighborhood of $\partial\Omega$ given by $(x, \varepsilon n)$ with $|\varepsilon| \le \varepsilon_0$, n the normal to $\partial\Omega$ and, if $H(x, t)$ is such that $H(x, 0) = F(x)$, $H(x, 1) = G(x) = x_0$, defining

$$\tilde{F}(x) = \begin{cases} F(x) & \text{if } x \in X \\ H(x, t) & \text{if } (x, \pm t\varepsilon_0) \text{ belongs to the tube} \\ x_0 & \text{in the complement of the tube.} \end{cases}$$

Clearly $\tilde{F}$ is an extension of F to $\mathbb{R}^N$.

2) If $F:\partial\Omega \longrightarrow \mathbb{R}^M-\{0\}$ is essential relative to Ω then any extension of F to Ω has a zero: this is the result we will be using.

3) From the proof of (2) it is clear that it is enough to have F inessential relative to the ball $\bar{B}$.

4) If $\partial\Omega = \partial B$, then F has a zero in B if F is not deformable to a constant map. Now the homotopy classes of maps from ∂B into $\mathbb{R}^M-\{0\}$ or, taking $F/|F|$, maps from $S^{N-1} \to S^{M-1}$, form the homotopy group $\Pi_{N-1}(S^{M-1})$.

In general we wish to study the homotopy classes of maps from $\partial\Omega$ into $\mathbb{R}^M-\{0\}$ and in particular those which are essential with respect to $\bar{\Omega}$, that is those which are not restrictions of mappings from $\bar{\Omega}$ into $\mathbb{R}^M-\{0\}$; in fact, if $[\bar{\Omega}, \mathbb{R}^M-\{0\}]$ denotes the set of equivalence classes of mappings from $\bar{\Omega}$ into $\mathbb{R}^M-\{0\}$ and i^* is the mapping: $[\Omega, \mathbb{R}^M-\{0\}] \xrightarrow{i^*} [\partial\Omega, \mathbb{R}^M-\{0\}]$ induced by restriction, we want that $[F] \notin$ Image i^*. This set of equivalence classes is called:

II-1) Cohomotopy groups

(This material is taken from E. H. Spanier: Borsuk's cohomotopy groups. Annals of Math. *50*, (1949), 203-248).

Definition. Let X be a compact topological space, A be a closed subset of X, then $\Pi^m(X, A)$ is the set of homotopy classes of maps $f: (X, A) \to (S^m, pt)$, where pt is a point in S^m.

Theorem: If the topological dimension of X is less than $2m-1$, then $\Pi^m(X, A)$ is an abelian group.

The *neutral element* is $(X, A) \to (pt, pt)$ that is the class of all inessential maps.

The *inverse* of $[F]$ is: $(X, A) \xrightarrow{f} (S^m, pt) \xrightarrow{i} (S^m, pt)$ where i is a change of orientation of S^m.

Properties.

a) If $f: Y \to X$, then one has $f^*: \Pi^m(X) \longrightarrow \Pi^m(Y)$ defined by: $Y \xrightarrow{f} X \xrightarrow{g} S^m$, that is $f^*[g] = [gf]$.

b) One defines $\delta: \Pi^m(A) \longrightarrow \Pi^{m+1}(X, A)$ as follows: $A \xrightarrow{f} S^m$, consider S^m as the equator of S^{m+1} and let E_+^{m+1} the north hemisphere. Since E_+^{m+1} is contractible one may extend f to $\tilde{f}: X \to E_+^{m+1}$. Let h be a deformation of S^m to p, the south pole of S^{m+1}, followed by an extension of E_+^{m+1} onto $S^{m+1}-\{p\}$ along meridians; then define $\delta[f] = [h\tilde{f}]: (X, A) \longrightarrow (S^{m+1}, pt)$.

c) This cohomotopy theory satisfies all Eilenberg-Steenrod axioms for cohomology, that is:

1) $(X, A) \xrightarrow{Id} (X, A)$: $Id^* = Id$ in $\Pi^m(X, A)$

2) $(X, A) \xrightarrow{f} (Y, B) \xrightarrow{g} (Z, C)$: $\Pi^m(Z, C) \xrightarrow{g^*} \Pi^m(Y, B) \xrightarrow{f^*} \Pi^m(X, A)$

that is $(gf)^* = f^*g^*$.

3) If $f:(X, A) \to (Y, B)$ and $f_o = f|_A$ then $f^*\delta = \delta f_o^*$ in

$$\begin{array}{ccc} \Pi^m(Y, B) & \xrightarrow{f^*} & \Pi^m(X, A) \\ \delta \uparrow & & \uparrow \delta \\ \Pi^{m-1}(B) & \xrightarrow[f_o^*]{} & \Pi^{m-1}(A) \end{array}$$

4) If $i:(X, p) \to (X, A)$, $j:A \to X$ are inclusions the sequence

$$\ldots \xrightarrow{\delta} \Pi^m(X, A) \xrightarrow{i^*} \Pi^m(X) \xrightarrow{j^*} \Pi^m(A) \xrightarrow{\delta} \Pi^{m+1}(X, A) \longrightarrow \ldots$$

is exact. (Recall that if one has $\xrightarrow{f} A \xrightarrow{g}$, exact means $\ker g = \operatorname{Im} f$), for $2m-2 < \dim X$, exactness is not completely true.

5) $(X, A) \underset{g}{\xrightarrow{f}} (Y, B)$, if f is homotopic to g, then $f^* = g^*$.

6) <u>Excision</u>. If $X = A \cup B$, then $\Pi^m(X, A) = \Pi^m(B, A \cap B)$.

7) <u>Normalization</u>. $\Pi^m(pt) = 0$.

In our case $F:\partial\Omega \to \mathbb{R}^M-\{0\}$ gives an element of $\Pi^{M-1}(\partial\Omega)$, F is inessential with respect to $\bar{\Omega}$ if and only if $[F]$ belongs to $\operatorname{Im} i^*$, that is if and only if $\delta[F] = 0$ in the exact sequence

$$\longrightarrow \Pi^{M-1}(\bar{\Omega}) \xrightarrow{i^*} \Pi^{M-1}(\partial\Omega) \xrightarrow{\delta} \Pi^M(\bar{\Omega}, \partial\Omega), \quad \text{if } N < 2M - 1.$$

<u>Examples</u>.

1) If X is a complex of dimension N, then $\Pi^m(X) = 0$ for $m > N$, since F is homotopic to a simplicial mapping whose image is contained in the N skeleton of S^m and hence deformable to a point.

2) If $X = S^{N-1}$ then $\Pi^{N-1}(S^{N-1}) = \Pi_{N-1}(S^{N-1}) = Z$ and each mapping

is characterized by its degree, that is the number of times $F(S^{N-1})$ covers S^{N-1}. $\Pi^{M-1}(S^{N-1}) = \Pi_{N-1}(S^{M-1})$ will be studied later.

3) If $X = \bar{B}$, ball of center x_0 and radius R in $\mathbb{R}^N$, then $\Pi^m(\bar{B}) = 0$ for all m, since if $F:\bar{B} \to S^m$ then F can be extended to $\mathbb{R}^N$ by $F(x) = F(x_0+R(x-x_0)/|x-x_0|)$ for $|x-x_0| \geq R$.

4) If Ω is homeomorphic to a ball B with k holes $B_1, \ldots, B_k$, then $\partial\Omega = \partial B \cup \partial B_1 \cup \ldots \cup \partial B_k$, disjoint sets, and clearly

$$\Pi^m(\partial\Omega) \overset{i^*}{\cong} \Pi^m(\partial B) \oplus \Pi^m(\partial B_i).$$

Now we shall see how the homeomorphism i^*, generated by the inclusion of $\partial B \cup (\cup \partial B_i)$ into $\partial\Omega$, behaves.

<u>Lemma 1</u>. $\Pi^m(\bar{\Omega}) = \Pi^m(\bar{B} - \cup B_i) \cong \oplus\, \Pi^m(\partial B_i) = \oplus\, \Pi_{N-1}(S^m)$.

<u>Proof</u>. By excision of $\cup B_i$, one has $\Pi^m(\bar{\Omega}, \cup\partial B_i) = \Pi^m(\bar{B}, \cup\bar{B}_i)$ but in the exact sequence

$$\longrightarrow \Pi^{m-1}(\cup\bar{B}_i) \xrightarrow{\delta} \Pi^m(\bar{B}, \cup\bar{B}_i) \longrightarrow \Pi^m(\bar{B}) \longrightarrow$$

both extreme groups are zero (being balls), then so is the middle one. From $\longrightarrow \Pi^m(\bar{\Omega}, \cup\partial B_i) \longrightarrow \Pi^m(\bar{\Omega}) \xrightarrow{j^*} \Pi^m(\cup\partial B_i) \longrightarrow \Pi^m(\bar{\Omega}, \cup\partial B_i) \longrightarrow$ we get that j^* is an isomorphism.

<u>Lemma 2</u>. If k^* is generated by the inclusion of ∂B into $\bar{B} - \cup B_i$, then $\Pi^m(\bar{B} - \cup B_i) \xrightarrow{k^*} \Pi^m(\partial B) \cong \Pi_{N-1}(S^m)$ is onto.

<u>Proof</u>. If $F:\partial B \longrightarrow \mathbb{R}^{m+1}-\{0\}$ represents an element of $\Pi^m(\partial B)$ we may define $F_j:\bar{B} - B_j \longrightarrow \mathbb{R}^{m+1}-\{0\}$ as a radial retraction towards the center of B_j. Clearly $F_j|\bar{\Omega}$ contributes only to $\Pi^m(\partial B_j)$ since on B_i, $i \neq j$, $F_j \neq 0$ is inessential.

Also F_j: $B_j \longrightarrow \mathbb{R}^{m+1}-\{0\}$ can be extended radially giving an element in $\Pi^m(\bar{\Omega})$ and in $\Pi^m(\partial B)$ but this extension does not contribute to

$\Pi^m(\partial B_i)$, $i \neq j$. This implies that the generators of $\Pi^m(\partial B_j)$ go into the generators of $\Pi^m(\bar{\Omega})$ (under j^*), with a direct algebraic sum, and into the generators of $\Pi^m(\partial B)$ (under $k^* j^{*-1}$).

Let then $F:\partial\Omega \to \mathbb{R}^{m+1}-\{0\}$ be an element of $\Pi^m(\partial\Omega)$. $F|\partial B$, $F|\partial B_i$ give elements $[F_o]$, $[F_i]$ in $\Pi^m(\partial B)$, $\Pi^m(\partial B_i)$; consider the diagram:

$$\begin{array}{ccccccccc}
 & & 0 & \longrightarrow & \Pi^m(\bar{\Omega}) & \xrightarrow{\;j^*\;} & \oplus\, \Pi^m(\partial B_i) & \xrightarrow{\;\delta_1\;} & 0 & & \\
 & & & & \uparrow \wr Id & & \uparrow i_1^* & & & & \\
 & & 0 & \longrightarrow & \Pi^m(\Omega) & \xrightarrow[\;i_o^*\;]{} & \Pi^m(\partial\Omega) & \longrightarrow & \Pi^{m+1}(\bar{\Omega}, \partial\Omega) & \longrightarrow & 0 \\
 & & & & \downarrow \wr Id & & \downarrow i_2^* & & & & \\
 & & \Pi^m(\bar{\Omega}, \partial B) & \longrightarrow & \Pi^m(\bar{\Omega}) & \xrightarrow[\;k^*\;]{} & \Pi^m(\partial B) & \xrightarrow{\;\delta_2\;} & 0 & &
\end{array}$$

i_o, i_1, i_2, j, k are induced by inclusions; from lemmas 1 and 2, δ_1 and δ_2 map into 0. Then one has $j^* = i_1^* i_o^*$, $k^* = i_2^* i_o^*$. Since j^* is one to one so is i_o^*: $\ker i_o^* = \operatorname{Im} i_3^* = 0$ (i_3 the inclusion of $\bar{\Omega}$ into $(\bar{\Omega}, \partial\Omega)$); $\ker\ i_3 = \operatorname{Im} \delta = \Pi^{m+1}(\bar{\Omega}, \partial\Omega)$.

Finally $i^* = i_1^* \oplus i_2^*$ is an isomorphism, $[F_o] = i_2^*[F]$, $\oplus\ [F_i] = i_1^*[F]$. We wish to know then $\delta[F] \neq 0$ or equivalently when $[F] \in \operatorname{Im} i_o^*$.

Let then $[G] = i_o^*[H]$ with $[H]$ in $\Pi^m(\bar{\Omega})$, then $i_1^*[G] = i_1^* i_o^*[H] = j^*[H]$ and $[H] = j^{*-1}\ i_1^*[G]$, $[G] = i_o^*\ j^{*-1}\ i_1^*[G]$ and $\operatorname{Im} i_o^* = \ker(I - i_o^*\ j^{*-1}\ i_1^*)$. Now $i_1^*\ i_o^*\ j^{*-1}\ i_1^* = i_1^*$, $i_2^*\ i_o^*\ j^{*-1}\ i_1^* = k^*\ j^{*-1}\ i_1^*$, then $i_1^*(I - i_o^*\ j^{*-1}\ i_1^*)[F] = 0$

$$i_2^*(I - i_o^*\ j^{*-1}\ i_1^*)[F] = [F_o] - \sum k^*\ j^{*-1}[F_i]$$

that is $i^*(I - i_o^*\ j^{*-1}\ i_1^*)[F] = 0 \oplus [F_o] - \sum k^* j^{*-1}[F_i]$

and, since i^* is an isomorphism, $[F]$ belongs to $\operatorname{Im} i_o^*$ if and only if

$$[F_o] - \sum k^* j^{*-1}[F_i] = 0 \quad \text{in} \quad \Pi^m(\partial B).$$

Definition. If Ω has the form $\bar{B} - \cup B_i$, $F:\partial\Omega \to \mathbb{R}^{m+1}-\{0\}$, if $[F_o]$, $[F_i]$ are the classes of $F|\partial B$, $F|\partial B_i$ in $\Pi_{N-1}(S^m)$, the *generalized degree* of F is

$$\deg_{N-1,m}(F, 0, \Omega) = [F_o] - \sum [F_i].$$

(Recall that the radial retraction $k^* j^{*-1}$ is the identity if one identifies $\Pi^m(\partial B_j)$ with $\Pi^m(\partial B)$). Although the above argument depends on the exactness, and hence on N, one may show that it remains valid for any N by using the results in [S. T. Hu: *Homotopy theory*, Academic Press, 1959].

5) If Ω is not of the form above it is more difficult to define a sum of invariants (this can be done using oriented framed bordisms but this is out of the scope of these notes). However if $[F] = 0$ in $\Pi^m(\partial\Omega)$, then F is inessential and extends to $\tilde{F}:\bar{B} \longrightarrow \mathbb{R}^{m+1}-\{0\}$. If we know that the zeroes of $F|\Omega$ are contained in $B_1, \ldots, B_k$, define G on $\bar{B} - \cup B_i$ as F on $\bar{\Omega}$, $\tilde{F}$ on $\bar{B} - \bar{\Omega}$, then G maps $\bar{B} - \cup B_i$ into $\mathbb{R}^{m+1}-\{0\}$ and $[G|\partial B] = 0$ since $\tilde{F}$ extends $G|\partial B$ to $\bar{B}$. From the above $\delta[G|\partial(\bar{B} - \cup B_i)] = o = [G_o]$ and $\sum[F_i] = \sum[F|\partial B_i] = 0$. This argument will be used in chapter III.

II-2) A particular case $N = M = m-1$: degree theory

There are many definitions of degree, some combinatorial (see [Pe] for its numerical computation) others generic (using Sard's lemma) and others differential (see [Ni]).

Here we shall use the Alexander-Pontrjagin duality between $\Pi^{N-1}(\partial\Omega)$ and $\sum_o(\mathbb{R}^N-\partial\Omega)$ the reduced homotopy group of $\mathbb{R}^N-\partial\Omega$ which is $\oplus Z$ with generators given by x_i, a point in the i-th bounded component of $\mathbb{R}^N-\partial\Omega$. This isomorphism enables one to write any element $[F]$ in $\Pi^{N-1}(\partial\Omega)$ as

$$[F] = \sum m_i \alpha_i$$

where m_i is an integer and α_i represents the mapping $x - x_i$ from $\partial\Omega$ into $\mathbb{R}^N - \{0\}$. Furthermore the isomorphism preserves the group applications, in particular if $\tilde{\Omega} = \Omega \cup \Omega_i$, Ω_i one of the bounded components of $\mathbb{R}^N - \partial\Omega$, then $\sum_0(\mathbb{R}^N - \partial\tilde{\Omega})$ has one generator less and $\Pi^{N-1}(\partial\tilde{\Omega}) \xrightarrow{i} \Pi^{N-1}(\partial\Omega)$ is one to one, sending α_j to α_j for $j \neq i$. Then δ sends $\Pi^{N-1}(\partial\Omega)$ onto $\oplus$ Z with as many generators as there are components in Ω.

<u>Consequences</u>

1) If $m_i \neq 0$ then any extension $\tilde{F}$ to F to the corresponding component Ω_i has a zero in Ω_i: if not F belongs to $\Pi^{N-1}(\partial(\Omega \cup \Omega_i))$, a group without the generator α_i.

2) Only a finite number of the m_i's are non-zero: if not take any extension $\tilde{F}$ to a ball $\bar{B}$, this extension will have an infinite number of zeroes, each one in a different component Ω_i. From this sequence one may extract a converging subsequence, also a zero of $\tilde{F}$ by continuity. But this zero cannot be in any Ω_i, since these are disjoint and open, nor in $\partial(\cup \Omega_i) = \partial(\Omega)$ since $F \neq 0$ on it.

3) <u>Definition</u>. Let $d(F, 0, \Omega) = \sum m_i$ where the sum is over the connected components of Ω. Also $d(F, p, \Omega) = d(F-p, 0, \Omega)$.

Note that it would have been more useful to define the degree as a direct sum, so that if each m_i is 0 then one has a non-zero extension to Ω (here this is the case only if Ω is connected), but then we wouldn't get the other definitions.

<u>Examples</u>.

1) $\Omega = B$, $F(x) = A(x-x_0)$ with x_0 in B and A an invertible

constant matrix: A is in $GL(N)$ which has two components according to the sign of the determinant. Taking a path in $GL(N)$ we may deform A to I^N or to $I^{N-1} \oplus \{-1\}$ and maintain $F|\partial\Omega \neq 0$. Then $[F] = \alpha = [x-x_0]$ in the first case and $[F] = -\alpha$ in the second (recall the definition of the inverse in $\Pi^m(X)$): $d(F, 0, B) = \text{sign} \det A$.

2) $\bar{\Omega} = \bar{B} - \cup B_i$ as in example 4 of the preceding section. Take x_0 in Ω, x_i in B_i and let $\alpha_0 = [x-x_0]$, $\alpha_i = [x-x_i]$, then

$$[F] = m_0\alpha_0 + \sum m_i\alpha_i \quad \text{in} \quad \Pi^{N-1}(\partial\Omega).$$

Since $\mathbb{R}^N-\bar{B}$ has only one component, then $\Pi^{N-1}(\partial B)$ is generated by α_0 and $\Pi^{N-1}(\partial B_i)$ by α_i with $k^* j^{*-1}\alpha_i = \alpha_0$.

Now $i_2^*[F] = m_0 i_2^* \alpha_0 + \sum m_i i_2^* \alpha_i = (m_0 + \sum m_i)\alpha_0$, since x_i and x_0 are in the same component of $\mathbb{R}^N-\bar{B}$. Similarly $i_1^*(\alpha_0) = 0$, since $x-x_0$ is inessential on B_i, $i_1^*(\alpha_i) = \alpha_i$.

Then the degree we have defined in last section, $[F_0] - \sum[F_i] = (m_0 + \sum m_i) - \sum m_i = m_0$, coincides with the new definition.

3) If F is C^1 on $\bar{\Omega}$, from Sard's lemma, there is a p close to 0 and regular value for F. Clearly $F-tp$ is a valid deformation ($F \neq 0$ on $\partial\Omega$ and p is arbitrarely small) and so $\deg(F, p, \Omega) = \deg(F, 0, \Omega)$.

But $F^{-1}(p) = \{x_1, \ldots, x_k\}$ and one can choose $B_1, \ldots, B_k$, balls centered at $x_1, \ldots, x_k$, such that $F(x)-p = DF(x_j)(x-x_j) + o(x-x_j)$ on ∂B_j. $[F(x)-p]$ is an element of $\Pi^{N-1}(\bar{\Omega} - \cup B_i)$, which is generated by α_i (with x_i in the component $\bar{\Omega}_i$ of $\mathbb{R}^N-\bar{\Omega}$) and $\beta_1, \ldots, \beta_k$ (corresponding to $B_1, \ldots, B_k$), while $\Pi^{N-1}(\partial\Omega)$ is generated by α_i and γ_j (γ_j corresponding to a component of Ω).

Now $\beta_j = [x-x_j] = \gamma_j$ if x_j belongs to Ω_j. Suppose, for simplicity, that $B_j \subset \Omega_0$ then

$$\Pi^{N-1}(\bar{\Omega} - \cup B_i) \xrightarrow{i^*} \Pi^{N-1}(\partial\Omega)$$

$$\alpha_i \longrightarrow \alpha_i$$

$$\beta_1, \ldots, \beta_k \longrightarrow \gamma_0$$

$$\gamma_j \quad j \geq 1$$

$$[F-p] = \sum m_i\alpha_i + \sum n_i\beta_i \longrightarrow \sum m_i\alpha_i + (\sum n_i)\,\gamma_0$$

thus $\deg(F, p, \Omega) = \sum n_i$, but by definition $n_j = \deg(F, p, B_j)$ (from the inclusion of ∂B_j in $\bar{\Omega} - \cup B_i$) and, by example 1, we may choose B_j so small that, on ∂B_j, $F(x)-p$ can be deformed to its linear part, that is $n_j = \mathrm{Sign}|DF(x_j)|$ and

$$\deg(F, p, \Omega) = \sum \mathrm{Sign}|DF(x_j)|$$

which is the usual definition.

As a matter of fact, all the definitions give an algebraic count of the number of solutions to $F(x) = p$ in Ω and are characterized uniquely by the following:

<u>Properties of degree</u>

i) *Invariance under homotopy:* $d(F, p, \Omega)$ depends only on the homotopy class of $F:\partial\Omega \to \mathbb{R}^N-\{0\}$.

ii) If p_1, p_2 are in the same component of $\mathbb{R}^N-\partial\Omega$ then $d(F, p_1, \Omega) = d(F, p_2, \Omega)$.

iii) If $F(x) \neq 0$ in $\bar{\Omega} - \cup\Omega_i$, Ω_i open disjoint subsets of Ω, then $d(F, p, \Omega) = \sum d(F, p, \Omega_i)$.

iv) *Product:* If $\Omega \subset \mathbb{R}^N$, $\Omega' \subset \mathbb{R}^{N'}$, $F:\partial\Omega \to \mathbb{R}^N-\{p\}$, $F':\partial\Omega' \to \mathbb{R}^{N'}-\{p'\}$ then

$$d(F\times F', (p, p'), \Omega\times\Omega') = d(F, p, \Omega)d(F', p', \Omega').$$

v) *Normalization:* $d(Id, p, \Omega) = 1$ if p is in Ω, 0 if not.

Exercise: Prove these properties.

Other properties

vi) *Composition:* $F:\partial\Omega \to \partial\Omega' \subset \mathbb{R}^N-\{0\}$, $G:\partial\Omega' \to \mathbb{R}^N-\{0\}$ then $d(G \circ F, 0, \Omega) = \sum d(F, \Omega_i, \Omega)d(G, 0, \Omega_i)$ where Ω_i is a bounded component of $\mathbb{R}^N-\partial\Omega'$ and $d(F, \Omega_i, \Omega)$ is the degree of F with respect to any point of Ω_i.

Exercise: Prove this property using the chain rule.

vii) *Borsuk's theorem:* If Ω is symmetric with respect to the origin and $F:\partial\Omega \to \mathbb{R}^N-\{0\}$ is odd then if $0 \in \Omega$ $d(F, 0, \Omega)$ is odd and even if $0 \notin \Omega$.

Proof. See [Ni, 1.7].

Exercises

1) If $F:\bar{B} \to \mathbb{R}^N$ is such that $(F(x), x) > 0$ on ∂B, then F has a zero in B (use $tF(x) + (1-t)x$).

2) Brower fixed point theorem: if $\phi:\bar{B} \to \mathbb{R}^N$ is such that $\phi(\partial B) \subset B$ then ϕ has a fixed point in B (use (1) on $F(x) = x - \phi(x)$).

3) If $F:\mathbb{R}^N \to \mathbb{R}^N$ is such that $(F(x), x)/|x| \to \infty$ when $|x| \to \infty$ then F is *onto* $\mathbb{R}^N$.

4) If Ω is symmetric in $\mathbb{R}^N$, $0 \in \Omega$, $F:\partial\Omega \to \mathbb{R}^k$, with $k < N$, then, if F is odd, F has a zero on $\partial\Omega$.

5) If $F: \Omega \subset \mathbb{C}^N \to \mathbb{C}^N-\{0\}$ is complex analytic, then $d(F, 0, \Omega)$ is non-negative.

6) If $F:\Omega \subset \mathbb{R}^n \times \mathbb{R}^m \to \mathbb{R}^{n+m}$ with $F|\partial\Omega \neq 0$, is of the form $F(x) = x + \phi(x)$ with $\phi(\Omega) \subset \mathbb{R}^n$, then $d(F, 0, \Omega) =$

$d(F|\Omega \cap \mathbb{R}^n, 0, \Omega \cap \mathbb{R}^n)$ (write $x = x_1 \oplus x_2$, use $\phi(x_1 + tx_2)$ changing Ω to $\Omega \cap \mathbb{R}^n \times B^m$ where B is a large ball in $\mathbb{R}^m$).

II-3) Different dimensions, homotopy of spheres

Let $\Omega \subset \mathbb{R}^N$ and $F: \partial\Omega \to \mathbb{R}^M - \{0\}$ representing an element of $\Pi^{M-1}(\partial\Omega)$.

If $N < M$, $\Pi^{M-1}(\partial\Omega) = 0$ since, after approximating F by a simplicial mapping, one may deform its image to a point without touching 0.

If $N = M$, we have the case studied in last section.

If $N > M$, the art of cohomotopy is not yet advanced enough to characterize the generators of the groups, except for special cases, like spheres, in which case $\Pi^{M-1}(S^{N-1}) = \Pi_{N-1}(S^{M-1})$, enough for a local bifurcation study. One has

•) $\Pi_n(S^k) = 0$, $n < k$; $\Pi_n(S^1) = 0$ for $n > 1$;

$\Pi_n(S^n) = Z$ characterized by degree; $\Pi_3(S^2) = Z$;

$\Pi_{n+1}(S^n) = Z_2$ if $n > 2$; $\Pi_{n+2}(S^n) = Z_2$ if $n > 3$;

$\Pi_{n+3}(S^n) = Z_{24}$ if $n > 4$; $\Pi_{n+p}(S^n) = \Pi_{n+p+1}(S^{n+1})$ if $n > p+1$.

•) $\Pi_3(S^2)$ is generated by the Hopf map defined as follows: considering S^3 as the unit sphere in $\mathbb{C}^2$, $\{(\lambda, z)/|\lambda|^2+|z|^2 = 1\}$ then

$$(\lambda, z) \xrightarrow{\eta} (\mathrm{Re}\ 2\bar{\lambda}z,\ \mathrm{Im}\ 2\bar{\lambda}z,\ |z|^2-|\lambda|^2) \text{ in } S^2.$$

•) *Suspension for spheres*

Let $\psi: S^{N-1} \to S^{M-1}$. Define $\sum\psi$, *the suspension of* ψ, as a mapping from S^N into S^M given by identifying first S^{N-1} with the equator of S^N, and likewise with S^{M-1} in S^M, then sending poles into poles and extending linearly over meridians.

That is if $\phi(x_1, \ldots, x_{n+1}) = (\psi(x_1, \ldots, x_N), x_{N+1})$ then $\sum\psi(x_1, \ldots, x_{N+1}) = \phi/|\phi|$. Since we shall study mappings from S^{N-1} into $\mathbb{R}^M-\{0\}$, we shall define

$$\sum F(x_1, \ldots, x_{N+1}) = (F(x_1, \ldots, x_N), x_{N+1})$$

so that $\sum\psi = \sum(F/|F|)$ and $\sum F/|\sum F|$ have the same homotopy class. (Exercise).

<u>Exercise</u>. Prove that if $F:S^{N-1} \to S^{N-1}$, then $d(F, 0, B^N) = d(\sum F, 0, B^{N+1})$.

Freudenthal suspension theorem [Sp. p. 458] says that, if $\ell \geq N-2M+3$, then $[\sum^\ell \psi]$ in $\Pi_{N-1+\ell}(S^{M-1+\ell})$ is equal to $[\sum^{\ell+1}\psi]$ in $\Pi_{N+\ell}(S^{M+\ell})$. (Note that these groups are equal).
In particular $\sum\eta$, the suspension of the Hopf map, generates $\Pi_4(S^3)$ and $\sum^{n-2}\eta$ generates $\Pi_{n+1}(S^n)$ $n > 2$.

•) *An important example: Whitehead's mapping*

Consider the mapping $F(x, \lambda) = (B(\lambda)x, |x|^2-|\lambda|^2)$ for x in $\mathbb{R}^d$, λ in $\mathbb{R}^k$, $F:\bar{B} = \{(x, \lambda)/|x|^2+|\lambda|^2 \leq 1\} \to \mathbb{R}^{d+1}$. We assume $B(0) = 0$, $B(\lambda)$ invertible for $\lambda \neq 0$. Note that the unique zero of F is $(0, 0)$ and that $F|\partial B$ defines an element of $\Pi_{k+d-1}(S^d)$.

<u>Exercise</u>. Prove that F is homotopic to $(B(\lambda)x, |x|^2-r^2)$, $0 < r < 1$ (use $|x|^2 - tr^2-(1-t)|\lambda|^2$). It is then clearly seen that one is looking for solutions with $|x| = r$; this will be used in the local bifurcation problem.

Consider now an element of $\Pi_{k-1}(GL(d))$, that is a mapping $\lambda \in S^{k-1} \to B(\lambda) \in GL(d)$ and the morphism

$$\Pi_{k-1}(Gl(d)) \xrightarrow{J} \Pi_{k+d-1}(S^d)$$

$$\lambda \in S^{k-1} \to B(\lambda) \longrightarrow \begin{cases} B(\lambda)\, x \\ |x|^2 - |\lambda|^2 \end{cases} \quad (\lambda,\, x) \in \partial B$$

Clearly the class of $F(x, \lambda)$ depends only on the class of $B(\lambda)$. This morphism is called Whitehead's J-homomorphism and we are interested in its image.

The left hand side group was studied by R. Bott and, if $k < d$, is periodic in k of period 8. The image of J was studied by Adams and we have

Theorem. $k < d$

$\Pi_{k-1}(GL(d)) = Z_2$ if $k \equiv 1,2[8]$, J is one to one

$\Pi_{k-1}(GL(d)) = 0$ if $k \equiv 3,5,6,7[8]$

$\Pi_{k-1}(GL(d)) = Z$ if $k \equiv 0,4[8]$, the image of J is a cyclic group of order $m(k)$, with $k = 4n$ and in the expansion

$$\frac{x}{e^x - 1} = 1 - \frac{x}{2} + \sum_{n=1} (-1)^{n-1} B_n \frac{x^n}{(2n)!}$$

(B_n are the Bernouilli numbers) then $m(k)$ is the denominator of B_n/k expressed in irreductible terms.

k = 1: $GL(d)$ has two components characterized by the sign of the determinant. $J(B(\lambda))$ will be nontrivial, that is F will have a nonzero degree in $\Pi_d(S^d)$, only if $\det B(\lambda)$ changes sign (in fact the degree is ± 2). This is the crucial fact for a one real parameter bifurcation result.

k = 2: $\Pi_1(Gl(d))$ is generated by

$$\begin{pmatrix} \mu & \nu & 0 \\ -\nu & \mu & \\ & 0 & I^{d-2} \end{pmatrix}$$

If $d > 2$, taking $z = x_1 + ix_2$, $\lambda = \mu + i\nu$, $F(x, \lambda)$ is the suspension of the Hopf map. If $B(\lambda)$ is analytic in λ, or in case of the Hopf bifurcation, it easy to compute $[B(\lambda)]$.

Particular case: If $B(\lambda) = \lambda_1 B_1 + \ldots + \lambda_k B_k$ with $|B(\lambda)x| \geq C|\lambda||x|$, that is $B(\lambda)$ is a nondegenerate quadratic form, then define $C_k = k$ for $k = 1, 2, 4, 8$; $C_k = 8$ for $k = 5, 6, 7$; $C_k = 4$ for $k = 3$; $C_{k+8} = 16C_k$ for $k \geq 8$, then

Theorem: (Alexander-Yorke)

If $B(\lambda)$ satisfies the above condition, then $d = r\, C_k$ with r an integer and $[B(\lambda)] \equiv r[2]$.

Note that if $k = d$ (as many unknowns as parameters) then $k = r\, C_k$ if and only if $k = 1, 2, 4, 8$. This comes from the fact that $B(\lambda)x$ represents a bilineal form from $\mathbb{R}^k \times \mathbb{R}^k$ into $\mathbb{R}^k$, that is a multiplication, but this is possible only in those dimensions: reals, complex numbers, quaternions and Cayley numbers. In general given $B(\lambda)$ one has not yet found simple criteria to compute its homotopy class.

Chapter 2

LOCAL BIFURCATION

In last chapter we have reduced the local study of

(2) $$Ax - T(\lambda)x - g(x, \lambda) = 0$$

to the d^*, real or complex, equations

(4) $$B(\lambda)x_1 - \tilde{g}(x_1, \lambda) = 0$$

with x_1 in $\mathbb{K}^d$, λ in $\mathbb{K}^k$, $B(0) = 0$ and

$\|\tilde{g}(x_1, \lambda) - \tilde{g}(x_1', \lambda')\| \le C[M(r)\|x_1-x_1'\| + |\lambda-\lambda'| + |\lambda|\|x_1-x_1'\|]$

$\|\tilde{g}(x_1, \lambda)\| \le C[\|x_1\|^2 + |\lambda|^{2n+1}]$ for λ, λ' small, $\|x_1\|$, $\|x_1'\| \le r$, $M(0) = 0$ and n will be given later.

We are interested in two sorts of solutions: trivial solutions with $\|x_1\|$ very small and λ fixed ($x_1 = 0$ if $\tilde{g}(0, \lambda) = 0$) and bifurcating solutions with $\|x_1\| = r$ and λ small. In this chapter we whall use the topological results to prove the existence of such solutions; in the first section we shall study the case of a one parameter (real or complex) bifurcation from the spectrum of a Fredholm operator A of zero index; in the second we shall consider the general case, with $T(\lambda)$ arbitrary, λ in $\mathbb{K}$ and A of non negative index; in the last section we shall give some results on multiparameter bifurcation.

§I A PARTICULAR CASE: $T(\lambda) = \lambda I$, $i(A) = 0$, λ IN $\mathbb{R}$ OR $\mathbb{C}$

Here B is included in E and, since this is a very frequent case in applications, we shall find first explicitely $B(\lambda)$ by choosing appropriate projections.

<u>Theorem</u>. Let A be Fredholm of index 0 then the following are equivalent:

i) For some finite α, $E = R(A^\alpha) \oplus \ker A^\alpha$,

ii) Ker $A^{\alpha+p} = \ker A^\alpha$, $R(A^{\alpha+p}) = R(A^\alpha)$ for all $p \ge 0$

iii) 0 is an eigenvalue of A isolated in the spectrum of A,

iv) The $d\times d$ matrix $B(\lambda)$ is invertible for $\lambda \ne 0$.

<u>Proof</u>. [See Iz.1, theorem 5.1]. Note that it can be shown that if $\alpha = \infty$ then $B(\lambda) \equiv 0$.

We shall suppose then that 0 is an isolated eigenvalue of A, in which case one can prove that A^α is also a Fredholm operator of zero index, $m = \dim \ker A^\alpha$ is the *algebraic multiplicity* of 0, $B = R(A^\alpha) \cap B \oplus \ker A^\alpha$, that if x is in B then

$$x = u + v \qquad u \in R(A^\alpha), \quad v \in \ker A^\alpha$$

and, if Q_1 is the projection of E onto $R(A^\alpha)$ parallel to $\ker A^\alpha$, equation (2) is equivalent to

$$Au - \lambda u = Q_1 g$$

$$Av - \lambda v = (I - Q_1) g$$

since $A - \lambda I$ sends $\ker A^\alpha$ into itself and $B \cap R(A^\alpha)$ onto $R(A^\alpha)$ (exercise).

Let K_1 be the inverse of A on $R(A^\alpha)$ then

$$u = (I - \lambda K_1)^{-1} K_1 Q_1 g(u+v, \lambda)$$

for λ small. This equation can be solved uniquely for u as a function of (v, λ), with the same Lipschitz and smallness conditions as g, see [Iz 1.P.16].

Since A is nilpotent on $\ker A^\alpha$, we may choose P and Q in such a way that A^α is in Jordan form on $\ker A^\alpha$, that is A has d blocks of dimension $k_1, \dots, k_d$ with $\sum k_i = m$, $\max k_i = \alpha$.
If $v = (y_1, \dots, y_m)$, we get on a typical block

$$y_2 - \lambda y_1 = g_1$$

$$y_2 - \lambda y_2 = g_2$$

$$y_k - \lambda y_{k-1} = g_{k-1}$$

$$-\lambda y_k = g_k$$

where $(g_1, \dots, g_k)$ are the components of $(I - Q_1) g(v + u(v, \lambda), \lambda)$ on the block. The equations can be written as

$$y_2 = g_1 + \lambda y_1$$
$$y_3 = g_2 + \lambda g_1 + \lambda^2 y_1$$
$$\text{--------------------}$$
$$y_k = g_{k-1}+\lambda g_{k-2}+\dots+\lambda^{k-2} g_1+\lambda^{k-1} y_1$$
$$\lambda^k y_1 + g_k + \lambda g_{k-1}+\dots+\lambda^{k-1} g_1 = 0.$$

Taking $n = \alpha$ in the growth condition on g, we may solve the first $k-1$ equations, in this and their similars in the other blocks, in a unique fashion in terms of $\lambda, y_1, y_{k_1+1}, y_{k_1+k_2+1}, \dots, y_{k_1+\dots+k_d+1}$, that is on the natural projection onto ker A. Calling this projection z we get the bifurcation equations

$$\lambda^{k_1} z_1 - \tilde{g}_1(z, \lambda) = 0$$
(4)
$$\text{----------------------}$$
$$\lambda^{k_d} z_d - \tilde{g}_d(z, \lambda) = 0$$

with $\|\tilde{g}\| \leq c(\|z\|^2 + |\lambda|^{2\alpha+1})$.

Here $B(\lambda)$ is diagonal.

Exercises

1) Verify the estimates in the contractions used.

2) If $M \equiv 2c+1$ show that there is an r_o small enough such that if $\|z\| = r \leq r_o$, $|\lambda| = (Mr)^{1/\alpha}$, then $\|B(\lambda)z\| \geq |\lambda|^{\alpha}\|z\| > \|\tilde{g}(z, \lambda)\|$.

3) Prove that the system

$$y_2 - \lambda y_1 = 0$$
$$\text{------------}$$
$$y_k - \lambda y_{k-1} = 0$$
$$- \lambda y_k = |y_1|^2+|\lambda|^{2k}$$

has no solution except $\lambda = z = 0$ (λ real or complex). This shows that the growth $|\lambda|^{2\alpha+1}$ is necessary.

4) If $M(r) = cr$, then there is a unique trivial solution $z = z(\lambda)$,

Lipschitz in λ.

Theorem:

i) For each $r \leq r_o$, there is a trivial solution with $\|Px\| < r$, $|\lambda| = (Mr)^{1/\alpha}$, λ in $\mathbb{R}$ or $\mathbb{C}$.

ii) If $m = \dim_{\mathbb{K}} \ker A^{\alpha}$ is odd, then, for each $r \leq r_o$, there is a nontrivial solution with $\|Px\| = r$, $|\lambda| < (Mr)^{1/\alpha}$.

iii) If $\lambda \in \mathbb{R}$, $\dim \ker A = 1$, m odd, then there are two bifurcating solutions with $Px = \pm r$.

iv) If $\lambda \in \mathbb{C}$, $\dim_{\mathbb{C}} \ker A = 1$, any m, then there is bifurcation in all directions Px with $\|Px\| = r$.

Proof.

i) Fix λ with $|\lambda| = (Mr)^{1/\alpha}$; from the above exercise, equation (4) has no zeroes on the sphere $\|z\| = r$, so that its degree is well defined and equal to the degree of $B(\lambda)z$ which is Sign λ^m for λ real and 1 for λ complex.

ii) Adding to equation (4) the side condition $\|z\|^2 - r^2$, this pair has no zeroes on the sphere $\partial B = \{(z, \lambda) \in \mathbb{K}^d \times \mathbb{K}, \|z\|^2 + |\lambda|^2 = r^2 + (Mr)^{2/\alpha}\}$, that is the generalized degree, as an element of $\Pi_d(S^d)$, in the real case, or of $\Pi_{2d+1}(S^{2d})$ in the complex case, is well defined. We may clearly deform $\tilde{g}$ to zero and get Whitehead's mapping.

Exercise

1) Using deformations in $B(\lambda)$ of the form

$$\begin{pmatrix} 1-t & t \\ -t & 1-t \end{pmatrix} \begin{pmatrix} (1-t)\lambda^i & -t \\ t\lambda^{i+j} & (1-t)\lambda^j \end{pmatrix}$$

replace $B(\lambda)$ by a diagonal matrix of the form $(\lambda^m, 1, \ldots, 1)$.

2) If $\lambda \in \mathbb{R}$, $m = 2n+\varepsilon$, $\varepsilon = 0, 1$, deform this matrix to $(\lambda^\varepsilon, 1, \dots, 1)$ via $((1-t)\lambda^{2n} + t)\lambda^\varepsilon$.

3) Deform $(z_1, \dots, z_d, |z|^2 - r^2)$ to $(0, \dots, 0, -1)$.

4) Solving $\lambda z_1 = \varepsilon^2$, $z_2 = \dots = z_d = 0$, $|z|^2 - r^2 = 0$, compute the index of each zero and prove that the degree with respect to the above ball is -2.

In the complex case, we get the suspension of the matrix $\lambda \to \lambda^m$, which defines an element of degree m in $\Pi_1(S^1)$ and, through the J-homomorphism a nontrivial element in $\Pi_{2d+1}(S^{2d})$ if and only if m is odd, if $d > 1$, always if $d = 1$. (In [Iz.1, Remark 2.9] there is a direct proof of this fact).

iii), iv) Here we have only one equation. Fixing $z = z_0$, with $|z_0| = r$, and considering the set $\{\lambda / |\lambda| \leq (Mr)^{1/\alpha}\}$, we may deform, on the boundary of this set, the equation to $\lambda^m z_0$ which, as a mapping from S^0 into $\mathbb{R}-\{0\}$ has a nonzero degree if and only if m is odd and, as a mapping from S^1 into $\mathbb{R}^2-\{0\}$, has degree m.

Notes

1) If $m = 1$ and $\tilde{g}$ is more regular one may use, in the real case, Morse theory to show the existence of two curves of solutions (see [Iz 1] and [Ni]).

2) In case (i) one may use the ball B in the space $\mathbb{K}^d \times \mathbb{K}$, adding to equation (4) the side condition $\lambda - \lambda_0 = 0$, with $|\lambda_0| = (Mr)^{1/\alpha}$. The same deformation leads to the computation of the degree of $(B(\lambda)z, \lambda-\lambda_0)$ which is, via $B(t\lambda+(1-t)\lambda_0)$ and via $\lambda - t\lambda_0$, the degree of $(B(\lambda_0)z, \lambda)$ obtaining, by suspension or product, the same degree as before.

Likewise in cases (iii) and (iv) we may consider (4) with the side condition $z-z_0 = 0$, compute the degree of $(B(\lambda)z, z-z_0)$ which, via

$B(\lambda)(tz + (1-t)z_0)$, gives $(B(\lambda)z_0, z-z_0)$ and, via $z-tz_0$, leads again to the suspension or product of $B(\lambda)z_0$. This enables one to consider equation (4) on the same set B, adding, according to ones wish, a suitable side condition.

<u>Corollary</u>:

1) If one of the $\tilde{g}$'s has the form $\lambda\tilde{h}$ with

$$|h(z, \lambda)| \leq C(\|z\|^2 + |\lambda|^{2\alpha})$$

then the result of the theorem is true for all m.

2) If one of the $\tilde{g}$'s is such that $\tilde{g}_\ell(z, 0) \neq 0$ for $\|z\| = r$, some $r > 0$ small, then if $\dim \ker A > 1$, there is always bifurcation with $\lambda \neq 0$.

<u>Exercises</u>

1) Prove the corollary.

2) If the multiplicity is even, we may find nonlinearities without bifurcation, that is giving an element of $\Pi^m(\bar{\Omega})$. For example if $B(\lambda)$ is the diagonal matrix $(\lambda^{k+2n}, \lambda^k)$ and, in the real case, $g(z_1, z_2, \lambda) = (tz_2, -tz_1)$ with $t = |z_1|^2 + |z_2|^2$, or, in the complex case,

$$g(z_1, z_2, \lambda) = (t[\lambda^n\bar{z}_2 - nt\bar{z}_1 + n\lambda^{k+n}|\lambda|^{-n-\varepsilon}z_2], -t[\bar{\lambda}^n|\lambda|^{-n+\varepsilon}\bar{z}_1 + nt\bar{z}_2 + n\lambda^{k+n}z_1])$$

with $\varepsilon = 0$ if $n = 0$, $\varepsilon = 1$ if $n > 0$ and $k \geq 2$, $\varepsilon = 1/2$ if $n > 0$ and $k = 1$.

Prove that the only solution is $z_1 = z_2 = 0$.

Note that in the complex case the nonlinearity is not complex analytic. In fact if $\tilde{g}$ is complex analytic in z there is always bifurcation, see [Da 2].

§II STUDY FOR $A-T(\lambda)$, $i(A) \geq 0$, λ IN $\mathbb{R}$ OR $\mathbb{C}$

Since the main idea of these notes is to suppose the minimum on the nonlinear part, it is clear, if $i(A)$ is negative, then the linear part $B(\lambda)x_1$ will be overdetermined and in general impossible to solve, unless we have a number of parameters large enough to compensate the index.

Since $T(\lambda)$ is analytic, one may show the following result:

Lemma. There is a $\lambda_0 > 0$ such that for $0 < |\lambda| < \lambda_0$

i) $A-T(\lambda)$ is Fredholm with $i(A-T(\lambda)) = i(A)$.

ii) $\dim \ker(A-T(\lambda)) = n_1 \leq d$, $\dim \operatorname{coker}(A-T(\lambda)) = n_2 \leq d^*$.

iii) If $U \equiv \{\lambda / A-T(\lambda)$ is Fredholm$\}$, then U is open; $i(A-T(\lambda))$, $\dim \ker(A-T(\lambda))$, $\dim \operatorname{coker}(A-T(\lambda))$ are constant on connected components of U, except at isolated points where the two last integers may be larger.

Proof. See lemmas 6.2 and 6.3 in [Iz.1].

Here we shall assume that for λ small, $\lambda \neq 0$, $\operatorname{codim} R(A-T(\lambda))=0$. This is equivalent to ask that $B(\lambda)$ is onto for $\lambda \neq 0$. Then, for each $\lambda \neq 0$, there is a $d^* \times d^*$ minor with nonzero determinant. Since $B(\lambda)$ is analytic, the determinants of the minors cannot accumulate zeroes near $\lambda = 0$, unless they are identically zero; this means that one of them has the form

$$\Delta(\lambda) = \lambda^m a(\lambda) \quad \text{with} \quad a(0) \neq 0$$

and we shall suppose that it corresponds to the first d^* coordinates, that is $x_1 = (y_1, \ldots, y_{d^*}, z_1, \ldots, z_i)$ with $i = i(A)$. m will be called the *algebraic multiplicity of* 0.

Theorem. Suppose g satisfies the usual conditions with $n = m-d^*+1$, then

i) $i(A) = 0$, the conclusions of the preceeding theorem are valid,

ii) $i(A) = 0$, the parity of m and the sign of $a(0)$ in the real case, m itself in the complex case, are independent of the projections P and Q;

iii) $i(A) > 0$, there is always bifurcation: in fact given constants $\varepsilon_1, \ldots, \varepsilon_i$ with $\sum|\varepsilon_j|^2 = o(r^2)$, (4) has bifurcating solutions with $\lambda^k z_1 = \varepsilon$, $z_2 = \varepsilon_2, \ldots, z_i = \varepsilon_i$ with $k = 0$ if m is odd and $k = 1$ if m is even.

<u>Proof</u>. As the theorem indicates we shall add, in case of positive index, the last i equations, obtaining thus a square matrix with determinant of the form $\lambda^{m+k}\, a(\lambda)$ that is with a leading term of odd power in λ.

<u>Exercise</u>. Prove that on the sphere $\|x_1\|^2 + |\lambda|^2 = r^2 + (Mr)^{2/n}$ one may deform the pair $(B(\lambda)x_1 - \tilde{g}(x_1, \lambda), \|x_1\|^2 - r^2)$ to $(B(\lambda)x_1, \|x_1\|^2 - r^2)$ where $B(\lambda)$ and $\tilde{g}$ represent the augmented equations in case of a positive index.

One can then deform $B(\lambda)$ to $(\lambda^m, 1, \ldots, 1)$ a diagonal matrix, see [Iz.1.p.31], or use in the real case the following

<u>Lemma</u>: $d(B(\lambda)x_1, \|x_1\|^2 - r^2, 0, B) = i_- - i_+$ where $i(\lambda) = d(B(\lambda)x_1, 0, \|x_1\| < r)$, $i_\pm = i(\lambda)$ with $\lambda \gtrless 0$.

(This implies that the degree is -2 Sign a(0) if m is odd, 0 if m is even).

<u>Proof of the lemma</u>. On ∂B deform the first mapping to $(B(\lambda)x_1, (Mr)^{2/n} - |\lambda|^2)$, using a linear deformation. This map has two

isolated zeroes $x_1 = 0$, $\lambda = \pm(Mr)^{1/n}$ with index equal to the sign of the Jacobian of the map, that is $-\mathrm{Sign}(2\lambda \det B(\lambda))$.

ii) [See Iz.1.p.33].

<u>Example</u>. If $A = I-\mu_0 T$, $\lambda = \mu-\mu_0$, $T(\lambda) = \lambda T$ with T compact, one has the classical bifurcation problem. Here A has zero index and we may define P and Q in two steps, taking first the decomposition on $R(A^\alpha) \oplus \ker A^\alpha$ and then taking A in Jordan form with blocks with 0 on the diagonal and 1 on the upper diagonal. Since $T = (I-A)/\mu_0$, T will have μ_0^{-1} on the diagonal and $-\mu_0^{-1}$ on the upper diagonal and $A-T(\lambda)$ will have a diagonal with $(\mu-\mu_0)/\mu_0$ and an upper diagonal of the form μ/μ_0. We are exactly in the situation of last section and we may reduce ourselves to the study on $\ker A$ with $B(\lambda)$ a diagonal matrix with elements of the form $(\mu-\mu_0)^{k_i}/\mu_0^{k_i}$, or consider the problem on $\ker A^\alpha$ and deform the off-diagonal elements to zero. The preceding lemma give then the classical Krasnosel'skii's result in the real case: bifurcation if the algebraic multiplicity of μ_0 is odd.

§III SEVERAL PARAMETERS BIFURCATION

In order to study the bifurcation equation (4) we still assume no particular knowledge of the nonlinear part but a well known behavior of the linear part $B(\lambda)$. Thus, in order to be able to use topological methods we need to deform the equation to its known part, that is to suppose $B(\lambda)x_1$, with some additional condition (like $\|x_1\|^2-r^2$ or $x_1-x_1^0$) which will imply the existence of bifurcating solutions, is different from zero on some sphere and dominates the nonlinear piece.

III-1) <u>A particular case</u> $d = d^* \leq k$

Suppose that there is a fixed x_1^0, with $\|x_1^0\| = r$, and fixed

$\lambda^o_{d+1}, \ldots, \lambda^o_k$ such that the only local solution of $B(\lambda)x^o_1 = 0$ is $\lambda_1 = \ldots = \lambda_d = 0$ and that $\| B(\lambda_1, \ldots, \lambda_d, \lambda^o_{d+1}, \ldots, \lambda^o_k)x^o_1\| > \|\tilde{g}(x^o_1, \lambda)\|$ for $|\lambda_1|^2 + \ldots + |\lambda_d|^2 = \rho^2$.

Then, on this sphere in $\mathbb{K}^d$, we may deform (4) to $B(\lambda_1, \ldots, \lambda^o_k)x^o_1$ and, if this field has a nonzero index, get a solution.

Note that, in order to use only one approach, we can consider the system

$$B(\lambda)x_1 - \tilde{g}(x_1, \lambda) = 0$$

$$x_1 - x^o_1 = 0$$

$$\lambda_j - \lambda^o_j = 0 \qquad j = d+1, \ldots, k$$

on the sphere $\{(x_1, \lambda)/|x_1|^2+|\lambda|^2 = r^2+\rho^2+|\lambda^o|^2\}$ in $\mathbb{K}^{k+d}$ with $2d+k-(d+1)+1 = k+d$ equations.

Under the above hypothesis, the degree of this system is well defined and equal to $d(B(\lambda)x_1, x_1-x^o_1, \lambda_j-\lambda^o_j, 0, B)$ and, deforming $B(\lambda)x_1$ to $B(\lambda_1, \ldots, \lambda_d, \lambda^o_{d+1}, \ldots, \lambda^o_k)x^o_1$ via a linear deformation, equal to $d(B(\lambda_1, \ldots, \lambda_d, \lambda^o_{d+1}, \ldots, \lambda^o_k)x^o_1, x_1, \lambda_j, 0, B)$ (deforming $(x_1 - x^o_1, \lambda_j - \lambda^o_j)$ to (x_1, λ_j)) obtaining thus the suspension of the first mapping and the same degree.

<u>Example</u>: <u>$k = d$</u> and $(B(\lambda)x^o_1)_i = P_i^{k_i}(\lambda) + Q_i^{k_i+1}(\lambda)$ where $P_i^{k_i}(\lambda)$ is a homogeneous polynomial of degree k_i and Q_i of higher order. If $(P_i^{k_i}(\lambda))_i$ has an isolated zero at $\lambda = 0$, then one has the following result.

<u>Theorem</u>. If $|\tilde{g}_i| \leq C(\|x_1\|^2 + |\lambda|^{\alpha+k_i+1})$, $\alpha = \max k_i$, $i=1, \ldots, d$, then

i) *Real case:* If all k_i are odd there is bifurcation.

ii) *Complex case:* There is always bifurcation.

Note: For $d = 1$ these are the cases (iii) and (iv) of the bifurcation theorem in section I.

Proof. Using the homogeneity of P^{k_i} in λ and the linearity in x_1, we may deform (4) to $(P_i^{k_i}(\lambda)x_1^o)_i$ (Exercise). The degree of this vector field is Πk_i in the complex case and its parity is the same as the parity of Πk_i in the real case, see [Cr. p.45-49], [St. p. 223-234]. In the real case this is a consequence of Borsuk's theorem, since the mapping defined by $(P_i^{k_i}(\lambda)x_1^o)_i$ is odd, hence with odd degree.

Examples.

1) If $\det B(\lambda) \neq 0$, λ small nonzero, and $(P_i^{k_i}(\lambda)x_1^o)_i$ has a nonzero degree for some x_1^o $(k = d)$, then we'll have bifurcation in all directions in ker A, since any x_1, with $\|x_1\| = \|x_1^o\|$, can be reached by a rotation from x_1^o, leaving the degree invariant. In fact the degree of $B(\lambda)x_1^o$ in $\Pi_{d-1}(S^{d-1})$ depends uniquely on the class of $B(\lambda)$ in $\Pi_{d-1}(GL(d))$; this group and its image in $\Pi_{d-1}(S^{d-1})$ under the mapping $B(\lambda) \xrightarrow{p} B(\lambda)x_1^o$ are known and appear in the exact sequence associated to the fiber bundle $O_{d-1} \xrightarrow{i} O_d \xrightarrow{p} S^{d-1}$:

$$\Pi_k(S^{d-1}) \longrightarrow \Pi_{k-1}(O_{d-1}) \xrightarrow{i_*} \Pi_{k-1}(O_d) \xrightarrow{p_*} \Pi_{k-1}(S^{d-1})$$

then the degree $p_*[B(\lambda)]$ is nonzero if and only if $B(\lambda)$ cannot be deformed, for $|\lambda| = \rho$, to a matrix of the form

$$\begin{pmatrix} C(\lambda) & 0 \\ 0 & I \end{pmatrix}$$

i.e. $[B(\lambda)]$ does not belong to Im i_*.

If $\det B(\lambda) \neq 0$ for $\lambda \neq 0$ and $k > d$, then, if one tries to fix $\lambda_{d+1}, \ldots, \lambda_k$, the degree is zero since one may deform $B(\lambda)$ to $B(0, \ldots, 0, \lambda_{d+1}, \ldots, \lambda_k)$, a constant matrix. However one may consider all the parameters and get an element in $\Pi_{k-1}(GL(d))$ and an invariant in $\Pi_{k+d-1}(S^{2d-1})$ if one considers $(B(\lambda)x_1, x-x_1^o)$, which is the suspension of the element $B(\lambda)x_1^o$ in $\Pi_{k-1}(S^{d-1})$; this last invariant is non-trivial if and only if $B(\lambda)$ cannot be deformed, as above, to a lower dimensional matrix. Here the topological results are known, their interpretation in terms of analysis is not complete.

2) If $\mathbb{K} = \mathbb{R}$, $d = k = 2$ one may use the results of [Kr.3] to compute the index of the vector field (See [Iz.1]).

III-2) Reduction to one parameter

As we have already seen, the local behavior of the bifurcation equation, at the level of topological invariants, is given by the linear part. It is then important to have a precise knowledge of the set of points λ where $\det B(\lambda) = 0$. Since this function is analytic in λ we have to use algebraic geometry to determine them. For instance if λ is in $\mathbb{C}^k$ this set is an algebraic variety whose irreducible components have local dimension $k-1$; in order to be able to perform the deformation to the linear part one has to move in a direction transversal to the variety, that is to reduce oneself to a one dimensional problem. In the real case the situation is more complicated, but we may say that if we want positive results by reduction to a one parameter problem it is necessary that the variety of zeroes of the determinant separates $\mathbb{R}^k$ locally in two connected components, since then the class of $(B(\lambda_t)x_1, \|x_1\|^2 - r^2)$, λ_t a path transversal to the variety, is determined by the sign of the determinant; if the variety does not separate the space, this sign does not change. The case $d = 1$ has already been studied.

Suppose then that $\det B(\lambda) = P_{k_o}(\lambda_1, \dots, \lambda_k) + \dots$ with $P_{k_o} \not\equiv 0$ (In case of a positive index, $B(\lambda)$ will represent a $d^* \times d^*$

minor with nontrivial determinant).

Theorem. If $g(x, \lambda)$ satisfies the usual conditions with a growth in λ of the form $|\lambda|^{2n+1}$, with $n = k_o - d^*+1$, then

i) $i(A) = 0$ and k_o odd, there is a k-1 dimensional family of bifurcating solutions.

ii) $i(A) = 0$, k_o even and there is a direction λ_o with $P_{k_o}(\lambda_o) = \ldots = P_{k_o+m-1}(\lambda_o) = 0$, $P_{k_o+m}(\lambda_o) \neq 0$, if m is odd and $n = k_o + m - d^* + 1$, the same conclusion holds.

iii) $i(A) > 0$, there is a $k-1+i(A)$ dimensional family of bifurcating solutions.

Proof. i) Consider a direction λ_o with $P_{k_o}(\lambda_o) \neq 0$ (without loss of generality suppose that this direction is the axis λ_1). Consider then the mapping

$$(B(\lambda)x_1 - \tilde{g}(x_1, \lambda), \ \|x_1\|^2 - r^2, \ \lambda_2 - \varepsilon_2, \ldots, \lambda_k - \varepsilon_k),$$

with $\sum \varepsilon_i^2 = o(r^2)$, on the sphere $\|x_1\|^2 + |\lambda|^2 = r^2 + (Mr)^{2/n}$, for some large M. We leave as an exercise to show that this mapping can be deformed to the suspension of $(B(\lambda_1, 0, \ldots, 0)x_1, \|x_1\|^2 - r^2)$ which has already been studied. iii) is analogous completing, as before, $B(\lambda)$ to a square system.

ii) It is enough to take $\varepsilon_2, \ldots, \varepsilon_k$ so small that, in $\det B(\lambda)$, the terms $P_{k_o}(\lambda_1, \varepsilon_2, \ldots, \varepsilon_k) + \ldots + P_{k_o+m-1}(\lambda_1, \varepsilon_2, \ldots, \varepsilon_k)$ are small compared to $P_{k_o+m}(\lambda_1, \varepsilon_2, \ldots, \varepsilon_k)$.

III-3) General case $k \leq d$

If the zeroes of $\det B(\lambda)$ form a variety of local dimension k_1, the previous analysis shows that the parameters must be moved in a transversal variety (so that the equation may be deformed to its linear part) of complementing dimension $k-k_1$, since if not we may deform $B(\lambda)$ to $B(\tilde{\lambda})$, with a constant $\tilde{\lambda}$ outside the variety, and get a trivial invariant.

Moving the transversal variety we'll have the possibility of a k_1-dimensional family of nontrivial solutions, putting additional conditions of the form $\lambda_i-\varepsilon_i = 0$, $i = 1, \ldots, k_1$. If we put ℓ conditions on (x_1, λ), in such a way that non trivial solutions are guaranteed, we'll get an invariant in $\Pi_{d+k-1}(S^{d^*+k_1+\ell-1})$ (d+k unknowns and $d^*+k_1+\ell$ equations on the sphere $\|x_1\|^2+|\lambda|^2 = r^2+\rho^2$, with r and ρ such that $|\lambda_{k_1+1}|^2 + \ldots + |\lambda_k|^2 \cong \rho^2$, $\det B(\varepsilon_1, \ldots, \varepsilon_{k_1}, \lambda_{k_1+1}, \ldots, \lambda_k) \neq 0$ and the conditions on (x_1, λ) give $\|x_1\| \cong r$).
If beforehand we fix $\lambda_1 = \ldots = \lambda_{k_1} = 0$ we'll get an invariant in $\Pi_{d+k-k_1-1}(S^{d^*+\ell-1})$ whose suspension is the first one; the meaning, from the analysis point of view, of the fact that the suspension may kill this last invariant is not clear yet.

In any case, we may reduce ourselves to the problem where $B(\lambda)$ has maximum rank for $\lambda \neq 0$ with an invariant in $\Pi_{d+k-1}(S^{d^*+\ell-1})$. Since those groups are zero if $d+k-1 < d^*+\ell-1$, we can't impose more than $\ell \leq i(A)+k$ conditions on (x_1, λ); for example if $d = d^* > k$ there is no sense to fix x_1 as in section III-1.
Note that we are considering here the real case; for the complex case d, d^*, k are even.
Note also that $i(A)$ may be negative provided $i(A)+k \geq 1$; for instance $d = 1$, $d^* = 2$, then $B(\lambda)x = (\lambda_1 x, \lambda_2 x)$, with a fixed $x = r$, has degree 1.

From now on, in order to simplify the study, we shall assume that $i(A) = 0$. With respect to the ℓ conditions on (x_1, λ), one could think that for each set of conditions, tayloring the solutions according to the particular needs of the problem, one would get a generalized Whitehead's mapping from $\Pi_{k+1}(GL(d))$ into $\Pi_{d+k-1}(S^{d+\ell-1})$ given by $B(\lambda) \longrightarrow (B(\lambda)x_1, F(\lambda, x_1))$, where F is an ℓ-vector of conditions, guaranteeing nontrivial solutions. To be more precise, suppose that for all (r, ρ) small enough and positive, one has the property that if $(B(\lambda)x_1-g(\lambda, x_1), F_r(\lambda, x_1)) = (0, 0)$ for $\|x_1\|^2+|\lambda|^2 \leq r^2+\rho^2$, then $x_1 \neq 0$. We are assuming that $B(0) = 0$, by reduction on ker A, and $g(0, \lambda) = 0$, so that $x_1 = 0$ is the trivial solution. Assume also that for some (r, ρ) the pair

$$\begin{matrix} B(\lambda)x_1 - tg(\lambda, x_1) \\ F_r(\lambda, x_1) \end{matrix}$$

is non-zero on the sphere $\|x_1\|^2+|\lambda|^2 = r^2+\rho^2$, for all t in $[0, 1]$, so that the pair represents an element of $\Pi_{k+d-1}(S^{d+\ell-1})$ for all t in the interval. We have then the rather disapointing result

<u>Theorem</u>. Given the above hypothesis, the above invariant is always trivial unless there is only one condition, $\ell = 1$, linearly deformable to $\|x_1\|^2-r^2$ or $[B(\lambda)x_1^o] \neq 0$ in $\Pi_{k-1}(S^{d-1})$ for x_1^o fixed, $\|x_1^o\| = r$ and $|\lambda| = \rho$. So if $k < d$, $B(\lambda)x_1^o$ is always deformable to a constant map, and the only J-homomorphism is Whitehead's mapping. If $B(\lambda)x_1^o$ is nontrivial we have bifurcation in all directions.

<u>Proof</u>. Note first that one has to study the class of the pair $(B(\lambda)x_1, F_r(\lambda, x_1))$ and, since $F_r(t\lambda, 0) \neq 0$ for $|t\lambda|^2 \leq r^2 + \rho^2$ ($x_1 = 0$ is solution of (4)), the class of $(B(\lambda)x_1, F_r(0, x_1))$, ($F_r(0, x_1) \neq 0$ for $\|x_1\|^2 = r^2+\rho^2$); so we may as well suppose that F_r does not depend on λ.

Secondly the deformation $(t+(1-t)|\lambda|)B(\lambda))$ is valid and the homotopy, defined by radial retraction onto the sphere of radius $|\lambda| = \rho$, $|\lambda|B(\lambda(t+(1-t)\rho/|\lambda|))$ is continuous $(B(\lambda)$ is uniformly bounded on the ball $|\lambda|^2 \leq \rho^2 + r^2$), so that one may replace $B(\lambda)$ by $|\lambda|(B(\rho\lambda/|\lambda|)$. This implies that if, on the sphere $|\lambda| = \rho$, $B(\lambda)$ is homotopic to $D(\lambda)$ then $(|\lambda|B(\rho\lambda/|\lambda|)x_1, F_r(x_1))$ has the class of $(|\lambda|D(\rho\lambda/|\lambda|)x_1, F_r(x_1))$; in particular if $D(\lambda)$ has the form

$$\begin{pmatrix} C(\lambda) & 0 \\ & \\ 0 & I \end{pmatrix}$$

(which is equivalent, as we have seen in section III-1, to $[B(\lambda)x_1^0] = 0$) then, with $x_1 = (\tilde{x}, x_d)$ in $\mathbb{R}^{d-1} \times \mathbb{R}$, $(B(\lambda)x_1, F_r(x_1))$ is homotopic to $(|\lambda|C(\rho\lambda/|\lambda|)\tilde{x}, |\lambda|x_d, F_r(\tilde{x}, x_d))$. One may replace $|\lambda|x_d$ by x_d, via $(t+(1-t)|\lambda|)$, and then $F_r(\tilde{x}, x_d)$ by $F_r(\tilde{x}, 0)$, via tx_d. Use then the deformation $F_r(\cos \pi t/2\tilde{x}, \sin \pi t/2\|\tilde{x}\|)$, which is valid since $F_r(0) \neq 0$ and $F_r(x_1) \neq 0$ for *all* x_1 in $\mathbb{R}^d$ with $\|x_1\|^2 = r^2+\rho^2$.

Now, if $\ell > 1$ or if $\ell = 1$ and $F_r(0, 0)$ and $F_r(0, (r^2+\rho^2)^{\frac{1}{2}})$ have the same sign, choose a point p in $\mathbb{R}^\ell-\{0\}$ such that $F_r(0, 0) \neq xp$ and $F_r(0, (r^2+p^2)^{\frac{1}{2}}) \neq xp$, for all $x \geq 0$. Use then the deformation $tF_r(0, \|\tilde{x}\|)-(1-t)p$ and then one may deform $(B(\lambda)x_1, F_r(x_1))$ to the constant map $(0, -p)$.

If $\ell = 1$ and $F_r(0, 0)\ F_r(0, (r^2+\rho^2)^{\frac{1}{2}}) < 0$, use the linear deformation $tF_r(0, \|\tilde{x}\|)-(1-t)$ Sign $F_r(0, 0)(\|\tilde{x}\|^2 - r^2)$.

Note that if $[B(\lambda)] = [I] = 0$ then $B(\lambda)x_1$ is deformable to x_1, as above for x_d, then to $x_1-x_1^0$ with $0 < \|x_1^0\| < r$, and the complete Whitehead's map to $(-x_1^0, 0)$ with class 0.

Examples.

1) If one has the condition $\|x_1\|^2 - r^2$ and $k < d$ then one may use the results in chapter one, see [A.Y.1], [Iz.3].

2) If λ is in $\mathbb{C}$, one recovers the results of the first section.

3) Several tricks to compute the class $B(\lambda)$ are given in [Iz.1].

4) For the case of the Hopf bifurcation problem, bifurcation of periodic solutions to an autonomous system (a problem with two parameters), the use of Whitehead's map leads to the Alexander-Yorke result. See [A.Y.], [Iz.1] for this particular approach in finite dimensions, [Iz.2] for the infinite dimensional version and applications to partial differential equations, in particular the Navier-Stokes equations. In fact in this problem there is a group action and one may use another invariant, the Fuller Index, to get finer results, see [C.M.Y.]; if the system is hamiltonian, then one may give an estimate on the number of solutions, see [F.R.].

5) The problem of the transition to turbulence for fluids might be a good example of several parameters bifurcation but there are still no definite results.

Chapter III

GLOBAL BIFURCATION

A natural question is to ask what happens to the bifurcating branches when they leave a bifurcation point: do they go to infinity, do they come back or do they stop somewhere? In this chapter we shall give, using only topological tools, several generalizations of the important Rabinowitz' theorem.

Several technical problems rise in the global study: if A is a Fredholm operator, in general $A-T(\lambda)$ is not a Fredholm operator for all λ and one runs into the essential spectrum, case which has been very little studied. Almost all known techniques ask for some compacity on the set of solutions and for some approximation scheme from finite dimensional mappings. For all these reasons and in order to cut down these technical difficulties, the equations we shall consider will be of the form

$$(5) \qquad (I - T(\lambda))x - g(x, \lambda) = 0$$

with x in E, some real or complex Banach space, $T(\lambda)$ a compact operator from $E \times \mathbb{K}^k$, analytic in λ, $g(x, \lambda)$ compact with $\|g(x, \lambda)\| = o(\|x\|)$ if $\|x\|$ goes to 0.

For example, if λ is in $\mathbb{K}$ and $I-T(\lambda)$ is invertible for some λ, then $I-T(\lambda)$ will be invertible for all λ except for a discrete number of λ's for which $I-T(\lambda)$ will be Fredholm of index 0 (see the lemma at the beginning of section II in last chapter).

In the first section we shall indicate how the invariants defined in the first chapter can be extended to infinite dimensional spaces, using stable cohomotopy, in the second we shall apply these invariants to the study of global solutions.

§1. COHOMOTOPY IN INFINITE DIMENSIONS

Let E^∞ be a real Banach space (in the complex case we forget its structure). Define an orientation in E^∞ by considering a sequence of n- dimensional subspaces E_n with $E_n \subset E_{n+1}$, $E^\infty = E_n \oplus E^{\infty-n}$, $E_{n+1} = E_n \oplus \mathbb{R}$, $E^{\infty-n} = E^{\infty-n-1} \oplus \mathbb{R}$. Set $P^{\infty-n} = E^{\infty-n} - \bar{\mathbb{R}}^+ = \{x \in E^{\infty-n},\ x = x_1 + r,\ x_1 \in E^{\infty-n-1},\ r \in \mathbb{R},\ r < 0\}$.

Definition. $LS(E^\infty)$, the Leray-Schauder category of E^∞ with

Objects: $X \subset E^\infty$, closed bounded sets,

Morphisms: Maps of the form I-compact

Homotopies: $x - F(x, t)$, $F: X \times I \to E^\infty$ compact.

Let $A \subset X$ be objects in $LS(E^\infty)$, define $\Pi^{\infty-n}(X, A)$ the set of homotopy classes of maps of the form $I-F$, F compact, with

$$I-F: \begin{cases} X & \to E^{\infty-n} - \{0\} \\ A & \to P^{\infty-n} \end{cases}$$

$\Pi^{\infty-n}(X) \equiv \Pi^{\infty-n}(X, \phi)$

Example. Let Ω be a bounded open subset of $E \times \mathbb{R}^k = E^\infty$, $E = E^{\infty-k}$. Let $F(x, \lambda)$ be compact on Ω with $x-F(x, \lambda) \neq 0$ for (x, λ) in $\partial\Omega$, a mapping from Ω into E; then $(x-F(x, \lambda), \lambda-\lambda)$ sends $\partial\Omega$ into $E-\{0\}$ and defines an element of $\Pi^{\infty-k}(\partial\Omega)$. In this case we shall consider only $x-F(x, \lambda)$.

Definition. $f = I-F: X \to E^{\infty-n}-\{0\}$ is *inessential with respect to* Y, $X \subset Y$, if f extends to $\tilde{f}: Y \to E^{\infty-n} - \{0\}$ in the category.

Note: As in the finite dimensional case (replacing Tietze's extension theorem by Dugundji's), if f is compactly homotopic to g and g is inessential, so is g.

Theorem. (Geba-Granas): $\Pi^{\infty-n}(X, A)$ is an abelian group.

Idea of the proof.

a) *Approximation by finite dimensional mappings*

Let $f = I-F:(X, A) \to (E^{\infty-n}-\{0\}, P^{\infty-n})$.

Since F is compact then $f(X)$ and $f(A)$ are closed in $E^{\infty-n}$ (Exercise). Let $\varepsilon < \min(\text{dist}(f(X), E_n), \text{dist}(f(A), E_n \oplus \bar{\mathbb{R}}^+))$, then F can be approximated by G, within an ε-distance, such that G maps X into a finite dimensional subspace L (we may suppose $E_{n+1} \subset L$ with $\dim L = n+1+m$). Then $g = I-G$ maps X into an ε-neighborhood of $E^{\infty-n}$. Let P_0 be the projection of E^∞ onto $E^{\infty-n}$, parallel to E_n, and define $\tilde{f}(x) = P_0 g(x) = x - (I-P_0)x - P_0 G(x) = x - \tilde{F}(x)$. Then $\tilde{F}(x)$ is in L since $P_0 G = G-(I-P_0)G$, $(I-P_0)G$ is

in $E_n \subset L$ and G is in L.
We may deform f to $\tilde{f}$ via $P_0((1-t)f(x) + tg(x))$ due to the definition of ε (Exercise). Let $\tilde{f}_L:(X \cap L, A \cap L) \xrightarrow{\tilde{f}} (E^{\infty-n} \cap L - \{0\}, P^{\infty-n} \cap L) = (E^{m+1} - \{0\}, P^{m+1})$, since $\dim (E^{\infty-n} \cap L) = m+1$. Since P^{m+1} is contractible, the last pair is homotopic to (S^m, pt), giving thus an element of $\Pi^m(X \cap L, A \cap L)$ which is a group if $\dim L = m + n + 1 \leq 2m - 2$, that is $m \geq n+3$. One can show that f and g are homotopic in $LS(E^\infty)$ if and only if $\tilde{f}_L$ and $\tilde{g}_L$ are homotopic.

b) *Finite range mappings*

Let $L \supset E_{n+1}$, of dimension $n+m+1$, be fixed. Let $\Pi_L^{\infty-n}(X, A)$ be the set of homotopy classes of maps $f:(X, A) \longrightarrow (E^{\infty-n}-\{0\}, P^{\infty-n})$ with $f = I-F$, $F(X) \subset L$ and homotopies $x-H(x, t)$, with $H(x, t)$ in L. Let $\tau:\Pi_L^{\infty-n}(X, A) \to \Pi^m(X \cap L, A \cap L)$ induced by restriction to $X \cap L:\tau(f) = \tilde{f}_L$.

Lemma. τ is one to one, onto.

Proof. τ *is onto:* If $\tilde{f}_L(x_1) = x_1-\tilde{F}_L(x_1)$ is an element of $\Pi^m(X \cap L, A \cap L)$, x_1 in L, let $x = x_1 \oplus x_2$ be in E^∞ and define $f(x) = x_2 \oplus x_1 - \tilde{F}_L(x_1)$: f extends $\tilde{f}_L$, choosing the complement of L in E^∞ contained in $E^{\infty-n}(E_{n+1} \subset L)$.

τ *is one to one:* If f and g give $\tilde{f}_L$ and $\tilde{g}_L$ homotopic via $\tilde{h}_L(x, t)$, let $Z = X \times \{0\} \cup X \cap L \times [0, 1] \cup X \times \{1\}$, closed subset of $X \times [0, 1]$.

Construct $H(x, t):Z \to L$ by

$$H(x, 0) = x - f(x)$$

$$H(x, 1) = x - g(x)$$

$$H(x, t) = x-\tilde{h}_L(x, t) \quad \text{for } x \text{ in } X \cap L.$$

From Dugundji's extension theorem, extend H to $\bar{H}: X \times [0, 1] \to L$ and let $h(x, t) = P_0(x - \bar{H}(x, t)) = x - (I - P_0)x - P_0 \bar{H}(x, t)$, P_0 the projection of E^∞ onto $E^{\infty-n}$ (if $h(x, t) = 0$, then $x - \bar{H}(x, t)$ belongs to $E_n \subset L$, x belongs to L and $h(x, t) = \tilde{h}_L(x, t)$).

From this lemma we may give to $\Pi_L^{\infty-n}(X, A)$ the group structure of $\Pi^m(X \cap L, A \cap L)$.

c) *The limit process*

Let $M \supset L$, $\dim M = k$. There is a Mayer-Vietoris homomorphism $\Delta: \Pi^m(X \cap L, A \cap L) \to \Pi^{k-1-n}(X \cap M, A \cap M)$ which induces, via τ, a homomorphism $\Pi_L^{\infty-n}(X, A) \to \Pi_M^{\infty-n}(X, A)$ commuting with all morphisms and coboundary operations. We get thus an inductive family of abelian groups and $\varinjlim_{L \to \infty} \Pi^{\dim L-1-n}(X \cap L, A \cap L) \equiv \sum^{\infty-n}(X, A)$ is called the *stable cohomotopy group* of (X, A). The limit of τ gives an isomorphism between $\sum^{\infty-n}(X, A)$ and $\Pi^{\infty-n}(X, A)$ which induces in this last set all the functorial properties of the stable group, in particular:

Theorem. (Geba-Granas): $\Pi^{\infty-n}$ defines a generalized cohomology functor on $LS(E^\infty)$, $\Pi^{\infty-n}(X, A)$ satisfies all Eilenberg-Steenrod axioms (see section II-1 of chapter I).

Examples.

1) The neutral element of $\Pi^{\infty-n}(X)$ represents all inessential mappings from X into $E^{\infty-n} - \{0\}$.

2) If $X = S$, a sphere in E^∞, the Mayer-Vietoris homomorphism is not necessary, one needs only the suspension $S \cap L = S^{n+m} \to S^m$. $\Pi^{\infty-n}(S) \equiv \Pi_n$ is the stable group for spheres, thus $\Pi^{\infty-0}(S) = Z$, $\Pi^{\infty-1}(S) = Z_2$, generated by the suspension of the Hopf map, $\Pi^{\infty-2}(S) = Z_2$,

$\Pi^{\infty-3}(S) = Z_{24} \dots$

3) $\Pi^{\infty-n}(\bar{B}) = 0$, B a ball in E^{∞} (Exercise).

4) If Ω is a bounded open set in E^{∞}, Ω connected, $\Pi^{\infty-0}(\partial\Omega)$ represents, as in the finite dimensional case, the Leray-Schauder degree.

Exercises: Extend all the finite dimensional results of section II in chapter I to the present situation, in particular the sum of invariants in the case of $\Omega = B - \cup B_i$, B, B_i balls in E^{∞}.

§II. GLOBAL SOLUTIONS

Suppose we want to study the set S of solutions to

$$x - F(x, \lambda) = 0 \tag{6}$$

with x in E, λ in $\mathbb{K}^k$, F compact on $E^{\infty} = E \times \mathbb{K}^k$.

Assume we know a set T of solutions, called trivial or stationary or "bad" solutions, and that we want to concentrate on a set B of nontrivial or periodic or "good" solutions.

Note first that if Ω is an open bounded subset of E^{∞} then $\bar{B} \cap \bar{\Omega}$ and $\bar{T} \cap \bar{\Omega}$ are compact (λ_n being bounded in a finite dimensional subspace has a converging subsequence, $F(x_n, \lambda_n)$ has also a converging subsequence to y; the x_{n_j} converge to y, λ_{n_j} to λ and, by continuity, $F(y, \lambda) = y$). Note also that $S \cap \bar{\Omega}$ is compact.

Suppose $S = B \cup T$, clearly $\bar{B} - B \subset T$.

Assume T is closed and call $\bar{B}$-B the *bifurcation set* of T (Note that everything may be defined relative to some set D in E^{∞} and work with induced topologies: this is the case for the Hopf bifurcation problem).

For all (x, λ) the function $d(x, \lambda)$, distance of (x, λ) to $T \cap \bar{\Omega}$ (Ω as above), is then a continuous function (Exercise).

Let (x_0, λ_0) be a point of $\bar{B}-B$ and let C be the connected component of $\bar{B}$ containing (x_0, λ_0) (Notice that C may be reduced to (x_0, λ_0) if one has a sequence of points in B converging to (x_0, λ_0) and not a continuum).

Suppose that C is bounded, hence compact from the above argument, and let Ω_1 be an open bounded subset of E^∞ containing C.

<u>Lemma</u>. There is an open set Ω such that $C \subset \Omega \subset \Omega_1$, $\bar{B} \cap \partial\Omega = \phi$.

<u>Proof</u>. (This is a particular case of a result by Whyburn).
Set $B_1 = \bar{B} \cap \partial\Omega$; B_1 and C are two disjoint compact subsets of $\bar{B} \cap \bar{\Omega}_1$ and hence at a positive distance one from the other.

It is easily seen that there is an $\varepsilon_0 > 0$, such that, if $\varepsilon < \varepsilon_0$, no ε-chain in $\bar{B} \cap \bar{\Omega}_1$ can join B_1 to C (an ε-chain is a finite number of balls, with diameter smaller than ε, $A_1, \ldots, A_m$ such that $\tilde{A}_i \equiv A_i \cap (\bar{B} \cap \bar{\Omega}_1)$ has the property that $\tilde{A}_i \cap \tilde{A}_j = \phi$ if and only if $|i-j| > 1$): in fact, if this is not true, there are a_n in B_1, b_n in C and $1/n$-chains joining a_n to b_n; by compacity there is a subsequence (a_{n_j}, b_{n_j}) converging to (a, b) in (B_1, C) and, for all $\varepsilon > 0$, there is an ε-chain joining a to b; let then C_a be the set of all z in $\bar{B} \cap \bar{\Omega}_1$ such that, for all $\varepsilon > 0$, a and z can be connected by an ε-chain : b is in C_a and clearly C_a is closed, hence compact. Furthermore C_a is connected since if not there are two open subsets X and Y with $C_a \cap X$, $C_a \cap Y$ disjoint, non-empty and covering C_a; take a continuous fonction λ from C_a into $[0, 1]$ with $\lambda(C_a \cap X) = 0$, $\lambda(C_a \cap Y) = 1$; now if x, y belong to C_a there is an ε-chain joining them (passing by a and eliminating multiple intersections) and so, if ε' is such that $|\lambda(x)-\lambda(y)| < \varepsilon' < 1$ for $|x-y| < \varepsilon$, taking x in $C_a \cap X$ and y in $C_a \cap Y$, there is a first step in the ε-chain from x to y where λ becomes 1, contradicting the continuity. Choose then $B_2 = \{z$ in $\bar{B} \cap \bar{\Omega}_1$ such that there is an x in B_1 and an ε_0-chain from z to $x\}$: $B_2 \cap C = \emptyset$, B_2 is closed since if z belongs to $\bar{B}_2$ the ball $B(z, \varepsilon_0)$ intersects B_2 and so z belongs to B_2; B_2 is relatively open since if z is in B_2, $B(z, \varepsilon_0) \cap \bar{B} \cap \bar{\Omega}_1$ is in B_2. Then if $C_2 = \bar{B} \cap \bar{\Omega}_1 - B_2$, $C_2 \cap B_2 = \emptyset$, $C_2 \cup B_2 = \bar{B} \cap \bar{\Omega}_1$, B_2 and C_2 are compact and hence at a positive distance ε_2 one from the other: Choose $\bar{\Omega}$ in Ω_1 as an $\varepsilon_2/2$ neighborhood of C_2, clearly $\partial\Omega \cap \bar{B} = \emptyset$. This ends the proof of the lemma.

Then if $x - F(x, \lambda) = 0$ on $\partial\Omega$, (x, λ) belongs to T and, if we add the condition $d(x, \lambda)^2 - r^2 = 0$ $(r > 0)$, the pair of equations is nonzero on $\partial\Omega$ and defines an element of $\Pi^{\infty-k+1}(\partial\Omega)$ (clearly we may impose any set of side conditions guaranteeing that the system is nonzero on $\partial\Omega$, defining thus an element in $\Pi^{\infty-k+\ell}(\partial\Omega)$ if there are ℓ conditions, $\ell \leq k$).

Note that the homotopy class is independent of r (r is the deformation); if r is very large, since $\bar{\Omega}$ is bounded, the second equation will be negative (distance of (x, λ) to $T \cap \bar{\Omega}$, (x, λ) in $\bar{\Omega}$); multiplying F and d by t, t in $[0, 1]$, the class of the pair will be the class of $(x, -r^2)$, which may be extended without zeroes to any Y in $LS(E^\infty)$, that is the pair is inessential and $[x-F(x, \lambda), d(x, \lambda)^2 - r^2] = 0$ in $\Pi^{\infty-k+1}(\partial\Omega)$, for all $r > 0$.

Take now r very small, so that zeroes of the pair are very close to T, i.e. of a bifurcation point. Suppose that all bifurcation points in C are isolated in T, then, since C is compact, they are in finite number and we may choose Ω_1 such that $\bar{\Omega}_1$ contains no other bifurcation points; if $P_0, \ldots, P_k$ are the bifurcation points of C, we can choose neighborhoods of these points such that there are no other bifurcation points in these neighborhoods and cover the rest of C by neighborhoods disjoint from T : the compacity of C enables then to construct Ω_1 such that, if r is small, one is, for $d(x, \lambda) = r$, in one of the neighborhoods of $P_0, \dots, P_k$; if $B_0, \dots, B_k$ are balls in these neighborhoods, the pair has no zeroes in $\Omega - \cup B_i$, if r is small, and we are exactly in the case of the definition of the generalized degree given in chapter I:

$$\sum[x - F(x, \lambda), d(x, \lambda)^2 - r^2] = 0$$

that is the sum of the local classes is 0 (the sum in $\Pi^{\infty-k+1}(\partial B)$, B a large ball containing $B_0, \dots, B_k$).

It remains thus to compute the local classes. We may anticipate that if (x_0, λ_0) is a bifurcation point with nontrivial local invariant, then C must connect it to another bifurcation point or else C cannot be bounded; this is the essence of Rabinowitz' theorem and we shall apply it to some examples.

II.1. One real or complex parameter

Here $F(x, \lambda) = T(\lambda)x + g(x, \lambda)$, where $T(\lambda)$ is a compact operator, $I-T(\tilde{\lambda})$ is invertible for some $\tilde{\lambda}$, $g(x, \lambda)$ is compact with $\|g(x, \lambda)\| = o(\|x\|)$. The set T of trivial solutions of (5) will be $\{(0, \lambda)\}$ and there is a chance of bifurcation only if $I-T(\lambda)$ is singular, that is for a discrete set of points, and the nontrivial solutions will be those of nonzero norm.

Now if λ_0 is a critical point for $I-T(\lambda)$, we may put $A = I-T(\lambda_0)$, $C(\lambda) = T(\lambda) - T(\lambda_0)$ and equation (5) is written as

$$Ax - C(\lambda)x = g(x, \lambda)$$

with $C(\lambda)$ a compact operator, analytic in λ, $C(0) = 0$, and A is a Fredholm operator of zero index; if K is the pseudo-inverse of A, P the projection onto ker A, Q onto $R(A)$, then (5) close to $(0, \lambda_0)$ is equivalent to

$$(I-P)x - (I-KQC(\lambda))^{-1} KQ(C(\lambda)Px-g(x, \lambda)) = 0$$

$$(I-Q) \; [C(\lambda)(I-KQC(\lambda))^{-1}Px+(I-C(\lambda)KQ)^{-1} g(x, \lambda)] = 0$$

If $B(\lambda) = (I-Q)\, C(\lambda)(I-KQC(\lambda))^{-1}P$, then $B(\lambda)$ has a determinant with dominant term $(\lambda-\lambda_0)^m$, where m was called the *algebraic multiplicity* of λ_0.

Theorem. Let C be the connected component of B containing $(0, \lambda_0)$, then i) C is not bounded

or ii) C is bounded, connects a finite number of critical points $(0, \lambda_j)$, with $I-T(\lambda_j)$ singular, and

Real case. If $\lambda_1 < \lambda_2 \dots < \lambda_p$ are the bifurcation points of C with odd algebraic multiplicity, let n_1 be the number of critical points λ, with odd multiplicity, such that $\lambda_1 < \lambda < \lambda_2$; let n_2 be their number with $\lambda_2 < \lambda < \lambda_3$, ..., n_{p-1} those with $\lambda_{p-1} < \lambda < \lambda_p$ (n_j and p may be 0) then

$$1 = (-1)^{n_1}-(-1)^{n_1+n_2} + \dots + (-1)^{k+1}(-1)^{\sum_{i=1}^{k} n_i} + \dots + (-1)^p\, (-1)^{\sum_{i=1}^{p-1} n_i}$$

Complex case: If $\lambda_1, \dots, \lambda_p$ are the bifurcation points of C with odd algebraic multiplicity, then p is even.

Consequences.

1) In the real case, p is even (exercise).

2) If $p = 0$, then λ_0 has even multiplicity; so, if λ_0 has odd multiplicity, then C connects $(0, \lambda_0)$ to an odd number of points $(0, \lambda)$, with λ of odd multiplicity or C is unbounded.

Proof. Clearly all hypothesis of the general discussion apply, so that, if C is compact, one may construct the open set Ω such that $x-F(x, \lambda) \neq 0$ on $\partial\Omega$, except if $x = 0$. Here $d(x, \lambda) = \|x\|$ since if (x, λ) is in Ω then $(0, \lambda)$ belongs to T.

We have thus $\sum[x-F(x, \lambda), \|x\|^2-r^2] = 0$, where the sum is over all bifurcation points of C and r is so small that the solutions of the pair appear in balls centered at these points and of the form $\|x\|^2 + |\lambda-\lambda_j|^2 < r^2 + \rho^2$.

We may deform, for r and ρ small enough, $g(x, \lambda)$ to 0 (if $|\lambda-\lambda_j| = \rho$, $I-T(\lambda)$ is invertible and $\|g(x, \lambda)\| = o(\|x\|)$ and the class of $[x-T(\lambda)x, \|x\|^2-r^2]$ is the same as the class of $[Ax_2 - QC(\lambda)(x_1+x_2) \oplus (I-Q)\, C(\lambda)(x_1+x_2), \rho^2-|\lambda-\lambda_j|^2]$, where $x_1 = Px$.

Exercise.

1) Prove that one can use a deformation on $(I-Q)\, C(\lambda)$ of the form $(I-Q)\, C(\lambda)\, (x_1+(1-t)x_2 + KQ\, C(\lambda)(x_1+x_2))$ (If the first part in the sum is zero then $x_2 = KQ\, C(\lambda)x)$.

2) Prove that one may repeat the process n times and obtain $(I-Q)\, C(\lambda)[I+KQC(\lambda) + \ldots + (KQC(\lambda))^n]x_1 + [KQC(\lambda)]^n\, x_2$.

3) If $n \geq m$, prove that one may deform this last expression to $B(\lambda)x_1$ and $Ax_2 - QC(\lambda)(x_1+x_2)$ to Ax_2.

We have thus to compute the class of $[(I-T(\lambda_j))x_2 + B(\lambda)x_1, \rho^2-|\lambda-\lambda_j|^2]$ on the sphere $\|x\|^2 + |\lambda-\lambda_j|^2 = r^2+\rho^2$.

In the complex case, since the characteristic values of $T(\lambda_j)$ are isolated, we may choose a path $f(t)$ from 1 to the origin avoiding all characteristic values; then $I-f(t)\, T(\lambda_j)$ is an admissible deformation and one obtains the suspension of $[B(\lambda)x_1, \rho^2-|\lambda-\lambda_j|^2]$, that is the suspension of $m_j\eta$ (η Hopf map), and so the sum of these suspension must be 0 in Z_2, that is $\sum m_j$ is even.

In the real case, consider the isomorphism of $E\times\mathbb{R}$ onto itself given by $(y_2, y_1, \lambda) = ((I-T(\lambda_j))x_2, Lx_1, \lambda)$, where L is an isomorphism from ker A onto coker A. This isomorphism, of the form I-compact, maps the sphere into the sphere $\{(y_2+y_1, \lambda)/\|Ky_2 \oplus L^{-1}y_1\|^2 + |\lambda-\lambda_j|^2 = r^2+\rho^2\}$ and the last mapping into $[y_2 \oplus B(\lambda)L^{-1}y_1, \rho^2-|\lambda-\lambda_j|^2]$ whose degree is ± 2 if m is odd, 0 if m is even. From the properties of degree (composition or Eilenberg-Steenrod axioms) this isomorphism induces also an isomorphism at the level of the groups $\Pi^{\infty-0}(S)$ so that the degree of $[(I-T(\lambda))x, \rho^2-|\lambda-\lambda_j|^2]$ is ± 2 if m is odd, 0 if m is even.

On the other hand, the zeroes of this map are $x = 0$, $\lambda = \lambda_j \pm \rho$, so that, as we have seen in last chapter, the degree is $i(\lambda_j-\rho) - i(\lambda_j+\rho)$, where $i(\lambda_j \pm \rho)$ is the index of $I-T(\lambda_j \pm \rho)$; then, since these indices are ± 1, $i(\lambda_j+\rho) = (-1)^m i(\lambda_j-\rho)$. Furthermore if λ is not a characteristic value, $I-T(\lambda+\varepsilon) = I-T(\lambda)+\varepsilon D(\lambda, \varepsilon)$, with $D(\lambda, \varepsilon)$ a bounded compact operator, so that the index may change only by crossing a characteristic value and only if its multiplicity is odd. Starting then from the first characteristic value on C of odd multiplicity it is an exercise to derive the formula of the theorem.

II-2) <u>Some examples</u>.

a) If $g(x, \lambda)$ is complex analytic, C is always unbounded [Da 2]; one uses the fact that in this case the degree is nonnegative and zero only if there are no solutions.

b) For the Hopf bifurcation problem a similar alternative holds: see [A.Y.I.], [Iz.2] and [C.M.Y.].

c) In all these notes the linear part was the only piece which was assumed to be known and so the topological invariants depend strongly on the class of $B(\lambda)$. However, if more is known on the nonlinear part, for example if the bifurcation equation has the form

(7) $$\lambda x - N(x) - R(x, \lambda) = 0$$

with $N(tx) = t^s N(x)$ for $t \geq 0$ and some $s > 1$, while

$$\|R(x, \lambda)\| \leq K|\lambda| \, \|x\|^s + o(\|x\|^s), \; x \text{ in } \mathbb{R}^d.$$

We say that U_0, of unit norm, is a *characteristic ray* of N if $N(U_0) = \mu U_0$.

Suppose U_0 is a characteristic ray then it is easy to see that $(tU_0, t^{s-1}\mu)$, $t \geq 0$, is solution of $\lambda x - N(x) = 0$ and that if U_0 is not a characteristic ray there are no solutions of (7) in any small enough conical neighborhood of U_0.

Let then U_0 be an isolated, in S^{d-1}, characteristic ray; it is not difficult to see that one may choose r and ρ small enough such that if one takes a small capped neighborhood of (U_0, μ) on the sphere $\|U\|^2 + |\tilde{\lambda}|^2 = 1 + |\mu|^2$ and one constructs a curved conical neighborhood of the form $x = tU$, $\lambda = t^{s-1}\tilde{\lambda}$, $(U, \tilde{\lambda})$ in the capped neighborhood and t such that $\|x\|^2 + |\lambda|^2 \leq r^2 + \rho^2$, then the degree of the pair $(\lambda x - N(x) - R(x, \lambda), \|x\|^2 + |\lambda|^2 - r^2)$ with respect to the curved conical neighborhood is defined and equal to the degree of $(\lambda x - N(x), \; \|x\|^2 + |\lambda|^2 - r^2)$.

If P is the projection on the tangent plane at U_0 to S^{d-1}, the first equation can be written

$$-PN\left(\frac{x}{\|x\|}\right)\|x\|^s \oplus \|x\| \left(\lambda - \left(N\left(\frac{x}{\|x\|}\right), \; \frac{x}{\|x\|}\right) \|x\|^{s-1}\right) \frac{x}{\|x\|}$$

which may be deformed to $-PN(U) \oplus \lambda - \mu \|x\|^{s-1}$ (exercise).

Then, since the index of $(\lambda - \mu \|x\|^{s-1}, \|x\|^2 + |\lambda|^2 - r^2)$ with respect to $(\|x\|, \lambda)$, at its only zero, is -1, the degree is

$$-\text{Index}\,(-PN(U), U_0)$$

(even if $\mu = 0$) see [Iz.1, chapter IV] for details.

If N is the gradient of $(N(x), x)/(s+1)$ and U_0 is an isolated relative extremum, then this last index is 1 for a maximum and $(-1)^{d-1}$ for a minimum and there is bifurcation along a curve $x(\lambda)$ with $x(0) = 0$, $x(\lambda)/\|x(\lambda)\| \to U_0$ when λ goes to zero.

Note that $(x = 0, \lambda = \pm r)$ are also isolated zeroes of the pair of equations and their degree with respect to a similar cone is the same as the degree of $(\lambda x, \|x\|^2+|\lambda|^2-r^2)$, equal to the degree of $(\text{Sign}(\lambda)\, x, \|x\|^2+|\lambda|^2-r^2)$; by computing the sign of the Jacobian, this index is 1, for $\lambda > 0$, and $(-1)^{d-1}$, for $\lambda < 0$. Now if all characteristic rays are isolated, we may add all these degrees to get the degree of $(\lambda x-N(x), \|x\|^2+|\lambda|^2-r^2)$ with respect to the ball of radius $(r^2+\rho^2)^{1/2}$: since, on the boundary of this ball, the second equation is never zero, we may deform the pair to $(1, \|x\|^2+|\lambda|^2-r^2)$ and then to $(1, 0)$, and hence the degree is 0: $\sum \text{Index}(-PN(U), U_i) = 1 + (-1)^{d-1} = \chi(S^{d-1})$ recovering thus the Lefchetz index formula.

Global results are easy to obtain, for example for

$$(I-\lambda T)x - N(x) - R(x, \lambda) = 0$$

with T, N, R compact, T and R as above, T self-adjoint; if U_0 is an isolated characteristic ray and C is the connected component of B containing the bifurcating branch in the direction U_0 then, if $\bar{C}$ is compact and meets all bifurcation points $(0, \lambda_j)$ along an isolated characteristic ray U_i^j, the following formula holds

$$\sum (-1)^{\beta_j} \text{Sign } \lambda_i \;\; \text{Index}(PN_i, U_i^j) = 0$$

where N_i denotes the projection of N restricted to $\ker(I-\lambda_i T)$ onto that kernel and β_j is the sum of the multiplicities of $I-\mu\lambda_i T$, $0 \leq \mu \leq 1$.

The proof, very similar to the previous results, is given in [Iz.1].

II-3) Concluding remarks

In these notes we have stressed the role of the linear part and we would like to give a quick review of some of the open questions suggested by the methods used.

1) The first problem is to understand, from the analysis point of view, the Bott periodicity theorem, that is to compute in terms of matrices the generators of $\Pi_{k-1}(GL(d))$, to be able to know the class of $B(\lambda)$ in these groups (this is difficult except in the special cases treated here) and to understand why these groups are trivial for $k \equiv 3, 5, 6, 7, [8]$. Another related question is the relation with other topological invariants such that K theory, Stiefeld manifolds, grassmanians, in particular with respect to the structure of the set of solutions. This first problem is mainly of translation from one field, topology, to another, analysis.

2) The second problem of translation is concerned with the topological dimension of the set of solutions. In this context see the recent article [A.A.].

3) A technical problem consist in extending the Geba-Granas theory of stable cohomotopy for mappings of the form $I-K$, where K is a k-set contraction. In this direction some work has been done when the codimension is zero or when this mapping sends the whole space into a fixed subspace.

4) The numerical implementation of the topological ideas for several parameters. For one parameter see [Pe].

5) If the equation has some special property, like analyticity or invariance under some group action, then one should not consider deformations which have not the same property. For one real parameter and complex analytic equations one has the property that the degree is

always non-negative; what happens, not only for the existence question, for a positive index or a several parameter problem? If the equation is odd or invariant under a finite group it is easy to define an index, for a continuous group one should look into equivariant cohomotopy groups, mainly if the action is not free. Not much is known either from the topological or from the analysis points of view, see however [Sa], [F.R.], [Co] ...

6) Find a "natural" conexion between these methods and the invariants used in variational problems: Morse index, genus, category...

7) Stability is a topological notion but very few results have been given in this direction, see [Sa.1] and the dynamical systems litterature.

8) Bifurcation for non compact perturbations as mentionned in the introduction and so on ...

REFERENCES

[A.A.] Alexander J.C., Antman S.S.: Global and local behavior of bifurcating multidimensional continua of solutions for multiparameter nonlinear eigenvalue problems. Univ. of Maryland Technical report. Oct. 1980.

[Ad.] Adams R. A: *Sobolev Spaces*. Academic Press. New-York (1975).

[Am] Amann H: Fixed point equations and nonlinear eigenvalue problems in ordered Banach Spaces. SIAM Review *18*, 4, (1976), 620-729.

[A.R.] Antman S. S., Rosenheld G.: Global behavior of buckled states of nonlinearly elastic rods - SIAM Review *20*, 3 (1978), 513-566.

[A.Y.] Alexander J. C., Yorke J.A.: Global bifurcation of periodic orbits. Amer. J. Math. *100* (1978), 263-292.

[A.Y.1] Alexander J.C., Yorke J.A.: Calculating bifurcation invariants as elements in the homotopy of the general linear group. J. Pure Appl. Algebra *13*, (1978), 1-8.

[Be] Berger M.S.: *Non linearity and functional analysis*. Academic Press, 1977.

[Bo] Böhme R.: Die Lösung der Verzweingungsgleichungen für nichtlineare Eigenvertprobleme. Math. Z. *127*, (1972), 105-126.

[Br] Brezis H.: *Operateurs maximaux monotones*. Math. Studies No. 5, North Holland, 1973.

[Ch] Chandrasekhar S.: *Hydrodynamic and hydromagnetic stability*. Oxford U. Press. 1961.

[C.M.Y.] Chow S.N., Mallet-Paret J., Yorke J.A.: Global Hopf bifurcation from a multiple eigenvalue. Nonlinear Anal. *2* (1978), 753-763.

[Co] Conley C.C. *Isolated invariant sets and the Morse index*. C.B.M.S. Regional Conf. Series in Math. No. 38. A.M.S. Providence R.I. 1978.

[C.R.] Ciarlet P. G., Rabier P.: *Les équations de von Karman*. Lect. Notes in Math. *826*, Springer Verlag, 1980.

[Cr] Cronin J.: *Fixed points and topological degree in nonlinear analysis*. A.M.S. Providence R.I. 1964.

[Da 1] Dancer E.N.: Bifurcation theory in real Banach space. Proc. London Math. Soc. *3*, 23 (1971), 699-734.

[Da 2] Dancer E.N.: Bifurcation theory for analytic operators. Proc. London Math. Soc. *3*, 26 (1973), 359-384.

[E.T.] Ekeland I., Temam R.: *Analyse convexe et problèmes variationnels*. Dunod, Paris, 1974.

[Fr] Friedman A. *Partial differential equations*. Holt, Rinehart and Winston. New-York, 1969.

[F.R.] Fadell F. R., Rabinowitz P.H.: Generalized cohomological index theories for Lie group actions with an application to bifurcation questions for Hamiltonian systems. Invent. Math *45* (1978), 134-174.

[Ge.Gr] Geba K., Granas A.: Infinite dimensional cohomology theories. J. Math. Pures et Appl. *52*, 2 (1973), 145-270.

[G.G.] Golubisky M., Guillemin V.: *Stable mappings and their singularities*. Graduate Texts in Math. No 14, Springer-Verlag, 1973.

[Go] Goldberg S.: *Unbounded linear operators*. Mc Graw-Hill, New-York, 1966.

[Io] Iooss G.: *Bifurcation of maps and applications*. North Holland Math. Studies No 36. Amsterdam, 1979.

[Iz 1] Ize J.: *Bifurcation theory for Fredholm operators*. Memoirs A.M.S. 7, 174, (1976).

[Iz 2] Ize J.: Periodic solutions of nonlinear parabolic equations. Comm. in P.D.E. *4*, 12, (1979), 1299-1387.

[Iz 3] Ize J.: Le problème de bifurcation de Hopf. Seminaire Brezis-Lions, 1975. A paraitre.

[Jo] Joseph D D. *Stability of fluid motions*, I. II. Springer-Verlag 1976.

[Kr 1] Krasnosel'skii M.A.: *Topological methods in the theory of non-linear integral equations*. Macmillan, New-York, 1964.

[Kr 2] Krasnosel'skii M.A.: *Positive solutions of operator equations*. P. Noordhoff, Groningen, 1964.

[Kr 3] Krasnosel'skii M.A.: *Plane vector fields*. Academic Press, 1966.

[L.M.] Lions J.L., Magenes E. *Nonhomogeneous boundary value problems and application, I. II*. Springer-Verlag, New-York, 1972.

[M.M.] Marsden J.E., Mc Cracken M.: *The Hopf bifurcation and its applications*. Appl. Math. Sciences n. 19. Springer-Verlag, New-York, 1976.

[Ni] Nirenberg L.: *Topics in nonlinear functional analysis*. Lecture Notes, Courant Inst., 1974.

[Ni 1] Nirenberg L.: Variational and topological methods in nonlinear problems. B.A.M.S. *4*, 3, (1981), 267-302.

[N.P.] *Nonlinear problems in the physical sciences and biology*. Proc. Battelle Summer Inst. Lect. Notes in Math. 322, Springer-Verlag, 1973.

[N.Y.] *Bifurcation theory and applications in scientific disciplines*. Ann. N.Y. Acad. Sci. *316*, 1978.

[Pe] Peitgen H.O., Walther H. O. Ed.: *Functional differential equations and approximation of fixed points*. Lect. Notes in Math. 730, Springer-Verlag, 1979.

[Pi] Pimbley G.H.: Eigenfunction branches of nonlinear operators and their bifurcations. Lect. Notes in Math. 104, Springer-Verlag, 1969.

[Ra] Rabinowitz P.H.: *Theorie du degré topologique et applications à des problèmes aux limites non linéaires*. Lect. Notes, Analyse Numérique Fonctionnelle, Univ. Paris VI, 1975.

[Ra 1] Rabinowitz P.H.: Free vibrations for a semilinear wave equations. C.P.A.M. *31* (1978), 31-68.

[Ra 2] Rabinowitz P.H.: Some aspects of nonlinear eigenvalue problems. Rocky Mountain J. of Math. *3*, 2, (1973), 161-202.

[Sa] Sattinger D.H.: Bifurcation and symmetry breaking in applied mathematics, B.A.M.S. *3*, 2, (1980), 779-819.

[Sa 1] Sattinger D.H.: *Topics in stability and bifurcation theory*. Lect. Notes in Math. 309, Springer-Verlag, 1973.

[St] Sather D.: Branching of solutions of nonlinear equations. Rocky mountain J. of Math. *3*, 2, (1973), 203-250.

[Sj] Sijbrand J.: *Studies in nonlinear stability and bifurcation theory*. Thesis, Utrecht, 1981.

[Sp] Spanier E.: *Algebraic topology*. Mc. Graw-Hill, 1966.

[Te] Temam R.: *On the theory and numerical analysis of the Navier-Stokes equations*. North Holland, 1977.

[Va] Vainberg M.M.: *Variational methods for the study of nonlinear operators*. Holden day series in Math. Physics, 1964.

SOBRE ESTABILIDAD TOPOLOGICA

Jorge Lewowicz

Universidad Simon Bolivar
Departamento de Matemática
Caracas, Venezuela

Un homeomorfismo f de un espacio métrico M es topológicamente estable (véase [7] y [8]) si para todo $\varepsilon > 0$ existe un C^0-entorno de f, N, tal que si $g \in N$, existe una función contínua $h: M \to N$ tal que $f(h(x)) = h(g(x))$ y $dist(x, h(x)) < \varepsilon$ para todo $x \in M$. Si ε es suficientemente pequenõ y M es una variedad, h resulta sobreyectiva.

Para dar una idea de alguno de los contenidos dinámicos de este concepto, mencionemos que en [8] Walters caracteriza la estabilidad topológica mediante las pseudo-órbitas de f (véase [2]). Dado $\delta > 0$, una sucesión $\{x_n: x_n \in M, n = \ldots -1, 0, 1, \ldots\}$ es una δ pseudo-órbita de f si $dist(f(x_n), x_{n+1}) < \delta$, $n \in Z$. En el artículo citado se prueba que f es topológicamdnte estable si y sólo si para todo $\varepsilon > 0$ existe $\delta > 0$ tal que si $\{x_n\}$ es una δ pseudo-órbita de f, entonces existe $x \in M$ tal que $dist(f^n(x), x_n) < \varepsilon$ para todo $n \in Z$. Si $x \in M$ es un punto no errante de f, es decir, si $x \in \Omega(f)$, es muy fácil construir una δ-pseudo órbita períodica por x, para cualquier $\delta > 0$. Si f es topológicamente estable existen entonces puntos y arbitrariamente cercanos a x tales que sus órbitas están cerca de las pseudo órbitas períodicas por x.

Si además f es expansivo, es decir, si existe $\alpha > 0$ tal que $dist(f^n(x), f^n(y)) < \alpha$ para todo $n \in Z$ implica $x = y$, entonces se puede concluir que las órbitas de f próximas a las pseudo órbitas períodicas por x mencionadas más arriba, son, en realidad, períodicas. Esto quiere decir que si f es topológicamente estable Ω es la adherencia del conjunto de sus puntos períodicos.

Como puede verificarse fácilmente, la siguiente definición de estabilidad topológica, que utilizaremos con frecuencia más adelante, es equivalente a la ya mencionada. f se dice topológicamente estable si para todo $\varepsilon > 0$, existe un C^0-entorno de f, N, tal que si $g \in N$, existe una relación de equivalencia R en M, invariante bajo g y tal que

1. $(x, y) \in R$ implica $dist(x, y) < \varepsilon$
2. f es conjugado con $\hat{g}$.

Aqui $\hat{g}$ denota el homeomorfismo inducido por g en el cociente M/R. Tal vez valga la pena recordar que dos homemorfismos $f_1: M_1 \to M_1$ y $f_2: M_2 \to M_2$ se dicen conjugados si existe un tercer homeomorfismo $h: M_1 \to M_2$ tal que $g\,h = h\,f$ y que un difeomorfismo $f: M \to M$ se dice C^1-estructuralmente estable si el conjunto de los difeomorfismos conjugados con f contiene un C^1-entorno de f. Desde este punto de vista puede decirse que un homeomorfismo topológicamente estable es C^0 estructuralmente estable salvo por una identificación de puntos (dada por R) que están, estuvieron y estarán muy cerca (a menos de ε). La estabilidad topológica de f expresa, de este modo, fuertes propiedades de persistencia de la dinámica de f.

Los difeomorfismos de Anosov son los primeros ejemplos de difeomorfismos estructuralmdnte estables que poseen gran riqueza dinámica. Si M es una variedad C^∞ riemanniana y $f: M \to M$ un difeomorfismo, decimos que f es de Anosov si existen subespacios E_x, I_x que varían continuamente con x, $E_x \oplus I_x = T_x M$, $f'(E_x) = E_{f(x)}$, $f'(I_x) = I_{f(x)}$ y tales que existen números k, λ $0 \leq k$, $0 < \lambda < 1$, con las propiedades:

i) $\|(f^n)'e_x\| \leq k\, \lambda^n \|e_n\|$, $e_x \in E_x$, $n \geq 0$

ii) $\|(f^n)'i_x\| \leq k\, \lambda^{|n|} \|i_x\|$, $i_x \in I_x$, $n \leq 0$.

En [7] Walters probó que los difeomorfismos de Anosov son topológicamente estables y Bowen mostró luego [1] que los subconjuntos invariantes hiperbólicos [6] de un difeomorfismo poseen propiedades de estabilidad análogas.

En lo que sigue, veremos algunas generalizaciones de estos resultados; para presentarlos adecuadamente convendrá recordar algunos resultados de [3].

Sea $f: M \to M$ un difeomorfismo y M una variedad compacta. En [3] se demuestra que f es de Anosov si y sólo si existe una forma cuadrática continua y no degenerada $A: TM \to R$ tal que $f^{\#} A - A > 0$. Una forma cuadrática A se dice no degenerada si $A_x = A/T_x M$ es no degenerada para todo $x \in M$ y se dice positiva si $A_x > 0$, $x \in M$. $f^{\#} A$ denota como siempre la forma definida por $f^{\#} A(X) = A(f'X)$ para toda sección X del fibrado TM. Sea $N \subset M \times M$ un entorno de la diagonal de $M \times M$ y $V: N \to R$ una función continua que se anula en la diagonal. Diremos que V es una función de Lyapunov para f si $V(f(x), f(y)) - V(x, y) > 0$ para todo (x, y) en un entorno de la diagonal $N' \subset N$, $x \neq y$. En términos de las funciones de Lyapunov, el resultado mencionado más arriba puede formularse de la siguiente manera:

<u>Teorema 1.</u> f es de Anosov si y sólo si existe una forma cuadrática no degenerada A y otra positiva B tales que definiendo $V(x, y)$

$= A_x(u)$ y $W(x, y) = B_x(u)$, $\exp_x u = y$, se tiene $V(f(x), f(y)) - V(x, y) > W(x, y)$, $x \neq y$.

El resultado que expondremos a continuación generaliza los resultados de [7] y de [3].

Sea V una función real y continua definida en un entorno de la diagonal en $M \times M$ y tal que $V(x, x) = 0$, $x \in M$. Decimos que V es no degenerada si existe $\rho > 0$ y para todo $x \in M$ subespacios $E_x, I_x \subset T_x M$, no necesariamente invariantes bajo f', dependiendo continuamente de x, tales que $\dim E_x = k$, $\dim I_x = \dim M - k$, $E_x \cap I_x = \{0\}$ y que

$$V(x, \exp_x u) < 0, \quad u \in E_x, \quad 0 < \|u\| \leq \rho$$

$$V(x, \exp_x v) > 0, \quad v \in I_x, \quad 0 < \|v\| \leq \rho.$$

Teorema 2. Sea f un homemorfismo de la variedad compacta M. Si f admite una función de Lyapunov no degenerada entonces es topológicamente estable.

Este resultado permite construir muchos homeomorfismos y difeomorfismos que no son de Anosov y sin embargo son topológicamente estables. El difeomorfismo de T^2 definido por

$$(x_1, x_2) \to (x_1 - \frac{\operatorname{sen} 2\pi x_1}{2\pi} + x_2, x_1)$$

admite la función de Lyapunov

$$(x_1 - y_1)(x_2 - y_2)$$

y no es de Anosov, como se verifica inmediatamente calculando su parte lineal en $(0, 0)$. Si $c: T^2 \to R$, $0 \leq c \leq 1$, es diferenciable, tiene primeras derivadas pequeñas, y $c(x) = 1$ si y sólo si x pertenece a algún minimal no trivial del isomorfismo de

$$T^2: (x_3, x_4) \to (2x_3 + x_4, \ x_3 + x_4),$$

entonces el difeomorfismo de T^4:

$$(x_1,x_2,x_3,x_4) \to (2x_1-c(x_3,x_4)f(x_1)+x_2, x_1-c(x_3,x_4)f(x_1)+x_2, 2x_3+x_4, x_3+x_4),$$

donde $f(x) = \frac{\operatorname{sen} 2\pi x}{2\pi}$, tiene todos sus puntos períodicos hiperbólicos, es topológicamente estable, y no admite descomposición no trivial de su fibrado tangente, por lo que no es de Anosov. La función de Lyapunov

$$V = (x_1-y_1)((x_1-y_1)-(x_2-y_2))+(x_3-y_3)((x_3-y_3)-(x_4-y_4))$$

permite verificar su estabilidad topológica.

Una noción estrechamente vinculada con la de estabilidad topológica es la de persistencia uniforme. Sea f un difeomorfismo de M; decimos que f es uniformemente persistente si para todo $\varepsilon > 0$ existe un C^0-entorno de f tal que si g pertenece a dicho entorno y $x \in M$, entonces existe $y \in M$ tal que

$$\operatorname{dist}(g^n(y), f^n(x)) < \varepsilon, \; n \in Z.$$

La estabilidad topológica implica la persistencia uniforme pero el recíproco no es cierto aún para difeomorfismos expansivos, como es fácil de verificar considerando, por ejemplo, levantamientos de difeomorfismos de Anosov del Toro a revestimentos con singularidades de T^2 (véase [5])

El teorema que sigue da una condición suficiente para la persistencia uniforme.

<u>Teorema 3</u>. Sea f un difeomorfismo expansivo de M y $P \subset M$ un conjunto denso de puntos períodicos hiperbólicos de f. Si para todo $x \in P$ o bien la variedad estable o bien la variedad inestable de x tienen dimensión 1, entonces f es uniformemente persistente.

La demostración del teorema 2 puede hacerse considerando un $\varepsilon > 0$ convenientemente pequeño y, dado un homeomorfismo $g: M \to M$, asociando a cada $x \in M$ el conjunto

$$R_x = \left\{ y \in M: \operatorname{dist}(f^n(x), g^n(y)) \leq \varepsilon, \quad n \in Z \right\}.$$

Si g está suficientemente cerca de f en la topología C^0, R_x es no vacío, y además para cada $y \in M$ existe $x \in M$ tal que

$$\operatorname{dist}(g^n(y), f^n(x)) \leq \varepsilon, \quad n \in Z.$$

La relación R puede definirse así

$$R = \left\{ (y, z) \in M \times M: y, z \in R_x \text{ para algún } x \in M \right\}.$$

Puesto que f es expansivo -*una condición necesaria y suficiente para que un difeomorfismo sea expansivo es que admita una función de Lyapunov*- si ε es suficientemente pequeño, R resulta una relación de equivalencia g-invariante. Si f y g fueran ambos expansivos con constante de expansividad α, y $\varepsilon < \frac{\alpha}{2}$, entonces R sería trivial y g sería conjugado con f.

Si f es de Anosov, existe un C^1-entorno de f y un $\alpha > 0$ tal que todo difeomorfismo de ese entorno tiene constante de expansividad α. Esto permite obtener como Corolario del Teorema 2 la estabilidad estructural de los difeomorfismos de Anosov.

A los efectos de dar una idea de algunos de los métodos empleados en la demostración del Teorema 2, expondremos una prueba esquemática (por más detalles véase [4]) del Teorema de Hartman sobre conjugación entre un campo vectorial y su parte lineal en un entorno de un punto de equilibrio hiperbólico.

Sean $\dot{x} = Ax$, A hiperbólica y $\dot{x} = Ax + f(x)$, $f(0) = f'(0) = 0$. El teorema de Hartman asegura que los flujos generados por ambas ecuaciones son localmente conjugados. Demostraremos primero el siguiente

<u>Lema</u>. Existe $\delta > 0$ tal que si para todo $x \in R^n$, $\|X(x) - A(x)\|$, $\|X'(x) - A\|$ y $\|(X'(x) - A)\, X(x)\|$ son menores que δ, entonces los flujos ϕ y ψ asociados respectivamente a $\dot{x} = Ax$ y $\dot{x} = X(x)$ son globalmente conjugados, es decir, existe un homeomorfismo h de R^n sobre R^n tal que $\phi(h(x), t) = h(\psi(x, t))$ para todo $x \in R^n$, $t \in R$.

Una vez demostrado lo anterior podrá concluirse la tesis del teorema de Hartman observando que $X(x) = Ax + \lambda(x)\,(x)$, donde λ vale 1 en un entorno del origen y 0 fuera de otro entorno, satisface las fuertes hipótesis del lema.

De la expansividad de ϕ y de las hipótesis, es fácil concluir la expansividad de ψ. Por consiguiente, si para $\varepsilon > 0$ conveniente mostramos que existe $\delta > 0$ tal que, si se satisfacen las hipótesis del lema, *para todo* $x \in R^n$ *existe* $y \in R^n$ *de modo que* $\|\phi(x, t) - \psi(y, t)\| < \varepsilon$ $t \in R$, podremos entonces asociar a cada $x \in R^n$ un y, único en virtud de la expansividad de ψ, con la propiedad arriba mencionada. La correspondencia así definida es la conjugación buscada h. Pasemos entonces a demostrar la afirmación en cursivas más arriba.

Sea $\dot{x} = Ax$, donde $A \in GL(n)$ y todos sus valores propios tienen parte real negativa, y sea $V(x) = (v\, x, x)$, v autoadjunta. La derivada de V respecto de $\dot{x} = Ax$, es $\dot{V}(x) = (\, w\, x, x)$, $w = vA + A^*v$.

La transformación $T = T_A \colon v \to w$ del espacio de matrices autoadjuntas de orden n en sí mismo es lineal y su núcleo lo constituyen la u tales que $U(x) = (u\, x, x)$ es integral primera de $\dot{x} = Ax$. Consecuentemente T es un isomorfismo.

Sea ahora A hiperbólica y $R^n = E \oplus I$, E e I invariantes bajo A, A/E (A/I) con valores propios de parte real negativa (positiva). Sean w_E, $w_I > 0$ en E e I respectivamente y

$v_E = T^{-1}_{A/E}(W_E)$, $u_E = T^{-1}_{A/E}(v_E)$ $v_I = T^{-1}_{A/I}(W_I)$, $u_I = T^{-1}_{A/I}(v_I)$.

Entonces, denotando con $P_E(P_I)$

$$U(x) = (u_E\, p_E\, x,\, P_E\, X) + (u_I\, P_I\, x,\, P_I\, X)$$

es positiva definida,

$$\dot{U}(X) = V(x) = (v_E\, P_E\, x,\, P_E\, x) + (v_I\, P_I\, x,\, P_I\, x)$$

es positiva en I y negativa en E, y

$$\ddot{U}(x) = W(x) = (w_E\, P_E\, x,\, P_E\, x) + (w_I\, P_I\, x,\, P_I\, x)$$

es positiva definida.

Consideremos, para todo $t \in R$, $K_t = \left\{y \in R^n\colon U(y-\phi(x,\, t)) \leq k\right\}$, donde $k > 0$ se escoge de modo que

$$U(x) \leq k \text{ implique } \|x\| \leq \varepsilon.$$

Teniendo en cuenta las observaciones del párrafo anterior, puede demostrarse fácilmente que es posible escoger el δ del Lema de modo que

$$\dot{U}(x,\, y) = \lim_{t \to 0} \frac{1}{t}\, (U(\psi(y,\, t) - \phi(x,\, t)) - U(y-x))$$

sea positiva (negativa) si $y-x \in I$ $(y-x \in E)$ y $U(y-x) = k$, y que

$$\ddot{U}(x,\, y) = \lim_{t \to 0} \frac{1}{t}\, (U(\phi(x,\, t),\, \psi(y,\, t)) - U(x,\, y))$$

sea positiva si $U(y-x) = k$. En estas condiciones, si para todo $y \in K_0$, $\psi(y,\, t) \notin K_t$ para algún $t \in R$, es posible definir $\tau\colon K_0 \to R^+$, continua, tal que

i) $\psi(y,\, t) \in K_t$, $|t| \leqq \tau(y)$,

ii) $\psi(y,\, t) \notin K_t$ a la derecha de $\tau(y)$, o a la izquierda de $-\tau(y)$.

La continuidad de τ es consecuencia del signo de $\ddot{U}$; más aún, de este hecho se desprende también que $\tau(\partial K_0) = 0$. De los signos de $\dot{U}(x, y)$ se deduce que

$$\alpha(y) = P_I(\psi(y,\tau(y))-\phi(x,\tau(y)))y \ \beta(y) = P_E(\psi(y_1,-\tau(y))-\phi(x_2,-\tau(y)))$$

$y \in K_0$, no son 0 a la vez; por lo tanto γ, $\gamma(y) = x + \alpha(y) + \beta(y)$ es una función continua, $x \in' \gamma(K_0)$ y $\gamma(y) = y$ para $y \in \partial K_0$. Como esto es absurdo, concluimos que existe $y \in K_0$ tal que $\psi(y, t) \in K_t$, $t \in R$, como queríamos demostrar.

Finalmente, tal vez valga la pena señalar que estos métodos pueden aplicarse también al estudio local de campos vectoriales con aproximaciones lineales no hiperbólicas, siempre que estos admitan funciones de Lyapunov apropiadas. En este sentido, un problema interesante consistiría en encontrar los campos polinomiales homogéneos de grado m, con la propiedad de que ninguna perturbación de grado mínimo mayor que m cambia su clase de conjugación local en el origen.

REFERENCIAS

[1] R. Bowen. w-limit sets for Axiom A diffeomorphisms. J. Differential Equations 18 (1975).

[2] R. Bowen. On Axiom A diffeomorphisms. CBMS 35 (1978).

[3] J. Lewowicz. Lyapunov functions and topological stability. J. Differential Equations 38 (1980), 192-209.

[4] J. Lewowicz, E. Lima de Sá and J. Tolosa. Lyapunov functions of two variables and a conjugacy theorem (A aparecer).

[5] T. O'Brien and W. Reddy. Each compact orientable surface of positive genus admits an expansive homeomorphism. Pacific J. Math. 35 (1970), 737-741.

[6] S. Smale. Differentiable Dynamical Systems. Bull. Amer. Math. Soc. 73 (1967), 747-817.

[7] P. Walters. Anosov diffeomorphisms are topologically stable. Topology 9 (1970), 71-78.

[8] P. Walters. On the pseudo orbit tracing property and its relationship to stability, Springer Lecture Notes Nº 668, pp. 231-244. Springer Verlag. Berlin/New York, 1978.

SOLVABILITY OF OPERATOR EQUATIONS INVOLVING NONLINEAR PERTURBATIONS OF FREDHOLM MAPPINGS OF NONNEGATIVE INDEX AND APPLICATIONS

P. S. Milojević *

Universidade Federal de Minas Gerais
Department of Mathematics - ICEx
Belo Horizonte, MG - Brazil

I. Introduction

Let X and Y be Banach spaces, $A:V \subset X \to Y$ a Fredholm map of index $i(A) \geq 0$ and $N:X \to Y$ a nonlinear map. Equations of the form

(1.1) $$Ax + Nx = f \qquad (x \in V, \quad f \in Y)$$

have been studied extensively in the last decade or so using a Liapunov-Schmidt splitting argument and/or various fixed point results and degree theories for compact or condensig displacement of the identity, coincidence degree, monotone operator theory, etc., (cf. [1, 3, 4, 5, 9, 19.]). Recently, in a number of papers [14-17], the author has studied Eq. (1.1) with N such that $A+N$ is either A - proper or strongly A-closed using the A-proper mapping approach, which is based only on the Brouwer degree. This approach yields constructive solvability results for Eq. (1.1) whenever $A+N$ is A-proper and at the same time unifies various above mentioned theories.

---

* The work was supported in part by CNP_q Grant 1980/81.

In this paper we extend various results from [15] to Eq. (1.1) with $i(A) > 0$ and from [22], and give some applications to BVP's for some semilinear elliptic equations. Our approach via the A-proper mapping theory is based only on the essentialness of maps acting between finite dimensional spaces, in contrast to the approaches of Nirenberg [18] and Mawhin [8] that are based on the essentialness [6, 18] of $I-C$, $C:X \to X$ compact, and of Borisovich, Sapronov and Zvyagin [2] and the author [12] based on the degree theory for nonlinear Fredholm maps. Moreover, when N is odd, we need only the Brouwer degree unlike the approach of Rabinowitz [22] based on the Leray-Schauder degree. It is essential in our approach to consider the A-properness or the strong A-closedness of $A+N$ with respect to projectionally complete schemes $\Gamma_i = \{X_n, P_n; Y_n, Q_n\}$ for (V, Y) with $\dim X_n - \dim Y_n = i > 0$ for each n and some given i (possibly $i = i(A)$). When X has a nice approximation property, such type of a scheme Γ_A induced by A exists and was widely used in [15-17]. We note that until our work [17] one has exclusively dealt in the A-proper mapping theory with schemes such that $\dim X_n = \dim Y_n$.

2. Zeros of odd A-proper maps and odd perturbations of Fredholm maps of positive index

In this section we prove a basic result on the existence of zeros of odd A-proper maps w.r.t. Γ_i, $i > 0$, and estimate the size of the set of zeros of such maps belonging to the boundary of a bounded symmetric neighborhood of zero in terms of its genus ([22]). Applications to odd pertubartions of Fredholm maps with positive index are also given.

Recall that if A is a nonempty closed and symmetric subset of $X\setminus\{0\}$, X is a real Banach space, we say the A has *genus* p, $\gamma(A)=p$, if there is an odd continuous map $\phi:A \to R^p\setminus\{0\}$ and p is the smallest integer having this property; $\gamma(\phi) = 0$. Let $D \subset X$. For a given $T: D \cap V \to Y$ we shall always assume that $Q_n T: X_n \cap D \to Y_n$ is continuous. Recall

Definition 2.1. (a) A map $T: D \cap V \to Y$ is *approximation-closed* (*A-closed*) w.r.t.Γ_i for (V, Y) if, whenever $\{x_{n_k} \in D \cap X_{n_k}\}$ is such that $x_{n_k} \to x$ and $Q_{n_k} T x_{n_k} - Q_{n_k} f \to 0$ for some f in Y, then $x \in D$ and $Tx = f$.

(b) T is *strongly* A-*closed* w.r.t.Γ_i if, whenever $\{x_{n_k} \in D \cap X_{n_k}\}$ is bounded and $Q_{n_k} T x_{n_k} - Q_{n_k} f \to 0$, then $Tx = f$ for some $x \in D$.

(c) T is A-*proper* w.r.t.Γ_i if, whenever $\{x_{n_k} \in D \cap X_{n_k}\}$ is bounded and $Q_{n_k} T x_{n_k} - Q_{n_k} f \to 0$, then some subsequence $x_{n_{k(i)}} \to x$.

(d) If f is given in advance in (a), (b), (c), resp., then T is A-closed, strongly A-closed, or A-proper at f w.r.t.Γ_i, resp.

The notion of an A-proper map w.r.t.Γ_0 of the form $I-T: X \to X$ was introduced by Kaniel [7, cond. IV] for studying fixed points of demicontinuous and bounded T. A-proper maps between X and X^* have been first studied by Pohožaev [21] and in more general setting by many authors (see [20] for the survey up to 1975 and, e.g. [10-17] for more recent development by the author). Various continuity conditions on T and/or scheme Γ_i imply its A-closedness (see [15-17] where one can also ind numereous examples of A-proper maps).

Our basic result for odd A-proper maps is the following

Theorem 2.1. *Let* $D \subset X$ *be open, bounded and symmetric relative to* 0 *and* $T: \bar{D} \to Y$ *A-proper at* 0 *w.r.t.*Γ_i, $i > 0$, *continuous and odd on* ∂D, *i.e.* $T(-x) = -T(x)$ *for* $x \in \partial D$. *Then* $Z = \{x \in \partial D \mid Tx = 0\} \neq \emptyset$ *and its genus* $\gamma(Z) \geq i$. *In particular, if* $i > 1$, *the equation* $Tx = 0$ *possesses infinitely many solutions on* ∂D.

Proof. If $Z = \emptyset$, then by the A-properness of T we can find an integer $n_o \geq 1$ such that $\Psi Q_n Tx \neq 0$ for $x \in \partial D \cap X_n$ and each $n \geq n_o$, where $\Psi: Y_n \to W_n \subset X_n$ is a linear homeomorphism. By the Borsuk theorem, $\deg(\Psi Q_n T, D \cap X_n, 0) \neq 0$ for each $n \geq n_o$, and the continuity of $\deg(\Psi Q_n T, D \cap X_n, a)$ in a implies that $\Psi Q_n T(D \cap X_n)$ contains a neighborhood of 0 in X_n, which is impossible since $\Psi Q_n(D \cap X_n) \subset W_n \neq X_n$. Therefore, $Z \neq \emptyset$ and is symmetric relative to 0 by the oddness of T.

Next, since T is continuous and A-proper at 0 and Γ_i is projectionally complete, it is easy to see that T is a proper mapping and consequently Z is a compact subset of X. Hence, by the properties of genus, $\gamma(Z) < \infty$ and there exists a uniform neighborhood $N_\delta(Z) = \{x \in X \mid d(x, Z) \leq \delta\}$ of Z such that $\gamma(N_\delta(Z)) = \gamma(Z)$.

Now we claim that there exists an integer $n_o \geq 1$ such that $N_\delta(Z) \supset Z_{1/n} \cap X_n$ for each $n \geq n_o$, where $Z_{1/n} = \{x \in \partial D \mid \|Q_n Tx\| \leq \frac{1}{n}\}$. If not, then there would exist $x_{n_k} \in \partial D \cap X_{n_k} \setminus N_\delta(Z)$ for each $k \geq 1$ such that $\|Q_{n_k} Tx_{n_k}\| \leq 1/n_k \to 0$ as $k \to \infty$. Then, since T is A-proper at 0, some subsequence $x_{n_{k(i)}} \to x$ with $Tx = 0$. But $x \in \overline{\partial D \setminus N_\delta(Z)}$ since $\partial D \cap X_n \setminus N_\delta(Z) \subset \partial D \setminus N_\delta(Z)$, which contradicts the fact that $x \in Z$. Therefore, such an n_o exists and by one of the properties of genus, $\gamma(Z) \geq \gamma(Z_{1/n} \cap X_n)$ for $n \geq n_o$. Since $\dim X_n - \dim Y_n = i$ for each n, by a result of Rabinowitz [22] we have that $\gamma(\{x \in \partial D \cap X_n \mid Q_n Tx = 0\}) \geq i$ for each $n \geq n_o$, and therefore, $\gamma(Z) \geq i$ by the monotonicity property of genus. ∎

That $Z \neq \emptyset$ was proven in [17]. We have

Corollary 2.1. *Let* $A: X \to Y$ *be a Fredholm map of index* $i > 0$ *and* $N: \bar{D} \subset X \to Y$ *a non linear map such that* $T \equiv A+N: \bar{D} \to Y$ *is A-proper at*

$0.w.r.t.\Gamma_A$ *and continuous. Then the conclusions of Theorem 2.1 hold.*

As N is Corollary 2.1 we can take e.g., a k-ball-contractive map with k suitably small (cf. [15]). In particular, when N is compact, Corollary 2.1 reduces to a result of Rabinowitz [22] whose proof is considerably more complicated and is based on constructing finite-dimensional odd approximations for T of Schauder type.

3. Solvability of operator equations involving nonlinear perturbations of Fredholm maps of pozitive index

In this section we prove a couple of general (approximation) solvability results for Eq. (1.1) with $i(A) > 0$ and quasibounded N such that $A+N$ is A-proper or strongly A-closed. They extend the corresponding results in [15] with $i(A) = 0$.

Throughout the section we assume that $\Gamma_i = \{X_n, P_n, Y_n, Q_n\}$ is such that $X_o = N(A) \subset X_n \subset V$, $Y_o \subset Y_n$, $Q_n(\tilde{Y}) \subset \tilde{Y}$ and $i = d - d^* = i(A) > 0$, where $\tilde{Y} = R(A)$, $Y = Y_o \oplus \tilde{Y}$, $d = \dim X_o$, $d^* = \dim Y_o$. An example of such a scheme is Γ_A. Let $\tilde{X}$ be such that $X = X_o \oplus \tilde{X}$ and each $x \in X$ has the unique representation $x = x_o + x_1$, $x_o \in X_o$, $x_1 \in \tilde{X}$. Let $Q_o: Y \to Y_o$ be a linear projection. We need

Definition 3.1. Let $D \subset R^d$ be bounded and closed. A continuous map $\phi: D \to R^{d^*}\setminus\{0\}$ is *inessential* relative to a bounded closed subset $B \subset R^d$ with $D \subset B$ if there is a continuous extension $T: B \to R^{d^*}\setminus\{0\}$ for ϕ.

We know that if $D = S^{d-1} \equiv \partial B(0, 1) \subset R^d$, then ϕ is inessential (relative to $\bar{B}(0, 1)$) if and only if the homotopy class of $\phi/|\phi|$ in $\Pi_{d-1}(S^{d^*-1})$ is trivial. In the case of more general D one needs to use framed cobordism classes of Pontrjagin. We also recall that if $\phi, \Psi: D \subset R^d \to R^{d^*}\setminus\{0\}$ are homotopic and ϕ is inessential (relative to $B \supset D$), then Ψ is inessential (relative to B). We impose the following conditions on N:

Assumption A. *For a given* $f_0 \in Y_0$ *there exist constants* $M > 0$, $K \geq 0$ *and* $\rho \geq 0$ *such that*

$$Q_0 N(x_0+x_1) \neq tf_0 \text{ for } \|x_1\| \leq r,\ r > K,\ \|x_0\| \geq r M + \rho,\ t \in [0, 1].$$

Assumption B. *For* $r > K$ *sufficiently large, the map* $Q_0 N : X_0 \to Y_0 \setminus \{0\}$ *for* $\|x_0\| = rM + \rho$ *has the nontrivial stable homotopy, i.e. all the suspensions of* $F : S^{d-1} \to S^{d^*-1}$ *are nontrivial, where* $F(\tau) = Q_0 N(r\tau)/\|Q_0 N(r\tau)\|$.

Remark 3.1. If $i(A) = 0$, then $Q_0 N : X_0 \to Y_0 \setminus \{0\}$ for $\|x_0\| = rM + \rho$ having the nontrivial stable signifies that $\deg_{BR}(Q_0 N, B(rM+\rho), 0) \cap \cap X_0, 0) \neq 0$. Regarding Assumptions A and B, the following result is useful in applications.

Lemma 3.1. (a) *Suppose that for some* $r > 0$ *large enough,* $0 \neq \phi(x_0) = \lim_{R\to\infty} Q_0 N(Rx_0+x_1)$ *exists uniformly for* $x_1 \in \tilde{X}$, $\|x_1\| \leq r$, *and* $x_0 \in X_0$, $\|x_0\| = 1$, *and* $\phi : \partial B(0, 1) \cap X_0 \to Y_0 \setminus \{0\}$ *has nontrivial stable homotopy. Then Assumptions* A *and* B *hold with* $f_0 = 0$.

(b) *Suppose that all the conditions of* (a) *hold and* $\|Q_0 N(Rx_0 + x_1)\| < \|\phi(x_0)\|$, $\|f_0\| < \|\phi(x_0)\|$ *for* $x_0 \in \partial(0, 1) \cap X_0$, $\|x_1\| \leq r$ *and* R *large. Then Assumptions* A *and* B *hold.*

Our first basic result is the following

Theorem 3.1. ([17]). *Let* $A : V \subset X \to Y$ *be a Fredholm map of index* $i(A) = d-d^* > 0$ *and* A*-proper w.r.t.* Γ_i, $f_0 \in Y_0$ *and* $N : X \to Y$ *such that Assumptions* A *and* B *hold. Suppose that* $S = (I-Q_0)N$ *is quasibounded, i.e.*

$$|S| = \limsup_{\|x\|_1\to\infty} \frac{\|Sx\|}{\|x\|_1} < \infty \text{ and } \delta|S| \max\{1, M\} < c, \tag{3.1}$$

where $\delta = \max \|Q_n\|$, $\|x\|_1 = \|x_o+x_1\|_1 = \max\{\|x_o\|, \|x_1\|\}$ *and* $c > 0$ *such that* $\|Q_nAx_1\| \geq c\,\|x_1\|$ *for each* $x_1 \in \tilde{X}_n = \tilde{X} \cap X_n$ *and all large* $n \geq n_o$. *Then*

(a) *If* $A+N$ *is A-proper and A-closed w.r.t.* Γ_i, Eq. (1.1) *is feebly approximation-solvable for each* $f \in f_o + \tilde{Y}$.

(b) *If* $A+N$ *is strongly A-closed w.r.t.* Γ_i, Eq. (1.1) *is solvable for each* $f \in f_o + \tilde{Y}$

Proof. Let $\varepsilon > 0$ be small. By (3.1) there exists an $R \geq K$ such that $\|Sx\| \leq (|S| + \varepsilon)\|x\|_1$ for all $\|x\|_1 \geq R$. We claim that there exists an $R_1 \geq R$ such that for each $r \geq R_1$ fixed, $Q_nAx + tQ_n(I-Q_o)Nx - tQ_nf_1 \neq 0$ for all $x = x_o + x_1$ with $\|x_o\| < rM + \rho$, $x_1 \in \tilde{X}_n$, $\|x_1\| = r$, $t \in [0, 1]$ and $n \geq n_o$, where $f_1 = (I-Q_o)f$. Indeed, if this were not the case, then for each $r \geq R$ with $r \to \infty$ there would exist $t \in [0, 1]$ and x with $\|x_o\| < rM + \rho$ and $x_1 \in X_{1n}$, $\|x_1\| = r$, for some $n \geq n_o$ such that $Q_nAx_1 + tQ_n(I-Q_o)Nx - tQ_nf_1 = 0$. Hence,

$$c\|x_1\| \leq \|Q_nAx_1\| \leq \delta\|Sx\| + \delta\|f_1\| \leq \delta(|S| + \varepsilon)\|x\|_1 + \delta\|f_1\|$$

or

$$c \leq \delta(|S|) + \varepsilon) \max\{1, M+\rho/r\} + \rho/r\|f_1\|.$$

Passing to the limit as $r \to \infty$ we obtain $c \leq \delta(|S|+\varepsilon)\max\{1, M\}$, in contradiction to (3.1). Hence, such an r exists.

Next, define $\bar{D} = \bar{D}_r = \{x = x_o+x_1 \mid \|x_o\| \leq rM+\rho, \|x_1\| \leq r\}$ with r chosen as above. For each $n \geq n_o$, define the homotopy $H_n : [0, 1]\times(\bar{D} \cap X_n) \to Y_n$ by $H_n(t,x) = Q_nAx+tQ_n(I-Q_o)Nx+Q_nQ_o^N(x_o+tx_1) - tQ_nf$. If $x \in \partial D$ is such that $\|x_1\| < r$, then $\|x_o\| = rM + \rho$ and in view of Assumption A and since $Y_o \subset Y_n$, $Q_nQ_oN(x_o+tx_1)-tQ_nQ_of = Q_oN(x_o + tx_1) - tQ_of \neq 0$ for each $t \in [0, 1]$. If $x \in \partial D \cap X_n$ is such that $\|x_o\| < rM + \rho$, then $\|x_1\| = r$ with $x_1 \in \tilde{X}_n$ and

$Q_nAx + tQ_n(I-Q_o)Nx - tQ_nf_1 \neq 0$ by our choice of r and belongs to $\tilde{Y}$. Consequently, $H_n(t, x) \neq 0$ for all $t \in [0, 1]$ and $x \in \partial(D \cap X_n)$ and therefore $H_n(1, \cdot): \partial D \cap X_n \to Y_n \backslash \{0\}$ is homotopic to $H_n(0, \cdot) = Q_nA + Q_oNP_o: \partial D \cap X_n \to Y_n \backslash \{0\}$ for each $n \geq n_o$. Therefore $H_n(1, \cdot)$ is essential relative to $\bar{D} \cap X_n$ if and only if $H_n(0, \cdot)$ is essential relative to $\bar{D} \cap X_n$.

Let us now show that $H_n(0, \cdot)$ is essential relative to $\bar{D} \cap X_n$. We have that $d_n = \dim \tilde{X}_n = \dim \tilde{Y}_n$, where $Y_n = Y_o \oplus \tilde{Y}_n$, $\tilde{Y}_n = \tilde{Y} \cap Y_n$ and $Q_nA: \tilde{X}_n \to \tilde{Y}_n$ is a bijection. Let $\{u_1, \ldots, u_d\}$, $\{v_1, \ldots, v_{d_n}\}$, $\{w_1, \ldots, w_{d^*}\}$ and $\{Q_nAv_1, \ldots, Q_nAv_{d_n}\}$ be fixed bases in X_o, $\tilde{X}_n$, Y_o and $\tilde{Y}_n$ respectively. Identify

$x_o + \tilde{x}_n = \sum_{i=1}^{d} a_iu_i + \sum_{i=1}^{d_n} b_iv_i$ with $[b, a] = (b_1, \ldots, b_{d_n}, a_1, \ldots, a_d) \in R^{d_n} \times R^d$, and call this identification by $C_n = B_n \times A_o$ where $A_o: X_o \to R^d$ and $B_n: \tilde{X}_n \to R^{d_n}$. Define a norm in $R^{d_n} \times R^d$ by $\|[b, a]\|_1 = \max\{\|a\|, (M + \rho/r) \cdot \|b\|\}$, where $\|a\| = \|\sum_{i=1}^{d} a_iu_i\|$ and $\|b\| = \|\sum_{i=1}^{d_n} b_iv_i\|$. Similarly, identify $y_o + \tilde{y}_n = \sum_{i=1}^{d^*} k_iw_i + \sum_{i=1}^{d_n} \ell_iQ_nAv_i$ with $[\ell, k] = (\ell_1, \ldots, \ell_{d_n}, k_1, \ldots, k_{d^*}) \in R^{d_n} \times R^{d^*}$, and call this identification by $D_n = L_n \times K_o$, where $K_o: Y_o \to R^{d^*}$ and $L_n: \tilde{Y}_n \to R^{d_n}$. Let a norm in $R^{d_n} \oplus R^{d^*}$ be that of $R^{d_n} \times R^d$ introduced above and $V \subset R^d$ such that $R^d = R^{d^*} \oplus V$.

Now, on the ball $\bar{B}_n(0, rM+\rho) = \{[b, a] \in R^{d_n} \times R^d \mid \|a\| \leq rM + \rho, \|b\| \leq r\}$, we define following map:

$$F_n([b, a]) = b + K_oQ_oN\left(\sum_{i=1}^{d} a_iu_i\right) \in R^{d_n} \oplus R^{d^*}.$$

We have that $H_{n,o}(x) = (D_n^{-1}F_nC_n)(x)$ and therefore $H_{n,o}$ is essential relative to $\bar{D} \cap X_n$ if and only if $D_n^{-1}F_nC_n$ is essential relative to $\bar{D} \cap X_n$. Since C_n is a homeomorphism from $\bar{D} \cap X_n$ onto $\bar{B}(0, rM+\rho)$

and D_n^{-1} is a homeomorphism of $F_n(\bar{B}(0, rM + \rho))$ onto its image, it is easy to check that $H_{n,o}$ is essential relative $\bar{D} \cap X_n$ if and only if F_n is essential relative to $\bar{B}_n(0, rM + \rho)$ (for $R^{d_n} \times R^d$).

Next we show that F_n is essential relative to $\bar{B}_n(0, rM + \rho)$. We first observe that for each $[b, a]$ we have that $F_n([a, b]) = b + K_oQ_oNA_o^{-1}(a)$ with $K_oQ_oNA_o^{-1}:\bar{B}(0, rM + \rho) = \{a\in R^d;\ \|a\| \leq rM+\rho\} \subset R^d \to R^{d^*} \subset R^d \oplus V$, and $K_oQ_oNA_o^{-1}:\partial B(0, rM+\rho) \to R^{d^*}\backslash\{0\}$, where $\dim V = d^* - d = i(A)$. Therefore, by Proposition 4.1.1 in [19], F_n is essential relative to $\bar{B}_n(0, rM+\rho)$ (for $R^{d_n} \times R^{d^*}$) if and only if the k - th suspension of $N_o \equiv K_oQ_oNA_o^{-1}$ is homotopically nontrivial, where $k = \dim R^{d_n} = \dim(\tilde{X} \cap X_n)$. But, since K_o and A_o^{-1} are homeomorphism (isometric) and $Q_oN:\partial B(0, rM+\rho) \subset X_o \to Y_o\backslash\{0\}$ has nontrivial stable homotopy class, it follows that N_o has a homotopically nontrivial k - th suspension.

Now, since $H_n(0, \cdot)$ is essential relative to $\bar{D} \cap X_n$ for each $n \geq n_o$ by the above discussion, there exists an $x_n \in \bar{D} \cap X_n$ such that $Q_nH(1, x_n) = Q_nAx_n + Q_nNx_n = Q_nf$ for each $n \geq n_o$. Therefore, if $A+N$ is strongly A-closed there exists an $x \in \bar{D}$ such that $Ax+Nx = f$. If $A+N$ is A-proper and A-closed, then some subsequence $x_{n_k} \to x$ and $Ax+Nx = f$.

■

<u>Corollary 3.1</u>. ([17]) *Let* $A:V \subset X \to Y$ *be as in Theorem 3.1,* $f_o \in Y_o$ *and* $N:X \to Y$ k *- ball contractive with* $k\delta < c$ *and continuous. Suppose that Assumptions* A, B *and condition* (3.2) *hold. Then Eq.* (3.1) *is feebly approximation-solvable for each* $f \in f_o \oplus \tilde{Y}$.

Theorem 3.1 was proved by the author [15] in the case of $i(A) = 0$ and we refer to [15] for a detailed discussion of this case and the

results of other authors. When N is compact, Corollary 3.1 reduces to a result of the author [12] which on the other hand extends the earlier results of Nirenberg [18] for compact and uniformly bounded N (i.e. $\|Nx\| \leq K$ for each $x \in X$ and some constant K) and of Borisovich-Sapronov-Zvyagin [2] for compact asymptotically linear N. The result of [18] was proven using a criteria ([6, 18]) for the essentialness of maps of the form $I-C$, $C:X \to X$ compact, while the authors of [2, 12] used the generalized degree theory for nonlinear Fredholm maps. In view of Lemma 3.1 (b), we see that Corollary 3.1 also includes a result of Berger [1] for N compact and uniformly bounded. Theorem 3.1 is also applicable to continuous bounded nonliner perturbations if $(A|_{\tilde{X}\cap V})^{-1}$ is compact.

Remark 3.2. Assumption A can be rephrazed as follows: for a given $f_0 \in Y_0$ there exist $\alpha \geq 0$ and $\rho > 0$ such that $Q_0N(x_0+x_1) = tf_0$ for $t \in [0, 1]$ implies $\|x_0\| < \alpha \|x_1\| + \rho$. Therefore, using this reformulation of Assumption A, we see that Theorem 3.1 remains valid (with the same proof) if (3.1) is replaced by (3.2) $|S| = \limsup_{\|x\|\to\infty} \frac{\|Sx\|}{\|x\|} < \infty$ and $(1 + \alpha)|S|\delta < c$.

When N has a (sub) linear growth, i.e., there are constants $\alpha > 0$, $\beta \geq 0$ and $\gamma \in [0, 1]$ such that $\|Nx\| \leq \alpha + \beta\|x\|^{\gamma}$, $x \in X$, then Assumption A can be relaxed to (cf. [15] when $i(A) = 0$):

Assumption A'. *For a given* $f_0 \in Y_0$ *there exist constants* $M > 0$, $K \geq 0$ *and* $\rho \geq 0$ *such that for* $\|x_1\| \leq r$, $r > K$, *and each* $\tau \geq r\tau(r)+\rho$ *for some* $\tau(r) \geq M$

$$Q_0N(\tau x_0+\tau^{\gamma}x_1) \neq tf_0 \quad \textit{for } x \in \partial B(0, 1) \cap X_0, \quad t\in [0, 1].$$

Our next result is the following continuation theorem proven in [14, 15] when $i(A) = 0$.

<u>Theorem 3.3</u>. [17] *Let* $\bar{D} = \{x = x_0 + x_1 \mid \|x_0\| \le r,\ \|x_1\| \le R\}$ *for some positive* r *and* R *and* $N:\bar{D} \to Y$ *such that* $Q_0Nx \neq 0$ *for* $x \in X_0 \cap \partial D$. *Suppose* $A+tN$ *is an A-proper and A-closed homotopy w.r.t.* Γ_i *and Assumption* B *holds for this* r. *Then,*

(a) *If* $Ax + t(I-Q_0)Nx \neq f$ *for* $x \in (X \setminus X_0) \cap \partial D$, $t \in [0, 1]$ *and some* $f \in \tilde{Y}$, *Eq.* (3.1) *is feebly approximation-solvable.*

(b) *If* $Ax + tNx \neq 0$ *for* $x \in (X \setminus X_0) \cap \partial D$, $t \in [0, 1]$, *Eq.* (3.1) *with* $f = 0$ *is feebly approximation-solvable.*

<u>Proof</u>. Define the homotopy $H:[0, 1] \times \bar{D} \to Y$ by $H(t, x) = = Ax + t(I-Q_0)Nx + Q_0Nx - tf$ with $f = 0$ in case (b). Then, in either case $H(t, x) \neq 0$, $x \in \partial D$ and $t \in [0, 1]$. Arguing by contradiction, it is easy to see, in view of the A-properness of $H(t, x)$, that there exists an $n_0 \geq 1$ such that for each $n \geq n_0$

$$(3.3) \qquad Q_nH(t, x) \neq 0 \quad \text{for} \quad x \in \partial D \cap X_n, \quad t \in [0, 1].$$

Moreover, for each fixed $n \geq n_0$, using the homotopy

$$F_n(t, x) = Q_nAx + Q_0N(x_0 + tx_1) \quad \text{for} \quad x_0 + x_1 \in (X_0 \oplus \tilde{X}_n) \cap \partial D =$$

$$= X_n \cap \partial D, \quad t \in [0, 1],$$ we obtain that $Q_nH_0 : \partial D \cap X_n \to Y_n \setminus \{0\}$ is homotopic to $Q_nA + Q_0NP_0 : \partial D \cap X_n \to Y_n \setminus \{0\}$. Now, in view of Assumption B, we obtain as in Theorem 3.1 that the last map is essential relative to $\bar{D} \cap X_n$ and therefore $Q_nH_1|_{\partial D \cap X_n}$ is essential relative to $\bar{D} \cap X_n$ for each $n \geq n_0$ by (3.3). The conclusions of the theorem now follow from the A-properness of $A+N$. ∎

When N is A-compact, Theorem 3.3 (b) includes the corresponding result of Mawhin [8] proven by him using his coincidence degree approach and a result about essential maps ([6, 18]).

4. Nonlinear perturbations of elliptic operators

In this section we shall look at some applications of the abstract results from Sections 2 and 3 to generalized boundary value problems for fully nonlinear perturbations of elliptic differential operators of nonnegative index.

A) Let $\Omega \subset R^n$ be a bounded domain of class C^{2m} and $L(x, D) = \sum_{|\alpha| \le 2m} a_\alpha(x) D^\alpha$ an elliptic operator with real coefficients $a_\alpha \in C^0(\bar{\Omega})$ for $|\alpha| = 2m$ and $a_\alpha \in L_\infty(\Omega)$ for $|\alpha| < 2m$. Let $B_j(x, D) = \sum_{|\alpha| \le m_j} b_{j,\alpha}(x) D^\alpha$ be m boundary operators with $m_j < 2m$ and $b_{j,\alpha} \in C^{2m-m_j}(\partial\Omega)$. Suppose that $L(x, D)$ and $B_j(x, D)$ satisfy the Lopatinski-Shapiro boundary conditions. Let $Y = L_2$ and $X = D(L) = W_2^{2m}(\Omega; \{B_j\}) = \{u \in W_2^{2m}(\Omega) \mid B_j u = 0,\ j = 1, \dots, m\}$. Let d be the dimension of the kernel of L and d^* the codimension of the cokernel of the range of L. Then $L: D(L) \to L_2$ is a Fredholm mapping of index $i(L) = d - d^*$.

We consider the boundary value problem

$$(4.1) \qquad \begin{aligned} Lu + F(x, u, Du, \dots, D^{2m}u) &= 0 \quad \text{in } \Omega \\ B_j u &= 0 \quad \text{on } \partial\Omega,\ j = 1, \dots, m \end{aligned}$$

where $F: \Omega \times R^{s_{2m}} \to R^1$ satisfies the Carathéodory conditions and s_{2m} is the number of multiïndices $\alpha = (\alpha_1, \dots, \alpha_n)$ with $|\alpha| \le 2m$. We impose inicially the following conditions on F:

(F.1) *For each* $v \in W_2^{2m}$ *the mapping* $u \to F(x, u, Du, \dots, D^{2m-1}u, D^{2m}v)$ *is continuous and bounded from* W_2^{2m-1} *into* L_2.

(F.2) $F: \Omega \times \mathbb{R}^{s_{2m}} \times \mathbb{R}^{s'_{2m}} \to \mathbb{R}$ *satisfies*

$$|F(x, \eta, \xi) - F(x, \eta, \xi')| \le \beta \sum_{|\alpha| \le 2m} |\xi_\alpha - \xi'_\alpha|$$

for all $(x, \eta) \in \Omega \times R^{s_{2m-1}}$, $\xi, \xi' \in R^{s_{2m}-s_{2m-1}}$ *and* β *small*.

Define $N: X \to Y$ by $Nu = F(x, u, Du, \ldots, D^{2m}u)$. We have shown in [14, 15] that N is k-ball-contractive and L and $L+N$ are A-proper and A-closed w.r.t. Γ_L for (X, Y). Hence, Theorem 2.1 implies

Theorem 4.1. *Suppose that* $i(L) > 0$, *Assumptions* (F1) *and* (F2) *hold and* F *is odd on* $R^{s_{2m}}$, *i.e.* $F(x, -\xi_1, -\xi_2, \ldots, -\xi_{2m}) = -F(x, \xi_1, \xi_2, \ldots, \xi_{2m})$ for $x \in \Omega$, $\xi \in {}^{s_{2m}}$. *Let* D *be an open, bounded and symmetric neighborhood of zero in* X *and* Z *the set of solutions of BVP* (4.1) *in* ∂D. *Then the genus* $\gamma(Z) \geq i$ *and, in particular, BVP* (4.1) *has infinitely many solutions in* ∂D *if* $i \geq 2$.

When F depends on the derivatives up to the order 2m-1, N is a compact mapping and consequently Theorem 4.1 includes the corresponding result of Rabinowitz [22].

B) In this part we give an extension of the result of Nirenberg [18] for BVP (4.1) with $f = 0$ and $i\ (L) \geq 0$ who assumed that F depended on the derivatives up to the order 2m-1. Suppose that X and Y are as in A), and $L: X \to Y$ is a Fredholm map of index $i = d - d^* \geq 0$ with smooth coefficients and satisfies the unique continuation condition, i.e. if $0 \neq u \in \text{Ker } L$, then meas $\{x \mid x \in \Omega, u(x) = 0\} = 0$. Let $w_1, \ldots, w_d$ (resp. $w_1', \ldots, w_{d^*}'$) be smooth functions spanning Ker L (resp. $Y_o = R(L)^{\perp}$). If $a = (a_1, \ldots, a_d) \in R^d$, we define

$$a \cdot w(x) = \sum_{i=1}^{d} a_i w_i(x), \quad x \in \Omega.$$

Suppose also

(F3) $|F(x, \xi)| \leq C$ *for all* $x \in \bar{\Omega}$, $\xi \in R^{s_{2m}}$ *and some constant* C.

(F4) $h(x, \xi) = \lim_{r \to \infty} F(x, r\xi)$ *exists uniformly for* $x \in \bar{\Omega}$ *and* $|\xi| = 1$.

Define a mapping $\phi: S^{d-1} \to R^{d^*}$, $\phi = (\phi_1, \ldots, \phi_{d^*})$ by

$$\phi_i(a) = (h(x, a\cdot w, \ldots, D^{2m}(a\cdot w), w_i^!)_{L_2}, \quad i = 1, \ldots, d^*.$$

Theorem 4.2. *Let* L *be as above and suppose that* (F1)-(F4) *hold and* $\phi: S^{d-1} \to R^{d^*}\setminus\{0\}$ *and has nontrivial stable homotopy. Then BVP* (4.1) *with* $f = 0$ *is feebly approximation-solvable w.r.t.* Γ_L.

Proof. We shall apply Theorem 3.1. Since X is a separable Hilbert space, the scheme Γ_L for (X, Y) induced by L is easy to construct. Let Q_o be the projection such that $R(Q_o) = Y_o$ and $\operatorname{Ker} Q_o = R(L)$. Define $N: X \to Y$ by $Nu = F(x, u, Du, \ldots, D^{2m}u)$. Then N is k-ball contractive as before and A and $A+N$ are A-proper and A-closed w.r.t. Γ_L.

Now, in view of Lemmas 1 and 2 in [18], we see that for $i = 1,\ldots,d$, $\lim_{r\to\infty} |\int_\Omega N(ra.w+u)w_i^! dx - \phi_i(x)| = 0$ uniformly for u in bounded subsets of X and $a \in S^{d-1}$. Therefore, for each bounded subset $B \subset R(L)$ there is $r_o > 0$ such that for each $r \geq r_o$, each $v \in B$, and each $a \in S^{d-1}$, one has $Q_oN(ra.w+v) \neq 0$. Hence, Assumption A holds with $f_o = 0$ as does Assumption B by Lemma 3.1. Since (3.1) clearly holds, we see that the conclusion follows from Theorem 3.1.

■

If L has some other properties, often (F.2) can be weaken or substituted by some other assumptions. For example, if $\partial\Omega$ is C^∞ and $a_\alpha \in C^{o,\lambda}(\bar{\Omega})$, $0 < \lambda < 1$, $|\alpha| \leq 2m$, $X = W_2^{2m}(\Omega) \cap \mathring{W}_2^m(\Omega)$ with $\|\cdot\|_{2m}$, $Y = L_2(\Omega)$, and $L: X \to Y$ is strongly uniformly elliptic, then there is a strongly uniformly elliptic operator $K = \sum_{|\alpha|\leq 2m} b_\alpha(x)D^\alpha$ and constants $c > 0$, $c_o > 0$ such that $(Lu, Ku)_o \geq c\|u\|_{2m}^2 - c_o\|u\|_o^2$ and $((-1)^m Ku, u)_o = \geq c_o\|u\|_m^2$, $u \in X$ (cf. [23]). Since $C = (-1)^m c_1 I: X \to Y$,

$c_1 > c_0$, is compact and $L+C$ is c-strongly K-monotone, it is bijective and therefore L is Fredholm of index zero. Now, if F is such that $(Nu-Nv, K(u-v)) \geq -c\,\|u-v\|^2$, then $L+tN$ is A-proper for $t \in [0, 1)$ and strongly A-closed for $t = 1$. Moreover, the size of β in (F2) dependes now on c as follows from (cf. [13, 17])

Proposition 4.1. (a) *Let* X *be densily and compactly embedded in a Banach space* X_0, $K:X \to X_0^*$, $(x, Kx) \geq \|x\|_0^2$, $x \in X$, *and* $A:V\subset X \to X_0$ *such that* $(Ax-Ay, K(x-y)) \geq c\|x-y\|^2 - c_0\|x-y\|_0^2$, $x,y \in V$, *and* $Q_nA:X_n \to Y_n$ *continuous. Then, if* $N:X \to X_0$ *is continuous and either k-ball contractive with* $k\delta < c$ *or ball-condensig if* $\delta = c = 1$ *and* $Q_n^*Kx = Kx$, $x \in X_n$, A *and* $A+N$ *are A-proper w.r.t.* Γ_0 *for* (V, X_0).
(b) *If* $A:X \to Y$ *is a-stable, i.e.* $\|Q_nA-Q_ny\| \geq c\|x-y\|$, $x,y \in X_n$, $n \geq 1$, $Q_nA:X_n \to Y_n$ *continuous and* $N:X \to Y$ *as in* (a), *then* A *and* $A+N$ *are A-proper. If* A *is demicontinuous, then* $A+N$ *is A-closed; when* A *is also continuous, it is a homeomorphism.*

We note that, in view of Proposition 4.1, the fixed point and surjectivity results for $A+N$ in [11] are valid uner the weaker demicontinuity condition on A.

REFERENCES

[1] M. S. Berger, *Nonlinearity and Functional Analysis*, Academic Press, N.Y., 1977.

[2] Yu. G. Borisovich, V. G. Zvyagin and Yu. I. Sapronov, *Nonlinear Fredholm mappings and Leary - Schauder Theory*, *Uspechi Mat.* Nauk 32 (1977), 3-74.

[3] L. Cesari, *Functional analysis, nonlinear differential equations and the alternative method*, *Lecture Notes in Pure and* Appl. Math., vol. 19, M. Dekker, N.Y. 1976, 1-197 (Editors: Cesari, Kannan, Shuur).

[4] S. Fucik, *Solvability of Nonlinear Equations and Boundary Value Problems*, D. Reidel Publ. Comp., 1980.

[5] G. E. Gaines and J. Mawhin, *Coincidence degree for nonlinear differential equations*, Lecture Notes in Math., vol. 568, Sprinver-Verlag, 1977.

[6] A. Granas, *The theory of compact vector fields and some of its applications to topology of functional spaces I*, Rozprawy Math. 20 (1962), 1-93.

[7] S. Kaniel, *Quasicompact nonlinear operators in Banach spaces and applications*, Arch. Rational Mech. Anal. 20 (1965), 259-278.

[8] J. Mawhin, *Stable homotopy and ordinary differential equations with nonlinear boundary conditions*, Rocky Mountain J. Of Math., 7(3) - (1977), 417-424.

[9] J. Mawhin, *Topological Degree Methods in Nonlinear Boundary Value Problems*, Regional Conf. Series in Math., Vol. 40, AMS, 1979.

[10] P. S. Milojević, *A generalization of Leray-Schauder theorem and surjectivity results for multivalued A-proper and pseudo mappings* J. Nonlinear Anal., TMA, 1(3) (1977), 263-276.

[11] P. S. Milojević, *Fixed point theorems for multivalued approximable mappings*, *Proc. Amer. Math. Soc.* 73(1) (1979), 65-72.

[12] P. S. Milojević, *The solvability of operator equations with asymptotic quasibounded nonlinearities*, Proc. Amer. Math. Soc. 76(2) (1979) - 293-298.

[13] P. S. Milojević, *Fredholm alternatives and surjectivity results for multivalued A-proper and condensing mappings with applications to nonlinear integral and differential equations*, Czechoslovak Math. J. 30(105) (1980), 387-417.

[14] P. S. Milojević, *Approximation-solvability of some nonlinear operator equations with applications*, Proc. Intern. Symp. Funct. Diff. Equat. and Bifurcation, July 1979, São Carlos, Brasil, Lecture Notes in Math. Vol. 799, Springer-Verlag, 289-316 (Ed. A. F. Ize).

[15] P. S. Milojević, *Approximation-solvability results for equations involving nonlinear perturbations of Fredholm mappings with applications to differential equations*, in Functional Analysis, Holomorphy and Approx. Theory, Proc. Int. Sem., Rio de Janeiro, August, 1979, Lecture Notes in Pure and Applied Math., M. Dekker, N.Y. (Ed. G. Zapata, to appear).

[16] P. S. Milojević, *Continuation theory for A-proper and strongly A-closed mappings and their uniform limits and nonlinear perturbations of Fredholm mappings*, in Functional Analysis, Holomorphy and Approx. Theory, Proc. Int. Sem., Rio de Janeiro, August 1980, North Holand Publ. Comp. (Ed. J. A. Barroso, to appear).

[17] P. S. Milojević, *Theory of A-proper and Pseudo A-closed Mappings*, Habilitation Memoir, Univ. Federal de Minas Gerais, Belo Horizonte, Brasil, December 1980, 1-208.

[18] L. Nirenberg, *An application of generalized degree to a class of nonlinear problems*, Troisième Colloque CBBRM d'analyse fonctionnelle, Vander, Louvain, 1971, 57-74.

[19] L. Nirenberg, *Topics in Nonlinear Functional Analysis*, New York University Lecture Notes, 1973/74.

[20] W. V. Petryshyn, *On the approximation-solvability of equations involving A-proper and pseudo A-proper mappings*, Bull. Amer. Soc. 81 (1975), 223-312.

[21] S. I. Pohozaev, *The solvability of nonlinear equations with odd operators*, Funct. Anal. i Priložen. 1 (1967), 66-73.

[22] P. H. Rabinowitz, *A note on a nonlinear elliptic equation*, Indiana Univ. Math. J., 22(1) - (1972), 43-49.

[23] I. V. Skrypnik, *On the coercivity inequality for pairs of linear elliptic operators*, Soviet Math. Doklady, 19(2) (1978), 324-327.

SOME REMARKS ON A WAVE EQUATION
WITH A NONLOCAL INTERACTION

Gustavo Perla Menzala[†]

Instituto de Matemática
Universidade Federal do Rio de Janeiro
Rio de Janeiro, RJ - Brazil

§.1) INTRODUCTION

In this note we shall prove some claims concerning the solutions of the semilinear wave equation

$$u_{tt} - \Delta u + m^2 u + |u|^{p-2} u \, (V*|u|^p) = 0 \tag{1.1}$$

where x runs in $\mathbb{R}^3$, $t \in \mathbb{R}$, m^2 is a positive constant and V is a spatial potential which belongs to $L^{3/5-p}(\mathbb{R}^3) + L^\infty(\mathbb{R}^3)$. In (1.1) the symbol $*$ goes for spatial convolution. We shall consider (1.1) with p in the interval $2 \leq p < 3$.

In our previous work, [3], [4], we studied the Cauchy problem for (1.1) as well as the asymptotic behavior of a class of solutions.

In this note we shall concentrate our attention on the following problems: First, let us suppose that the potential V is negative in a small neighborhood of a point in $\mathbb{R}^3$. Then we shall prove that if $V(-x) = V(x)$ for all x, we can choose initial data at $t = 0$ for (1.1) in such a way that the solution u *does not exist* for all time. In the last section we prove that the regular solutions of (1.1) enjoy

---

† This work was supported by FNDCT and CEPG-UFRJ (Brazil)

the *causality* property. This property of the solutions is very important when we consider the group $\{U(t), -\infty < t < \infty\}$ of nonlinear operators which takes the pair of initial data at time $t = 0$, to the pair (u, u_t) where u is the solution of (1.1) at time t. Due to causality we know that if the initial data f at time $t = 0$ is smooth with compact support then each component of $U(t)f = (u, u_t)$ also has compact support at each instant t.

This paper is an extended version of an invited lecture given by me at the Latin American School on Differential Equations (June 29, July 17) at the University of São Paulo. We would like to thank the organizers for their kind invitation and financial support during the meeting.

We use standard notation: We denote by $L^r(\mathbb{R}^3)$ $1 \leq r < \infty$ the space of functions (class of) in $\mathbb{R}^3$ whose rth powers are integrable, with the norm $\|f\|_r^r = \int |f(x)|^r dx$ and by $L^\infty(\mathbb{R}^3)$ the space of measurable essentially bounded functions in $\mathbb{R}^3$ with the norm $\|f\|_\infty = \text{ess sup } |f(x)|$. $H^j(\mathbb{R}^3)$ $(j=1,2,\ldots)$ denotes the usual Sobolev space of square integrable functions from $\mathbb{R}^3$ into $\mathbb{R}$ which together with their partial derivatives up to order j, in the sense of distributions belong to $L^2(\mathbb{R}^3)$. The norm in $H^j(\mathbb{R}^3)$ will be denoted by $\| \ \|_{H^j}$. From now on, an integral sign to which no domain is attached will be understood to be taken over all space $\mathbb{R}^3$. The Laplacian operator will be denoted by Δ and grad u denotes the gradient of u (in space variables).

§.2) NONEXISTENCE OF GLOBAL SOLUTIONS IF $V < 0$

Let us consider the semilinear wave equation

$$u_{tt} - \Delta u + u + |u|^{p-2} u \, (V*|u|^p) = 0 \tag{2.1}$$

$$u(x,0) = u_0(x), \quad u_t(x,0) = u_1(x)$$

where $x \in \mathbb{R}^3$, $t \in \mathbb{R}$, $2 \le p < 3$, $*$ denotes spatial convolution, $V: \mathbb{R}^3 \to \mathbb{R}$ belongs to $L^{3/5-p}(\mathbb{R}^3) + L^{\infty}(\mathbb{R}^3)$ and $V(-x) = V(x)$ for all x. Let $X = H^1(\mathbb{R}^3) \times L^2(\mathbb{R}^3)$ and define

$$A = \begin{bmatrix} 0 & I \\ \Delta - I & 0 \end{bmatrix} \text{ with } D(A) = H^2(\mathbb{R}^3) \times H^1(\mathbb{R}^3).$$

It is well known that A generates a strongly continuous group $U(t)$ of unitary operators on X. Let $N\binom{u}{v} = \begin{pmatrix} 0 \\ -|u|^{p-2}u(V*|u|^p) \end{pmatrix}$, so we we can write (2.1) in the form

$$\frac{dz}{dt} = Az + N(z), \quad z(0) = Z_0 \tag{2.2}$$

where $z = \binom{u}{u_t}$ and $z_0 = \binom{u_0}{u_1}$. Let us assume that z_0 belongs to X. We shall call a weak solution z of (2.1) a solution of the integral equation

$$z(t) = U(t)z_0 + \int_0^t U(t-s)N(z(s))ds. \tag{2.3}$$

The first component of z satisfies (2.1) in the sense of distributions. It was proved in [4] that under the above hypothesis on V, there exists a unique weak solution $z = \binom{u}{u_t}$

$z \in C(T_{min} < 0 < T_{max}; X)$ of (2.1), where $I = (T_{min}, T_{max})$ is the maximal interval of existence. If $T_{max} < +\infty$ (respec. $-\infty < T_{min}$), then $\lim_{t \nearrow T_{max}} \|z(t)\|_X = +\infty$ (respect. $\lim_{t \searrow T_{min}} \|z(t)\|_X = \infty$). Now, if if $V(x) \ge 0$, a simple argument shows that the above solution exists globally, that is $I = (-\infty, \infty)$. We would like to show in this section that without this hypothesis we may not have global existence:

<u>Theorem 1</u>. Let $V: \mathbb{R}^3 \to \mathbb{R}$ be an even function such that $V \in L^{3/5-p}(\mathbb{R}^3) + L^{\infty}(\mathbb{R}^3)$ with $2 \le p < 3$. If $V(x) < 0$ near the

origem, then there exists initial data (u_0, u_1) for which the interval I of existence (found in [4]) is finite.

Proof: Let $r > 0$ such that $V(x) < 0$ for x in $\{x \in \mathbb{R}^3, |x| < r\}$. Choose a test function $\phi(x)$ with support in $\{x \in \mathbb{R}^3, |x| < \frac{r}{2}\}$. Thus, we have

$$\iint |\phi(x)|^p \, V(x-y) |\phi(y)|^p \, dxdy < 0.$$

Let $\varepsilon > 0$ and choose $u_0(x) = \varepsilon^{-1}\phi(x)$ and $u_1(x)$ any test function such that $\int u_1(x)\phi(x)dx > 0$. Now, we consider the equation (2.1) with initial data (u_0, u_1) at $t = 0$. It is easy to show (and it was proved in [4]) that "energy" E is conserved, that is

$$\begin{aligned} E(t) &= \frac{1}{2} \int \left[|u_t|^2 + |\mathrm{grad}\ u|^2 + u^2\right] dx + \frac{1}{2p} \iint |u|^p V(x-y) |u|^p \, dxdy \\ &= E(0) = \frac{1}{2} \int \left[u_1^2(x) + \frac{1}{\varepsilon^2} |\mathrm{grad}\ \phi|^2 + \frac{1}{\varepsilon^2} |\phi|^2\right] dx + \\ &+ \frac{1}{4\varepsilon^4} \iint |\phi(x)|^p \, V(x-y) |\phi(y)|^p \, dxdy. \end{aligned} \tag{2.4}$$

Observe that $E(0) < 0$ if ε is chosen sufficiently small. Let us consider $g(t) = \frac{1}{2}\int u^2(x,t)dx$. Clearly, we have

$$\frac{d}{dt} g^{-1/2}(t)\Big|_{t=0} = -\frac{1}{2} g^{-3/2}(0) \int u_1(x) u_0(x) dx < 0 \tag{2.5}$$

and

$$\frac{d^2}{dt^2} g(t) = -\left[\int (|\mathrm{grad}\ u|^2 + |u|^2 + |u|^p (V * |u|^p)) dx\right] + \int |u_t|^2 dx.$$

Thus

$$\begin{aligned} g(t)\frac{d^2}{dt^2} g(t) - \frac{3}{2}\left(\frac{dg(t)}{dt}\right)^2 &= -E(0) \int u^2 dx + \frac{1-p}{2p} \int u^2 dx \int |u|^p \, V * |u|^p \\ &+ \int |u_t|^2 \int u^2 - \frac{3}{2}\left(\int u u_t\right)^2 \end{aligned} \tag{2.6}$$

By using the Cauchy-Schwarz's inequality we know that $\int u_t^2 \, dx \int u^2 dx - \left(\int u u_t dx\right)^2 \geq 0$ and $-\frac{1}{2}\left(\int u u_t dx\right)^2 \geq -E(0) \int u^2 dx$.

Therefore, from (2.6) we obtain

$$g(t)\frac{d^2}{dt^2}g(t) - \frac{3}{2}\left(\frac{dg(t)}{dt}\right)^2 \geq$$

$$\geq \int u^2 dx \left[-2E(0) + \frac{1-p}{2p}\int |u|^p (V*|u|^p)dx\right]. \tag{2.7}$$

Since $E(t) = E(0)$ implies that

$$\frac{1-p}{2p}\int |u|^p\, (V*|u|^p)dx \geq (1-p)E(0)$$

then, it follows from (2.7) that

$$g(t)\frac{d^2}{dt^2}g(t) - \frac{3}{2}\left(\frac{dg}{dt}(t)\right)^2 \geq \int u^2 dx \left[(-p-1)E(0)\right] \geq 0 \tag{2.8}$$

because $E(0) < 0$ and $2 \leq p < 3$.

Observe that (2.8) is equivalent to saying that $\frac{d^2}{dt^2}g^{-1/2}(t) \leq 0$ for all t.

From (2.5) and (2.8) we see that the graph of $g^{-1/2}(t)$ looks like picture 1

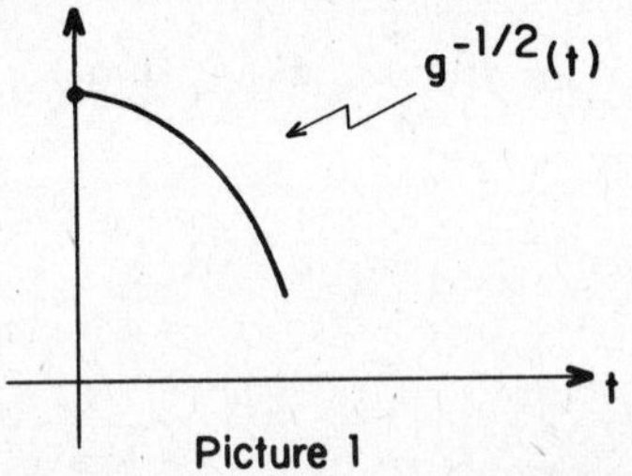

Picture 1

Hence we should have a finite $T > 0$ such that

$$\lim_{\substack{t\to T\\ t<T}} g^{-1/2}(t) = 0, \quad \text{thus} \quad \lim_{\substack{t\to T\\ t<T}} \|u(\cdot, t)\|_{L^2} = \infty.$$

This proves the theorem

§.3) FINITE PROPAGATION SPEED

In this section we shall prove a property of the solutions of (1.1) known as causality and which says essentially the following:

If the initial data at $t = 0$ is smooth and vanishes in the ball $\{x \in \mathbb{R}^3, |x-x_o| \leq r\}$, then the corresponding solution of (1.1) vanishes on the double cone

$$\left\{(x,t) \in \mathbb{R}^3 \times \mathbb{R},\ |x-x_o| < r - |t|,\ \ t \in I\right\} \tag{3.1}$$

where I is the maximal interval of existence mentioned in §.2):

Theorem 2 (Causality). Let $V: \mathbb{R}^3 \to \mathbb{R}$ belong to $L^{3/5-p}(\mathbb{R}^3)+L^\infty(\mathbb{R}^3)$ with $2 \leq p < 3$. Let $u_o \in H^1(\mathbb{R}^3)$ and $u_1 \in L^2(\mathbb{R}^3)$. Then, the values of the solution in $\{(x,t), |x-x_o| < r, t = t_o\}$ depend only on the values of the initial data in $\{(x,0), |x-x_o| < r + |t_o|\}$ provided that t_o belongs to the interval of existence I and r is a positive number.

Proof: Clearly, the above assertion is equivalent to the statement: If the initial data vanishes at $\{(x,0), |x-x_o| < r\}$, then the solution vanishes in the double cone (3.1). Just for convenience we consider only positive times. We may assume solution u is smooth (by approximating u_o, u_1 by smooth functions). We know that the solution has bounded energy in $0 \leq t \leq T_1$ if $[0,T_1] \subseteq I$. Let us multiply the equation by u_t to get identity

$$\frac{1}{2}\frac{\partial}{\partial t}\left[u_t^2+|\text{grad } u|^2 + m^2u^2\right] - \text{div}\left[u_t \text{ grad } u\right] = -u_t|u|^{p-1}\ (V*|u|^p). \tag{3.2}$$

Let $0 < T < T_1$ and

$$C_T = \left\{(x,t),\ |x-x_o| < r-t,\ 0 < t < T\right\}.$$

Let

$$\Omega_t = \left\{x,\ |x-x_o| < r-t\right\}.$$

Integrating the above identity (3.2) over C_T and using the divergence theorem, we obtain

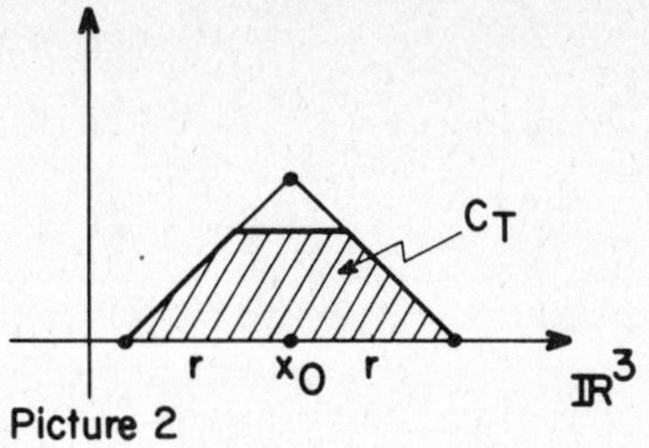

Picture 2

$$e(T)-e(0)+\frac{1}{\sqrt{2}}\int_{K_T}\left[\frac{1}{2}(u_t^2+|\text{grad }u|^2+m^2u^2)-u_t(\eta\cdot\text{grad }u)\right]dS$$

$$=-\int_0^T\int_{\Omega_t} u_t|u|^{p-1}\,(V*|u|^p)dxdt \tag{3.3}$$

where

$$e(t)=\frac{1}{2}\int_{\Omega_t}(u_t^2+|\text{grad }u|^2+m^2u^2)dx$$

K_T denotes the lateral boundary of C_T and η is the unit normal to Ω_t. Since the integrand of the third term of the left hand side of (3.3) is positive, we obtain the inequality

$$e(T)\le e(0)+\left|\int_0^T\int_{\Omega_t} u_t|u|^{p-1}(V*|u|^p)dxdt\right|. \tag{3.4}$$

By using Hölder's inequality we obtain

$$\left|\int_{\Omega_t} u_t|u|^{p-1}(V*|u|^p)dxdt\right| \le \int_{\Omega_t} u_t^2dx\int_{\Omega_t}|u|^{2(p-1)}(V*|u|^p)^2dx. \tag{3.5}$$

Since $V=V_1+V_2$ with V_1+V_2 with $V_1\in L^{3/5-p}$ and $V_2\in L^\infty$ then by interpolation and Sobolev's inequality we obtain

$$\int_{\Omega_t}|u|^{2(p-1)}(V_2*|u|^p)^2dx \le \|V_2\|_\infty^2\int|u|^{2(p-1)}dx\left[\int|u|^pdx\right]^2$$

$$\le \|V_2\|_\infty^2\,\|u\|_2^{10-2p}\,\|u\|_6^{6(p-2)} \le$$

$$\le C\,\|V_2\|_\infty^2\,\|u\|_2^{10-2p}\,\|\text{grad }u\|_2^{6(p-2)}.$$

Thus, from (3.6) and (3.5) we obtain

$$\left|\int_{\Omega_t} u_t|u|^{p-1}(V_2*|u|^p)dx\right| \le \|V_2\|_\infty \|u\|_2^{4-p} \|\text{grad } u\|_2^{3(p-2)} e(t)$$
$$\le C(V_2, E(0),p)e(t).$$

Similarly

$$\left|\int u_t|u|^{p-1}(V_1*|u|^p)dx\right| \le \|V_1\|_{3/5-p} \|u\|_6^{2p-1}\left(\int u_t^2\, dx\right)^{1/2}$$
$$\le C(V_1, E(0),p)e(t).$$

Finally, this implies that

$$e(T) \le e(0) + C\int_0^T e(t)dt$$

thus

$$e(T) \le e(0)e^{ct}.$$

Since $e(0) = 0$, it follows that $e(t) = 0$. In case $m = 0$, we write the equation in the form

$$u_{tt} - u + u + \left[|u|^{p-1}(V*|u|^p)-u\right] = 0$$

and we add to estimate (3.4) the following estimate

$$\int_{\Omega_t} u_t u\, dx \le C\int_{\Omega_t} (u_t^2+|\text{grad } u|^2+m^2u^2)dx.$$

This proves the theorem.

REFERENCES

[1] Glassey, R.T.; Blow-up theorems for nonlinear wave equations, Math. Z. 132, 183-203, (1973).

[2] Perla Menzala, G. On regular solutions of a nonlinear equation of Choquard's type, Proc. Royal Soc. Edinburgh 86 A, 291-301 (1980).

[3] Perla Menzala, G. and Strauss, W.A. On a wave equation with a cubic convolution, J. of Differential equations, (to appear).

[4] Perla Menzala, G., On a semilinear wave equation: The Cauchy problem and the asymptotic behavior of solutions, J. Math. Analy. Appl. (to appear).

[5] Strauss, W.A., Nonlinear invariant wave equations, Lecture Notes in Physics, Vol. 73, Edited by G. Velo and A. Wightman, 197-249, Springer-Verlag (1978).

[6] Strauss, W.A., Nonlinear scattering at low energy, J. of Functional Analy. 41, 1, 110-133 (1981).

THE MOUNTAIN PASS THEOREM: THEME AND VARIATIONS

Paul H. Rabinowitz[†]

Department of Mathematics
University of Wisconsin
Madison, Wisconsin, USA

Introduction

This paper contains a somewhat expanded version of a series of lectures presented at the First Latin American School on Differential Equations held in São Paulo during June 29 - July 17, 1981. The goal of these lectures was twofold: First to survey applications of minimax methods to critical point theory which have been carried out during the past few years. The second goal was to study in more detail the so-called Mountain Pass Theorem together with some related results and give some applications to partial differential equations, in particular semilinear elliptic boundary value problems.

In §1 we will carry out our survey of minimax methods in critical point theory. The Mountain Pass Theorem as well as the tools required to prove it will be treated in §2 and some applications to

---

[†] This research was sponsored in part by the National Science Foundation under Grant No. MCS-8110556 and by the United States Army under Contract No. - DAAG29-80-C-0041. Reproduction in whole or in part is permitted for any purpose of the United States Government.

elliptic partial differential equations will be given in §3. In order to obtain a "symmetric" version of the Mountain Pass Theorem (in §4), first index theories, in particular the notion of genus, will be discussed in §4. Lastly in §5 we will indicate how the ideas of §2 and §4 can be applied to bifurcation problems in which variational structure is present.

We thank the organizers of the school, especially Professor De Figueiredo and Hönig for their hospitality and for making these lectures possible.

§1: Minimax methods in critical point theory

This section surveys some contributions to the above subject made during the last few years. See also the lectures of A. Castro [1], Nirenberg [2], Palais [3], and the author [4] for some other aspects of this subject. We will focus on results that are spiritual descendents of Ljusternik and Schnirelman rather than Morse. Below $C^k(X, Y)$ denotes the set of k times continuously differentiable mappings from X to Y.

To begin suppose $I \in C^1(\mathbb{R}^n, \mathbb{R})$. A *critical point* of I is a point at which the derivative of I vanishes. More generally, let E be a real Banach space and $I \in C^1(E, \mathbb{R})$. For any $u \in E$, $I'(u)$, the Frechet derivative of I at u, is a continuous linear functional on E, i.e. $I'(u) \in E^*$. We say u is a critical point of I if $I'(u) = 0$, i.e.

$$I'(u)\phi = 0 \tag{1.1}$$

for all $\phi \in E$. If E is a Hilbert space, there is a natural identification between E and E^* and then the mapping $u \to I'(u)$ can be interpreted as a mapping of E onto itself. If u is a critical point of I, we say $I(u)$ is a *critical value* of I.

In applications to differential equations, verifying (1.1)

corresponds to finding a weak solution of the differential equation. Indeed one of the main uses of critical point theory is as a tool for obtaining the existence of solutions of differential equations.

The critical points we are most familiar with are local or global maxima or minima. Here, however, our primary concern is with indefinite functionals. We will not attempt a precise definition of such objects but what we have in mind are functionals which are not bounded from above or below, even modulo sub-spaces or submanifolds of finite dimension or co-dimension. As an example, consider

$$(1.2) \qquad I(u) = \int_0^1 \left[\frac{1}{2}\left|\frac{du}{dx}\right|^2 - F(x, u)\right] dx$$

for u in the Sobolev space $W_0^{1,2}[0, 1]$ where $F(x, \xi) \to \infty$ as $\xi \to \infty$ at a more rapid rate than quadratic, e.g. $F(x, \xi) = |\xi|^p$, $p > 2$. Then $I(u) \to -\infty$ as $u \to \infty$ along any ray while if we choose $u_k = \sin k\pi x$, $I(u_k) \to \infty$ as $k \to \infty$. Indefinite functionals occur naturally in mechanics where one treats Lagrangians, i.e. integrands of the form kinetic energy minus potential energy. One could try to give a Morse theoretic type of classification of C^2 indefinite functionals in terms of measuring the degree of indefiniteness of $I''(u)$ but we shall not pursue this point since we will usually only require $I \in C^1$ for what follows.

Since we are searching for critical points of $I \in C^1(E, \mathbb{R})$, in the large, in general some compactness must be built into the problem. One useful way to do this is via the *Palais-Smale conditions* (PS). We say I satisfies (PS) if any sequence (u_m) such that (i) $I(u_m)$ is bounded and (ii) $I'(u_m) \to 0$ is precompact, i.e. possesses a convergent subsequence. Actually in applying minimax methods to obtain critical points, one usually get a minimax characterization of a corresponding critical value c of I. Thus one only needs a local version of (PS) to show c is a critical value, namely any sequence $(u_m) \subset I^{-1}(c-\varepsilon, c+\varepsilon)$ for some $\varepsilon > 0$ and satisfying (ii) is precompact. Examples where (PS) is satisfied will be given in §3.

Now we will describe some critical point theorems. Below B_ρ denotes an open ball of radius ρ about 0. We begin with the following version of the Mountain Pass Theorem:

<u>Theorem 1.3</u>. Let E be a real Banach space, $I \in C^1(E, \mathbb{R})$ and satisfy (PS). Suppose

(I_1) $I(0) = 0$ and there are constants $\rho, \alpha > 0$ such that $I|_{\partial B_\rho} \geq \alpha$.

(I_2) There is an $e \in E \backslash B_\rho$ such that $I(e) \leq 0$.

Then I has a critical value $c \geq \alpha$. Moreover c can be characterized as follows: If

$$\Gamma = \left\{ g \in C([0, 1], E) \mid g(0) = 0, \ g(1) = e \right\},$$

then

$$c = \inf_{g \in \Gamma} \max_{t \in [0, 1]} I(g(t)). \tag{1.4}$$

Theorem 1.3 is due to Ambrosetti and the author [5]. Note that if $E = \mathbb{R}^3$, we can interpret (I_1), (I_2) as saying that the origin lies in a valley surrounded by a mountain range and (1.4) tells us that the lowest pass through the range contains a critical point. Note that I need not be bounded from above as in (1.2).

In applications of the Mountain Pass Theorem, usually $I > 0$ in $B_\rho \backslash \{0\}$. Thus $u = 0$ is a local minimum of I so we have an obvious critical point to begin with and obtain a second one via Theorem 1.3. In the following "Saddle Point Theorem" however, there is no evident initial critical point.

<u>Theorem 1.5</u>. Let E be a real Banach space such that $E = E_1 \oplus E_2$ with E_1 finite dimensional and let $I \in C^1(E, \mathbb{R})$ satisfying (PS).

Suppose

(I_3) There is a $b_1 \in \mathbb{R}$ and a bounded neighborhood Ω of 0 in E_1 such that $I|_{\partial\Omega} < b_1$

(I_4) There is a $b_2 > b_1$ such that $I|_{E_2} \geq b_2$.

Then I has a critical value $c \geq b_2$. Moreover c can be characterized as follows: If

$$S = \left\{h \in C(\Omega, E) \mid h(u) = u, \quad u \in \partial\Omega\right\},$$

then

$$c = \inf_{h \in S} \max_{u \in \bar{\Omega}} I(h(u)). \tag{1.6}$$

Theorem 1.5 is due to the author [6]. In applications, it seems to be useful only for functionals which grow at most quadratically [6, 7] unlike Theorem 1.3 where much more rapid growth is possible [5, 7]. There are several variants of Theorem 1.5 in the literature including situations where one gets a unique "saddle point" without (PS). See e.g. the lectures of A. Castro [1], Lazer, Landesman, and Meyers [8], Ahmad, Lazer, and Paul [9], Castro and Lazer [10], Amann [11], Amann and Zehnder [12], and Coron [13]. Although Theorems 1.3 and 1.5 seem rather different, in fact in a more restrictive setting, they are actually both special cases of a general critical point theorem in which E_1 may also be infinite dimensional. A common feature of (I_1) - (I_2) and (I_3) - (I_4) is that both sets of hypotheses imply (again in a more restrictive setting) a certain topological linking must occur. See Benci-Rabinowitz [7] for more details. See also Nirenberg [2] for some other remarks on linking and other results related to Theorems 1.3 and 1.5 and Hofer [14] for some generalizations of [7].

If one is dealing with a functional, I, which is invariant under a group of symmetries, then one can hope to obtain multiple critical points of I. Thus suppose we have a group G of mappings of E to E. We say I is invariant under G if $I(g(u)) = I(u)$ for all $g \in G$ and $u \in E$. For example if $G = \{id, -id\} \cong \mathbb{Z}_2$ where id denotes the identity map and I is even, I is invariant under G. The groups that have been studied thus far in applications are $\mathbb{Z}_2$ and S^1. To minimize technicalities, we will restrict ourselves to $\mathbb{Z}_2$ in our discussion. The earliest result we know of in this direction is due to Ljusternik and Schnirelman [15] who, unlike the cases given above, treated a constrained variational problem. However their work is in a sense the forerunner of all of the results we are discussing here. They proved in particular

Theorem 1.7. Suppose $g \in C^1(\mathbb{R}^n, \mathbb{R})$ and g is even. Then $g|_{S^{n-1}}$ possesses at least n distinct pairs of critical points.

Note that critical points of g occur in pairs since g is even. A special case of Theorem 1.7 that is familiar to all is $g(x) = (Ax, x)$ where $(.,.)$ denotes the usual inner product in $\mathbb{R}^n$ and A is a symmetric $n \times n$ matrix. Then the critical points correspond to eigenvectors of A and there are n linearly independent ones. A minimax characterization of the critical values of g will be given in §4. Infinite dimensional versions of Theorem 1.7 which give an infinite number of critical points of g restricted to a ball or more general manifold have been obtained by several authors: Ljusternik [16], Berger [17], Browder [18], Amann [19] and many others.

Perhaps the first results for unconstrained symmetric variational problems are due to D. Clark [20]. The following is a variant of his main result:

Theorem 1.8. Let E be a real Banach space and $I \in C^1(E, \mathbb{R})$ satisfying (PS). Suppose I satisfies

(I_5) I is even.

(I_6) I is bounded from below.

(I_7) $I(0) = 0$ and there exists $K \subset E$ with K homeomorphic to S^{j-1} by an odd homeomorphism.

Then I possesses at least j distinct pairs of critical points.

Again there is a minimax characterization of corresponding critical values [20, 4].

Next we will give a symmetric version of the Mountain Pass Theorem. This is also due to Ambrosetti and the author [5].

Theorem 1.9. Let E be a real infinite dimensional Banach space and $I \in C^1(E, \mathbb{R})$ satisfying (PS). Suppose I satisfies (I_1), (I_5), and

(I_8) For all finite dimensional subspaces $\tilde{E} \subset E$, there exists $R(\tilde{E}) > 0$ such that $I(u) \leq 0$ for $u \in \tilde{E} \backslash B_{R(\tilde{E})}$.

Then I possesses an unbounded sequence of critical values (c_m).

If E is n dimensional, we can simplify (I_8) and need not assume (PS) to conclude that I possesses at least n distinct pairs of critical points with corresponding positive critical values. As with our previous results, the critical values c_m possess a minimax characterization — see §4. Hypothesis (I_1) can be weakened as we shall also show in §4.

We will complete our survey with some bifurcation results. Consider the equation

$$Lu + H(u) = \lambda u \tag{1.10}$$

where L is a continuous linear map from E to E, $H \in C(E, E)$ with $H(u) = o(\|u\|)$ at $u = 0$, and $\lambda \in \mathbb{R}$. A *solution* of (1.10) is a pair $(\lambda, u) \in \mathbb{R} \times E$. Thus (1.10) possesses the family of trivial solutions $\{(\lambda, 0) \mid \lambda \in \mathbb{R}\}$. We say $(\mu, 0)$ is a *bifurcation point* for (1.10) if every neighborhood of $(\mu, 0)$ contains nontrivial solutions of (1.10). It is easy to see that a necessary condition for $(\mu, 0)$ to be a bifurcation point is $\mu \in \sigma(L)$, the spectrum of L. Simple examples show this necessary condition is not sufficient [4]. However when E is a Hilbert space and we have variational structure, i.e. $Lu + H(u) = I'(u)$, roughly speaking $\mu \in \sigma(L)$ is also sufficient for bifurcation to occur. More precisely:

<u>Theorem 1.11</u>. Let E be a real Hilbert space and $I \in C^2(E, \mathbb{R})$ where $I'(u) = Lu + H(u)$, L is linear, and $H(u) = o(\|u\|)$ at $u = 0$. If $\mu \in \sigma(L)$ is an isolated eigenvalue of finite multiplicity, then $(\mu, 0)$ is a bifurcation point. Moreover there is an $r_0 > 0$ such that for each $r \in (0, r_0)$, there are at least two distinct points $u(r) \in \partial B_r$ with corresponding $\lambda(r)$ such that $(\lambda(r), u(r))$ satisfy (1.10) and $(\lambda(r), \phi(r)) \to (\mu, 0)$ as $r \to 0$.

Theorem 1.11 is due independently to Böhme [21] and Marino [22]. (Actually Marino stated a less general result but uses essentially the same arguments as Böhme). Aside from smoothness considerations, Theorem 1.11 generalizes an earlier result of Krasnoselski [16]. If I is even, the resulting symmetry can be used to strengthen the conclusions of the theorem:

<u>Corollary 1.12</u>. [21-22] Let I be even and satisfy the hypotheses of Theorem 1.11. If μ is of multiplicity n, then for each $r \in (0, r_0)$, (1.10) has at least n distinct pairs of solutions $(\lambda_j(r), \pm u_j(r))$, $1 \le j \le n$ with $u_j(r) \in \partial B_r$ and $(\lambda_j(r), u_j(r)) \to$ $\to (\mu, 0)$ as $r \to 0$ for each j.

One can also study the set of solutions of (1.10) as a function of λ rather than $\|u\|$. Indeed the author has shown [23]:

<u>Theorem 1.13</u>. Under the hypotheses of Theorem 1.11, either

(i) $(\mu, 0)$ is not an isolated solution of (1.10) in $\{\mu\} \times E$, or

(ii) There is a one sided neighborhood Λ of μ in $\mathbb{R}$ such that for each $\lambda \in \Lambda\backslash\{\mu\}$, (1.10) has at least two nontrivial solutions, or

(iii) there is a neighborhood X of μ in $\mathbb{R}$ such that for each $\lambda \in X\backslash\{\mu\}$, (1.10) has at least one nontrivial solution.

Note that alternative (i) of Theorem 1.13 occurs e.g. if $H = 0$, i.e. (1.10) is a linear eigenvalue problem. Once again in the even case a stronger result due to Fadell and the author prevails [24]:

Corollary 1.14. Under the hypotheses of Corollary 1.12, either (i) holds or there exists integers $k,m \geq 0$ such that $k+m \geq n$ and left and right neighborhoods Λ_ℓ, Λ_r of μ such that for each $\lambda \in \Lambda_\ell$ (resp Λ_r), (1.10) possesses at least k (resp. m) distinct pairs of nontrivial solutions.

The proofs of Theorems 1.13 and Corollary 1.14 are based in part on ideas that go into the proof of the Mountain Pass Theorem and will be sketched in §5.

§2. The Mountain Pass Theorem

In this section we will prove the Mountain Pass Theorem. An important tool in its proof as well as all of the results stated in §1 is a so-called "Deformation Theorem" which we shall describe next. For $I \in C^1(E, \mathbb{R})$ and $c \in \mathbb{R}$, let

$$K_c = \left\{u \in E \mid I(u) = c \text{ and } I'(u) = 0\right\} \text{ and } A_c = \left\{u \in E \mid I(u) \leq c\right\}.$$

Thus K_c is the set of critical points of I having critical value c. Note that it is compact via (PS).

Theorem 2.1. Let E be a real Banach space, $I \in C^1(E, \mathbb{R})$, and satisfy (PS). For any $c \in \mathbb{R}$, $\bar{\varepsilon} > 0$, and neighborhood $\mathcal{O}$ of K_c, there exists $\varepsilon \in (0, \bar{\varepsilon})$ and $\eta \in C([0, 1] \times E, E)$ such that

1° $\eta(0, u) = u$ for all $u \in E$

2° $\eta(t, u) = u$ if $I(u) \notin [c-\bar{\varepsilon}, c+\bar{\varepsilon}]$ for each $t \in [0, 1]$

3° $\eta(1, A_{c+\varepsilon} \setminus \mathcal{O}) \subset A_{c-\varepsilon}$

4° If $K_c = \phi$, $\eta(1, A_{c+\varepsilon}) \subset A_{c-\varepsilon}$

5° If I is even, $\eta(t, u)$ is odd in u for each $t \in [0, 1]$.

The proof of Theorem 2.1 is not difficult but is lengthy. See e.g. [4] or [20]. We will indicate some of the main ideas behind the proof. Suppose e.g. $E = \mathbb{R}^n$ and $I \in C^2(\mathbb{R}^n, \mathbb{R})$. Consider the Cauchy problem

$$\frac{d\eta}{dt} = -I'(\eta), \quad \eta(0, u) = u, \tag{2.2}$$

i.e. we study the negative gradient flow corresponding to I. Since $I \in C^2$, (2.2) has a unique solution, $\eta(t, u)$, for each $u \in \mathbb{R}^n$, the solution existing for some time interval depending on u and satisfying 1° and 5° (modulo existence up to $t = 1$). Observe that

$$\frac{d}{dt} I(\eta(t, u)) = -|I'(\eta(t, u))|^2 \tag{2.3}$$

so $I(\eta(t, u))$ strictly decreases on solution orbits of (2.2) provided that $I'(u) \neq 0$. This fact suggests that we can satisfy 3° - 4° (again aside from problems with existence of the solution up to $t = 1$). Unfortunately 2° will not hold. If however we modify the right hand side of (2.2) by including a factor $\chi(u)$ where $\chi(u) = 0$ if $|I(u) - c| \geq \bar{\varepsilon}$ and $\chi(u) = 1$ if $|I(u) - c| \leq \delta$ for sufficiently small δ, we localize matters with respect to A_c and satisfy 2°. Finally by also rescaling $I'(u)$ away from $\mathcal{O}$ and using (PS) we can construct an appropriately modified right hand side of (2.2) for which 1° - 5° are verified.

The same ideas are used in the general case of a Banach space E. However two additional technical problems must be confronted:

(i) $\frac{d\eta}{dt} \in E$ while $I'(u) \in E^*$ so (2.2) doesn't make sense; even if it did, (ii) $I \in C^1$ so I' is merely continuous and there need not exist a solution of (2.2) let alone a unique one. Both of these difficulties can be overcome with the aid of the notion of a pseudogradient vector field due to Palais [τ]. See e.g. [3, 4, 18, 20].

Remark 2.4. The above ideas carry over to cases where E is not a linear space but a manifold such as e.g. in Theorem 1.9 and its generalizations. See [3, 18].

With the aid of Theorem 2.1, the proof of the Mountain Pass Theorem is immediate.

Proof of Theorem 1.3. Suppose c as defined in (1.4) is not a critical value of I so $K_c = \emptyset$. Let $\bar{\varepsilon} = \frac{\alpha}{2}$ and invoke Theorem 2.1 obtaining $\varepsilon \in (0, \bar{\varepsilon})$ and η. Choose $g \in \Gamma$ such that

$$\max_{t \in [0,1]} I(g(t)) \leq c + \varepsilon. \tag{2.5}$$

Consider $\eta(1, g(t))$. By (2.5) and 4° of Theorem 2.1,

$$\max_{t \in [0,1]} I(\eta(1, g(t)) \leq c - \varepsilon. \tag{2.6}$$

On the other hand, $\eta(1, g(t)) \in C([0, 1], E)$. Moreover by 2° of Theorem 2.1, $\eta(1, g(0)) = \eta(1, 0) = 0$ and $\eta(1, g(1)) = \eta(1, e) = e$ since $I(0) = 0$, $I(e) \leq 0$, and $c - \bar{\varepsilon} \geq \frac{\alpha}{2} > 0$. Therefore $\eta(1, g(t)) \in \Gamma$ so

$$\max_{t \in [0,1]} I(\eta(1, g(t))) \geq c, \tag{2.7}$$

contrary to (2.6). Thus c is a critical value of I and the proof is complete.

Remark 2.8. Suppose $I \in C^1(E, \mathbb{R})$ and satisfies (PS). Let η be the mapping given by Theorem 2.1 (so η depends on c, $\bar{\varepsilon}$, and $\mathcal{O}$).

Let S be a family of subsets of E and set

$$c = \inf_{K \in S} \sup_{u \in K} I(u). \tag{2.9}$$

Assume c is finite and with e.g. $\bar{\varepsilon} = 1$ and $O = \emptyset$, $\eta(1, \cdot): S \to S$. Then c is a critical value of I. Indeed the same proof as in Theorem 1.3 works here. Of course in practice the main difficulties are to find S and to verify (PS). There are even further pitfalls, e.g. I may have some known critical values (e.g. possibly a local minimum at $u=0$ as in Theorem 1.3) and one then has to avoid them to get anything new.

Remark 2.10. For a simple example of a situation where Remark 2.8 applies, suppose $I \in C^1(E, \mathbb{R})$, satisfies (PS), and is bounded from below. Then taking $S = \{\{x\} \mid x \in E\}$, we see c as defined by (2.9) is simply

$$c = \inf_{\{x\} \in S} \sup_{u \in \{x\}} I(u) = \inf_{E} I$$

and c is a critical value of I.

Remark 2.11. For some variations on and generalizations of Theorem 1.3, see e.g. Nirenberg [2], Benci-Rabinowitz [7] and Hofer [14].

§3. Some Applications

In this section we will sketch some applications of Theorem 1.3 to semilinear elliptic-boundary value problems. Let $\Omega \subset \mathbb{R}^n$ be a bounded domain with a smooth boundary and consider the boundary value problem:

$$-\Delta u = p(x, u), \quad x \in \Omega; \; u = 0, \quad x \in \partial\Omega. \tag{3.1}$$

We assume:

(p_1) p is locally Lipschitz continuous in $\bar{\Omega} \times \mathbb{R}$,

(p_2) there are constants $a_1, a_2 \geq 0$ and $s < \frac{n+2}{n-2}$ such that

$$|p(x, \xi)| \leq a_1 + a_2|\xi|^s$$

for all $(x, \xi) \in \bar{\Omega} \times \mathbb{R}$.

If $n = 1$ or 2, (p_2) can be weakened. Here and for the remainder of this section, let $E = W_0^{1,2}(\Omega)$ where we employ the usual Sobolev space notation. As norm in E we can take

$$\|u\| = \left(\int_\Omega |\nabla u|^2 dx\right)^{\frac{1}{2}}.$$

For $u \in E$, set

$$I(u) \equiv \frac{1}{2}\|u\|^2 - \int_\Omega P(x, u)dx \tag{3.2}$$

where

$$P(x, \xi) = \int_0^\xi p(x, t)dt.$$

Then (p_1)-(p_2) imply $I \in C^1(E, \mathbb{R})$ (see [4]). Moreover for $u, \phi \in E$,

$$I'(u)\phi = \int_\Omega [\nabla u \cdot \nabla\phi - p(x, u)\phi]dx \tag{3.3}$$

so critical points of I in E are weak solutions of (3.1). Hypotheses (p_1)-(p_2) and standard results from elliptic regularity theory further imply weak solutions of (3.1) are classical solutions. Therefore in what follows we will mainly focus on the existence of critical points of I. Note that under the canonical identification of E and E^*, $u \in E$ corresponds to the map

$$\phi \to \int_\Omega \nabla u \cdot \nabla\phi dx \in E^*.$$

Similarly using (p_1), (p_2), the second term on the right hand side of (3.3) is a compact operator $\mathcal{P}(u)$ on E [4, 5]. Thus we can write (3.3) in the equivalent form

$$I'(u) = u - \mathcal{P}(u). \tag{3.4}$$

Now further suppose p satisfies

$$p(x, \xi) = o(|\xi|) \text{ at } \xi = 0, \tag{p_3}$$

(p_4) there are constants $\mu > 2$ and $\bar{r} > 0$ such that

$$0 < \mu P(x, \xi) \leq \xi p(x, \xi)$$

for $|\xi| > \bar{r}$.

By (p_3), $P(x, \xi) = o(|\xi|^2)$ at $\xi = o$ and this fact combined with (p_2) and the Sobolev inequality imply

(3.5) $$\int_\Omega P(x, u)dx = o(\|u\|^2) \text{ at } u = o.$$

Thus (3.5) and the form of I in (3.2) imply I satisfies (I_1). Integrating the differential inequality (p_4) shows there are constants $a_3, a_4 \geq 0$ such that

(3.6) $$P(x, \xi) \geq a_3|\xi|^\mu - a_4$$

for all $(x, \xi) \in \bar{\Omega} \times \mathbb{R}$. Since $\mu > 2$, (3.2) and (3.6) show $I(\beta u) \to -\infty$ as $\beta \to \infty$ for any $u \in E\backslash\{0\}$. Hence I satisfies (I_2). For future reference note that in a similar fashion for any finite dimensional subspace $\tilde{E}$ of E, $I(u) \to -\infty$ as $u \to \infty$ uniformly for $u \in \tilde{E}\backslash\{0\}$ so I further satisfies (I_8). Lastly we claim I satisfies (PS). Once that fact has been established, all of the conditions of Theorem 1.3 will have been verified and we will have proved.

<u>Theorem 3.7</u>. If p satisfies (p_1)-(p_4), (3.1) possesses a nontrivial (classical) solution.

To check (PS), suppose $(u_m) \subset E$ with $I(u_m)$ bounded, say $|I(u_m)| \leq M$ and $I'(u_m) \to 0$. Therefore for large m,

(3.8) $$|I'(u_m)\phi| \leq \|\phi\| \quad \text{for all } \phi \in E.$$

Choosing $\phi = u_m$ and dropping subscripts, we have

(3.9) $$\begin{aligned} M + \mu^{-1}\|u\| &\geq |I(u) - \mu^{-1}I'(u)u| \geq \\ &\geq (2^{-1}-\mu^{-1})\|u\|^2 + \int_\Omega (\mu^{-1}p(x, u)u-P(x, u))dx \geq \\ &\geq a_5\|u\|^2 - a_6 \end{aligned}$$

via (p_4) for some constants a_5, a_6 independent of m. It follows from (3.9) that (u_m) is bounded in E. But then $I'(u_m) \to 0$, (3.4), and the compactness of P implies that (u_m) has a convergent subsequence. Hence (PS) is satisfied and Theorem 3.7 is proved.

Remark 3.10. Theorem 3.5 is due to Ambrosetti and the author [5]. The Maximum Principle and a slight modification of the above argument can be used to further prove (3.1) possesses both a positive and a negative solution [5].

For our second application, consider the nonlinear eigenvalue problem:

$$(3.11) \qquad -\Delta u = \lambda e^u, \quad x \in \Omega \subset \mathbb{R}^2; \quad u = 0, \quad x \in \partial\Omega.$$

More generally we could replace e^u by $f(x, u)$ where $f(x, 0) > 0$ in $\bar{\Omega} \subset \mathbb{R}^n$ and f is monotone increasing and strictly convex in u. Let

$$F(\lambda, u) \equiv -\Delta u - \lambda e^u$$

where $\lambda > 0$ and $u \in C_0^{2,\alpha}(\Omega)$ for some fixed $\alpha \in (0, 1)$. ($C_0^{2,\alpha}(\Omega)$ is the Banach space of functions which vanish on $\partial\Omega$, are twice continuously differentiable in $\bar{\Omega}$, and whose second derivatives are Hölder continuous with exponent α). Then $F: \mathbb{R}\times C_0^{2,\alpha}(\Omega) \to C^\alpha(\Omega)$, F is continuously differentiable, $F(0, 0) = 0$ and

$$(3.12) \qquad F_u(0, 0)v = -\Delta v$$

where F_u denotes the Frechet derivative of F with respect to u. By the Schauder theory of linear elliptic boundary value problems, (3.12) is an isomorphism of $C_0^{2,\alpha}(\Omega)$ to $C^\alpha(\Omega)$. Hence by the Implicit Function Theorem F possesses a curve of zeroes $(\lambda, u_\lambda(x))$ for λ near 0 which passes through $(0, 0)$. A continuation argument allows us to extend this curve as parametrized by λ to a maximal interval $[0, \bar{\lambda})$ in which $F_u(\lambda, u_\lambda)$ is an isomorphism between $C_0^{2,\alpha}(\Omega)$ and $C^\alpha(\Omega)$. The convexity of e^u forces $\bar{\lambda}$ to be finite [25]. Moreover the Maximum Principle implies any solution of (3.11) for $\lambda > 0$ is positive in Ω and in particular $u_\lambda(x) > 0$ in Ω for $\lambda \in (0, \bar{\lambda})$. (Alternatively we could have obtained u_λ for $\lambda \in (0, \bar{\lambda})$ as the minimal positive solution of (3.11) via a monotone iteration scheme).

<u>Theorem 3.13</u> [25]: For each $\lambda \in (0, \bar{\lambda})$, (3.11) possesses a second positive solution $\bar{u}_\lambda(x)$ with $\bar{u}_\lambda(x) > u_\lambda(x) > 0$ for $x \in \Omega$.

We will show how the existence of $\bar{u}_\lambda$ follows from the Mountain Pass Theorem. First a change of coordinates is required. Define

$$p(x, \xi) = e^{\xi + u_\lambda} - e^{u_\lambda} \quad \text{if} \quad \xi \geq 0; \quad = 0 \quad \text{if} \quad \xi < 0.$$

Let $v = u - u_\lambda$ where u is a solution of (3.11) with $u \geq u_\lambda$ so v satisfies

$$-\Delta v = \lambda p(x, v) \text{ in } \Omega; \quad v = 0, \ x \in \partial\Omega. \tag{3.14}$$

The functional corresponding to (3.14) is

$$I(v) = \int_\Omega \left[\frac{1}{2} |\nabla v|^2 - \lambda P(x, v)\right] dx \tag{3.15}$$

where P is the primitive of p. With the aid of the Sobolev Embedding Theorem for $n = 2$ we have $I \in C^1(E, \mathbb{R})$ and I satisfies (PS) since p satisfies (p_4). Actually p only satisfies (p_4) for $\xi > 0$ while for $\xi < 0$, the strict inequality is lost. Thus (3.6) does not hold for negative ξ but the weaker inequality suffices for us to verify (PS) and also (I_2) is satisfied by choosing e to be a sufficiently large multiple of any positive function in Ω. Lastly (I_1) follows since for $\lambda \in (0, \bar{\lambda})$, $\sigma(F_u(\lambda, u_\lambda)) \subset \mathbb{R}^+$ which implies $v = 0$ is a strict local minimum for I. Thus Theorem 1.3 yields a weak solution of (3.14) and regularity arguments and the Maximum Principle then imply v is smooth and $v > 0$ in Ω. Hence $\bar{u}_\lambda = v + u_\lambda > u_\lambda$ in Ω and our sketch of the proof of Theorem 3.13 is complete. For more details see Crandall-Rabinowitz [25].

As our final application, consider

(3.16) $$-\Delta u = \lambda p(u), \quad x \in \Omega; \quad u = 0, \quad x \in \partial\Omega$$

where Ω is a bounded domain in $\mathbb{R}^n$ with a smooth boundary. We assume p satisfies (p_1), (p_3), and

(p_5) there is a $\bar{\xi} > 0$ such that $p(\xi) > 0$ in $(0, \bar{\xi})$.

Note that $u \equiv 0$ is then a trivial solution of (3.16). Define $\bar{p}(\xi) = p(\xi)$ for $\xi \in (0, \bar{\xi})$ and $\bar{p}(\xi) = 0$ otherwise. Easy applications of the Maximum Principle show if u is a solution of

(3.17) $$-\Delta u = \lambda \bar{p}(u), \quad x \in \Omega; \quad u = 0, \quad x \in \partial\Omega,$$

then

$$\left\{x \in \Omega \mid u(x) > \bar{\xi} \text{ or } u(x) < 0\right\} = \phi.$$

Thus $0 \le u(x) \le \bar{\xi}$ and u satisfies (3.16). More careful use of the Maximum Principle proves $u > 0$ in Ω if $u \not\equiv 0$. Consequently to find solutions of (3.16), it suffices to obtain critical points of

(3.18) $$I_\lambda(u) = \int_\Omega \left[\frac{1}{2} |\nabla u|^2 - \lambda \bar{P}(x, u)\right] dx,$$

$\bar{P}$ being the primitive of $\bar{p}$. Since $\bar{p}$ satisfies (p_1) - (p_2), $I_\lambda \in C^1(E, \mathbb{R})$ as earlier and (p_3) implies I_λ satisfies (I_1). To verify (PS), note that for some constants K, K_1 we have:

(3.19) $$0 \le \int_\Omega P(u) dx \le K \int_\Omega |u| dx \le K_1 \|u\|$$

for all $u \in E$ via the definition of $\bar{p}$ and the Schwarz and Poincaré inequalities. Thus if $(u_m) \subset E$ and $I_\lambda(u_m)$ is bounded, then (3.18) - (3.19) imply that (u_m) is bounded. If further $I'_\lambda(u_m) \to 0$ as $m \to \infty$, then (3.4) shows (u_m) has a convergent subsequence and (PS) holds.

To apply the Mountain Pass Theorem, it remains to verify (I_2). But in fact this condition need not hold and $u \equiv 0$ may be the only solution of (3.17) unless we impose some restrictions on λ. Indeed the bounds we have for u $(0 \le u(x) \le \bar{\xi})$, (p_3), and the Schauder estimates for linear elliptic equations show $u \equiv 0$ for λ near 0. However if e.g. $v \in E\setminus\{0\}$ with support in $(0, \bar{\xi})$, (3.18) shows $I_\lambda(v) < 0$ for all large $\lambda > 0$. Hence by Remark 2.10, for such λ the functional I_λ has a negative minimum at e.g. $\underline{u}_\lambda$. The corresponding solution of (3.16) is not $u \equiv 0$ since $I_\lambda(\underline{u}_\lambda) < 0 = I_\lambda(0)$. Consequently $\underline{u}_\lambda > 0$ in Ω by above remarks. Moreover we can now satisfy (I_2) with $e = \underline{u}_\lambda$. Hence by Theorem 1.3 and our above remarks there is a second positive solution $\bar{u}_\lambda$ of (3.16) with $I_\lambda(\bar{u}_\lambda) > 0 > I_\lambda(\underline{u}_\lambda)$ so $\bar{u}_\lambda \neq \underline{u}_\lambda$. Thus we have shown

Theorem 3.20 [5]: Let p satisfy (p_1), (p_3), and (p_5). Then for all large $\lambda > 0$, (3.16) possesses at least two distinct positive solutions.

§4. Index Theories and the Symmetric Mountain Pass Theorem

When dealing with a functional, I, which is invariant under a group of symmetries, one can often exploit these symmetries to obtain multiple critical points of I. In this section we will indicate some of the main ideas that go into the proofs of such results, give a version of the Mountain Pass Theorem for even functionals, and apply it to (3.1).

To begin, we will discuss index theories. For simplicity, we will only consider the important special case of a $\mathbb{Z}_2$ action. For more general index theories, see [26-28]. Let E be a real Banach space and $\mathcal{E}$ the family of closed subsets K of $E\setminus\{0\}$ which are symmetric with respect to the origin, i.e. $x \in K$ implies $-x \in K$. An *index theory* for $\mathcal{E}$ is a mapping $\text{Ind}: \mathcal{E} \to \mathbb{N} \cup \{\infty\}$ which possesses the following properties:

(4.1) 1° (*Normalization*) If $x \in E\setminus\{0\}$, $\text{Ind}\{x, -x\} = 1$

2^0 (*Mapping Property*): If $A, B \in E$ and $\phi \in C(A, B)$ is odd, $\text{Ind } A \leq \text{Ind } B$.

3^0 (*Subadditivity*): If $A, B \in E$, $\text{Ind } A \cup B \leq \text{Ind } A + \text{Ind } B$.

4^0 (*Continuity Property*): If $K \in E$ is compact and $N_\delta(K) \equiv \{x \in E \mid \|x-K\| \leq \delta\}$, there is a $\delta > 0$ such that $\text{Ind } N_\delta(K) = \text{Ind } K$.

We will deduce a few easy consequences of $1^0 - 4^0$ which will be useful in the sequel. We always assume $A, B \in E$.

5^0 If $A \subset B$, $\text{Ind } A \leq \text{Ind } B$.

6^0 If $\phi \in C(A, E)$ is an odd homeomorphism, $\text{Ind } \phi(A) = \text{Ind } A$.

7^0 If $\text{Ind } B < \infty$, $\text{Ind } \overline{A \backslash B} \geq \text{Ind } A - \text{Ind } B$

8^0 If A is compact, $\text{Ind } A < \infty$.

Proofs. Taking $\phi = \text{id}$ in 2^0 gives 5^0; taking $B = \phi(A)$ in 2^0 and also interchanging the roles of A and B yields 6^0; and 7^0 follows from 5^0, 3^0 and the fact that $A \subset \overline{A \backslash B} \cup B$. To get 8^0, note that by 1^0 and 4^0, each point $x \in E \backslash \{0\}$ lies in a closed ball A such that $\text{Ind}[A \cup (-A)] = 1$. By compactness, finitely many such balls cover K so by 3^0, $\text{Ind } K < \infty$.

The earliest index theory we know of is due to Ljusternik and Schnirelman [29]. It is valid in a much more general setting than the above one. Other such index theories have been introduced by Yang [24], Conner and Floyd [30], Krasnoselski [16] and others — see [26]. Perhaps the simplest one and easiest to work with for our purposes is the notion of *genus*. If $A \in E$, we say A has genus n, denoted by $\gamma(A) = n$, if there is an $f \in C(A, \mathbb{R}^n \backslash \{0\})$ with f odd and n the smallest integer with this property.

Let us verify that $\gamma: E \to \mathbb{N} \cup \{\infty\}$ satisfies $1^0 - 4^0$ of (4.1). It is clear that 1^0 is satisfied. For 2^0, suppose that $\gamma(B) < \infty$ or there is nothing to prove. If $\gamma(B) = n$, there is an $f \in C(B, \mathbb{R}^n\setminus\{0\})$ with f odd. Consequently $f\circ\phi \in C(A, \mathbb{R}^n\setminus\{0\})$ and is odd so $\gamma(A) \leq n = \gamma(B)$. To get 3°, let $\gamma(A) = m$, $\gamma(B) = n$, and again $\gamma(A)$, $\gamma(B) < \infty$. Therefore there are functions $\phi \in C(A, \mathbb{R}^m\setminus\{0\})$, $\psi \in C(B, \mathbb{R}^n\setminus\{0\})$ with ϕ, ψ odd. By the Tietze Extension Theorem, ϕ and ψ can be continuously extended to $\hat{\phi} \in C(E, \mathbb{R}^m)$, $\hat{\psi} \in C(E, \mathbb{R}^n)$. Replacing $\hat{\phi}$ and $\hat{\psi}$ by their odd parts, it can be assumed that $\hat{\phi}$, $\hat{\psi}$ are odd. If $f \equiv (\hat{\phi}, \hat{\psi})$, $f \in C(A \cup B, \mathbb{R}^{m+n}\setminus\{0\})$ and is odd. Hence $\gamma(A \cup B) \leq m+n = \gamma(A) + \gamma(B)$ and 3° holds. Lastly if K is compact and $\gamma(K) = j$, there is a $\phi \in C(A, \mathbb{R}^j\setminus\{0\})$ with ϕ odd. As with 3°, ϕ has an odd extension $\hat{\phi} \in C(E, R^j)$. Since K is compact and $\phi \neq 0$ on K, $\hat{\phi} \neq 0$ on $N_\delta(K)$ for all small $\delta > 0$. Thus γ provides us with an index theory.

For later applications one needs to know the genus of finite dimensional spheres in E. The following proposition carries out this calculation.

Proposition 4.2. Let Ω be a symmetric bounded neighborhood of 0 in $\mathbb{R}^k$ and let $A \in E$ be homeomorphic to $\partial\Omega$ by an odd homeomorphism, h. Then $\gamma(A) = k$.

Proof. Clearly $\gamma(A) \leq k$. If $\gamma(A) = j < k$, there is a $\phi \in C(A, \mathbb{R}^j\setminus\{0\})$ with ϕ odd. Therefore $\phi\circ h^{-1} \in C(\partial\Omega, \mathbb{R}^j\setminus\{0\})$ and is odd. Since $\Omega \subset \mathbb{R}^k$ and $j < k$, one of the versions of the Borsuk-Ulam Theorem [29] implies $\phi \circ h^{-1}$ must vanish somewhere on $\partial\Omega$, a contradiction.

Corollary 4.3. If V is a closed subspace of E of co-dimension j and $A \in E$ with $\gamma(A) > j$, then $A \cap V \neq \emptyset$.

Proof. If $A \cap V = \emptyset$ and P denotes the projector of E onto $V^\perp$, a complementary subspace to V, then $P \in C(A, V^\perp\setminus\{0\})$ and is odd. By 2° of (4.1), $\gamma(A) \leq \gamma(P(A))$. Since $P(A) \subset V^\perp\setminus\{0\}$, radial projection of $P(A)$ to $\partial B_1 \cap V^\perp$ is a continuous odd map. Therefore by 2° of (4.1) again, $\gamma(A) \le \gamma(\partial B_1 \cap V^\perp)$. Since $V^\perp$ is j dimensional, $\partial B_1 \cap V^\perp$ is homeomorphic to S^{j-1} in $\mathbb{R}^j$ by an odd map. Hence by Proposition 4.2 $\gamma(A) \le j$, contrary to hypothesis. Thus $A \cap V \neq \phi$.

As an exercise in using an index theory to get multiple critical points of a functional, we will prove the result of Ljusternik and Schnirelman (Theorem 1.7) mentioned in §1.

Theorem 1.7. If $g \in C^1(\mathbb{R}^n, \mathbb{R})$ and is even $g|_{S^{n-1}}$ possesses at least n distinct pairs of critical points.

Proof. We will define n families of sets Γ_j, $1 \le j \le n$, and obtain n critical points by minimaxing g over these families. Set

$$\Gamma_j = \left\{ K \subset S^{n-1} \mid \gamma(K) \ge j \right\}, \quad 1 \le j \le n. \tag{4.4}$$

The sets Γ_j possess the following properties:

(4.5) (g_1) $\Gamma_j \neq \phi$, $1 \le j \le n$.

(g_2) (*Monotonicity*): $\Gamma_{j+1} \subset \Gamma_j$.

(g_3) (*Excision*): If $B \in \Gamma_j$ and $Y \in E$ with $\gamma(Y) \le s < j$, then $\overline{B \setminus Y} \in \Gamma_{j-s}$.

(g_4) (*Invariance*): If $f \in C(S^{n-1}, S^{n-1})$ is odd, then $f: \Gamma_j \to \Gamma_j$, i.e. $B \in \Gamma_j$ implies $f(B) \in \Gamma_j$.

Indeed (g_1) is a consequence of Proposition 4.2, (g_2) is trivial, (g_3) follows from 7° of (4.1), and (g_4) from 2° of (4.1). Now define

$$c_j = \inf_{B \in \Gamma_j} \max_{x \in B} g(x) \qquad 1 \le j \le n. \tag{4.6}$$

We will show c_j is a critical value of $g|_{S^{n-1}}$, $1 \le j \le n$. This is not sufficient to prove Theorem 1.7 because some of the c_j's may coincide resulting in fewer than n pairs of critical points of

$g|_{S^{n-1}}$. Consequently to complete the proof, it suffices to prove the stronger assertion that if

$$K_c = \left\{ x \in S^{n-1} \mid g(x) = c \text{ and } g|'_{S^{n-1}}(x) = 0 \right\},$$

(4.7) $$c_j = \dots = c_{j+k-1} \equiv c \quad \text{implies} \quad \gamma(K_c) \geq k.$$

If $k > 1$ in (4.7), i.e. if c is a *degenerate critical value*, K_c contains infinitely many distinct critical points since the definition of genus shows $\gamma(A) = 1$ if $A \in E$ consists of finitely many points. Now to prove (4.7), suppose $\gamma(K_c) < k$. By 4^0 of (4.1), there is a $\delta > 0$ such that $\gamma(N_\delta(K_c)) = \gamma(K_c)$. Note that $N_\delta(K_c) \not\subset S^{n-1}$. Let $O = N_\delta(K_c) \cap S^{n-1}$. Then $K_c \subset O \subset N_\delta(K_c)$ so by 5^0 of (4.1), $\gamma(O) < k$. By Remark 2.4 and Theorem 2.1 with e.g. $\bar{\varepsilon} = 1$, there is an $\varepsilon \in (0, \bar{\varepsilon})$ and $\eta \in C([0, 1] \times S^{n-1}, S^{n-1})$ such that

(4.8) $$\eta(1, A_{c+\varepsilon} \backslash O) \subset A_{c-\varepsilon}$$

where $A_s = \{x \in S^{n-1} | g(x) \leq s\}$. Choose $B \in \Gamma_{j+k-1}$ such that

(4.9) $$\max_{x \in B} g(x) \leq c + \varepsilon$$

i.e. $B \subset A_{c+\varepsilon}$. Therefore by (4.8) - (4.9),

(4.10) $$\max_{x \in \eta(1, \overline{B \backslash O})} g(x) \leq c - \varepsilon.$$

But by (g_3) of (4.5), $\overline{B \backslash O} \in \Gamma_j$ and by (g_4) of (4.5), $\eta(1, \overline{B \backslash O}) \in \Gamma_j$. Consequently the definition of $c = c_j$ shows

(4.11) $$\max_{x \in \eta(1, \overline{B \backslash O})} g(x) \geq c$$

contrary to (4.1). The proof is complete.

<u>Remark 4.12</u>. Note that $c_1 \leq c_2 \leq \dots \leq c_n$ via (g_2) of (4.5). Clearly c_1 is the minimum of g over S^{n-1} and it is not difficult to show

that c_n is the maximum of g over S^{n-1} [4].

Next we will show how variants of the ideas used in the above proof yield Theorem 1.9, a symmetric version of the Mountain Pass Theorem. For convenience we restate it.

<u>Theorem 1.9</u>. Let E be a real infinite dimensional Banach space and $I \in C^1(E, R)$ satisfying (PS). If I satisfies:

(I_1) $I(0) = 0$ and there are constants $\rho, \alpha > 0$ such that $I|_{\partial B_\rho} \geq \alpha$.

(I_5) I is even.

(I_8) For all finite dimensional subspaces $\tilde{E} \subset E$, there exists $R(\tilde{E})$ such that $I(u) \leq 0$ for $u \in \tilde{E} \setminus B_{R(\tilde{E})}$,

then I possesses an unbounded sequence of critical values.

<u>Proof</u>. We will produce a sequence of sets Γ_j which satisfy (g_1)-(g_3) of (4.5) as well as a modified form of (g_4). These properties imply the numbers c_j as defined in (4.6) are critical values of I for each $j \in \mathbb{N}$ and (c_j) must be unbounded.

To begin we choose any sequence of unit vectors $(e_m) \subset E$ such that $e_{m+1} \notin \text{span}\{e_1, \ldots, e_m\} \equiv E_m$. Let $R_m \equiv R(E_m)$, $D_m \equiv \bar{B}_{R_m} \cap E_m$, and

$$G_m = \left\{h \in C(D_m, E) \mid h \text{ is odd and } h = id \text{ on } \partial B_{R_m} \cap E_m\right\}.$$

Since $id \in G_m$, $G_m \neq \emptyset$. Finally set

$$\Gamma_j = \left\{h(\overline{D_m \setminus Y}) \mid m \geq j,\ h \in G_m,\ Y \in \mathcal{E}, \text{ and } \gamma(Y) \leq m - j\right\}.$$

Note that $h(\overline{D_m \setminus Y}) = \overline{h(D_m \setminus Y)}$ and is compact. Clearly $\Gamma_j \neq \emptyset$ for all $j \in \mathbb{N}$ so (g_1) of (4.5) holds. Likewise (g_2) of (4.5) is easy to verify while (g_3) and

$(\bar{g}_4)$ If $\phi \in C(E, E)$ is odd and $\phi = id$ on $\partial B_{R_m} \cap E_m$ for all $m \in \mathbb{N}$, then $\phi: \Gamma_j \to \Gamma_j$ for each $j \in \mathbb{N}$, follow in a straightforward fashion.

Now we define c_j via (4.6) for each $j \in \mathbb{N}$. Certainly $c_j < \infty$ since each $B \in \Gamma_j$ is compact. We claim

$$c_1 \geq \alpha > 0. \tag{4.13}$$

Assuming (4.13) for now, it follows that c_j is a critical value of I for each $j \in \mathbb{N}$ and the stronger multiplicity assertion

$$c_j = \dots = c_{j+\ell-1} \equiv c \quad \text{implies} \quad \gamma(K_c) \geq k \tag{4.14}$$

holds. Indeed to prove (4.14), note first that $c = c_j \geq c_1 \geq \alpha$ by (g_2) of (4.5) and (4.6) and therefore $0 \notin K_c$ since $I(0) = 0$. Hence $K_c \in E$ and $\gamma(K_c) < \infty$ via (PS) and 8° of (4.1). Now repeating an argument from the proof of Theorem 1.7, if $\gamma(K_c) < k$, there is a $\delta > 0$ such that $\gamma(N_\delta(K_c)) < k$ by 4° of (4.1). Theorem 2.1 with $\bar{\varepsilon} = \frac{\alpha}{2}$ and $\mathcal{O} = N_\delta(K_c)$ provides an $\varepsilon \in (0, \bar{\varepsilon})$ and $\eta \in C([0,1] \times E, E)$ such that

$$\eta(1, A_{c+\varepsilon} \backslash \mathcal{O}) \subset A_{c-\varepsilon}. \tag{4.15}$$

Choose $B \in \Gamma_{j+k-1}$ such that

$$\max_B I \leq c + \varepsilon. \tag{4.16}$$

Therefore by (4.15)

$$\max_{\eta(1, \overline{B \backslash \mathcal{O}})} I \leq c - \varepsilon. \tag{4.17}$$

But by (g_3), $\overline{B \backslash \mathcal{O}} \in \Gamma_j$ and by our choice of $\bar{\varepsilon}$ and 2° and 6° of Theorem 2.1, $\eta(1, \cdot) \in G_m$ for all $m \in \mathbb{N}$. Consequently $\eta(1, \overline{B \backslash \mathcal{O}}) \in \Gamma_j$ by $(\overline{g_4})$ and the definition of c_j then gives

$$\max_{\eta(1, \overline{B \backslash \mathcal{O}})} I \geq c \tag{4.18}$$

contrary to (4.17). Thus (4.14) is valid.

To verify (4.13), let $B \in \gamma_1$. Therefore $B = h(\overline{D_m \backslash Y})$ where $Y \in E$ and $\gamma(Y) \leq m-1$. Since $I \geq \alpha$ on ∂B_ρ by (I_1) and $I(u) \leq 0$ outside D_m in E_m via (I_8), $\rho < R_m$. Let $\hat{\Omega} = \{x \in D_m \mid h(x) \in B_\rho\}$

and let Ω denote the component of $\hat{\Omega}$ containing 0. Since h is odd and $h = \mathrm{id}$ on $\partial B_{R_m} \cap E_m$, Ω is a symmetric bounded open neighborhood of 0 in E_m so making the obvious identification between E and $\mathbb{R}^m$, we have $\gamma(\partial\Omega) = m$ via Proposition 4.2. Moreover if $x \in \partial\Omega$, $h(x) \in \partial B_\rho$. Hence if $W = \{x \in D_m \mid h(x) \in \partial B_\rho\}$, $\gamma(W) \geq \gamma(\partial\Omega) = m$ via 5^0 of (4.1) and $\gamma(\overline{W\setminus Y}) \geq m-(m-1) = 1$ by 7^0 of (4.1). Thus $\overline{W\setminus Y} \neq \emptyset$ and there exists $\xi \in W$ such that $h(\xi) \in B \cap \partial B_\rho$. Consequently by (I_1),

$$\max_B I \geq I(h(\xi)) \geq \inf_{\partial B_\rho} I \geq \alpha. \tag{4.19}$$

Since (4.19) is valid for all $B \in \Gamma_1$, $c_1 \geq \alpha$.

To complete the proof of Theorem 1.9, we must show $c_j \to \infty$ as $j \to \infty$. If not, since c_j is a monotone nondecreasing sequence by (g_2) of (4.5), there is a $\bar{c} < \infty$ such that $c_j \to \bar{c}$ as $j \to \infty$. Note that $\bar{c} > c_j$ for all j for otherwise $\gamma(K_{\bar{c}}) = \infty$ via (4.14). However $K_{\bar{c}} \in E$ and is compact so $\gamma(K_{\bar{c}}) < \infty$. Let

$$H = \left\{u \in E \mid c_1 \leq I(u) \leq \bar{c} \text{ and } I'(u) = 0\right\}.$$

By (PS), H is compact and by (I_5) and (4.19), $H \in E$. Again applying 4^0 and 8^0 of (4.1), there is a $\delta > 0$ such that $\gamma(N_\delta(H)) = \gamma(H) < \infty$. Suppose $\gamma(H) = n$. Invoking the Deformation Theorem with $\bar{\varepsilon} = \bar{c} - c_1$, $c = \bar{c}$, and $O = N_\delta(H)$, there is an $\varepsilon \in (0, \bar{\varepsilon})$ and $\eta \in C([0,1]\times E, E)$ with $\eta(1, u)$ odd such that

$$\eta(1, \overline{A_{\bar{c}+\varepsilon}\setminus O}) \subset A_{\bar{c}-\varepsilon}. \tag{4.20}$$

Choose the smallest value of m such that

$$c_m > \bar{c} - \varepsilon. \tag{4.21}$$

(Since $\bar{c} - \varepsilon > \bar{c} - (\bar{c} - a) = c_1$, $m > 1$). Choose $B \in \Gamma_{n+m}$ such that

$$\max_B I \leq \bar{c} + \varepsilon. \tag{4.22}$$

Note that $\eta(1, u) = u$ for $u \in \partial B_{R_n} \cap E_k$ for all $k \in \mathbb{N}$ via the definition of R_k, our choice of $\bar{E}$, and 2° and 6° of Theorem 2.1. Hence by (g_3) and $(\bar{g}_4)$, $\overline{B\setminus\mathcal{O}}$ and $\eta(1, \overline{B\setminus\mathcal{O}}) \in \Gamma_m$. But then by (4.20)-(4.22),

$$c_m \le \max_{\eta(1, \overline{B\setminus\mathcal{O}})} I \le \bar{c} - \bar{\varepsilon} < c_m$$

which is impossible. The proof is complete.

The next result shows how we can weaken the hypotheses of Theorem 1.9 somewhat. We will pay a price, however, since c_j as defined by (4.6) will no longer be a critical value of I unless j is sufficiently large. Let $E_j^{\perp}$ be a complementary subspace to E_j.

<u>Theorem 4.23</u>. Suppose E is a real Banach, $I \in C^1(E, \mathbb{R})$ and satisfies (PS). If I satisfies (I_5), (I_8), and

(I_9) there are constants $\rho, \alpha > 0$ and $n \in \mathbb{N}$ such that

$$I\Big|_{\partial B_\rho \cap E_n} \ge \alpha,$$

then I possesses an unbounded sequence of critical values.

<u>Proof</u>. Define c_j as in (4.6) and argue exactly as in the proof of Theorem 1.9 with one exception. Namely we can no longer show that $c_1 \ge \alpha$ but merely that $c_{m+1} \ge \alpha$. The proof of this fact is similar to the earlier case and will be omitted.

Now we can apply Theorem 1.9 to the setting of Theorem 3.7.

<u>Theorem 4.24</u>. If $p(x, \xi)$ satisfies (p_1) - (p_4) and is odd in ξ, (3.1) possesses an unbounded sequence of solutions.

<u>Proof</u>. From the proof of Theorem 3.7 we know I defined in (3.2) lies in $C^1(E, \mathbb{R})$, satisfies (PS) and (I_1). The remarks following (3.6) also show (I_8) holds and $p(x, \xi)$ being odd in ξ implies (I_5). Hence Theorem 1.9 gives an unbounded sequence of critical values (c_j)

of I. The form of I in (3.2), (p_2), and the Sobolev inequality shows I is bounded on bounded sets in E. Hence any sequence of critical points u_j corresponding to c_j must be unbounded.

Actually we can get a much better version of Theorem 4.23 instead of Theorem 1.9.

Theorem 4.25. If p satisfies (p_1), (p_2), (p_4), and $p(x, \xi)$ is odd in ξ, then (3.1) possesses an unbounded sequence of solutions.

Proof. The only role (p_3) plays in the proofs of Theorem 3.7 and Theorem 4.24 is in verifying (I_1). Thus it suffices to use (p_2) and various calculus inequalities valid for $u \in E = W_0^{1,2}(\Omega)$ to establish (I_9). We take E_j to be the span of the first j eigenfunctions of $-\Delta$ (with respect to o boundary conditions in Ω) ordered by increasing eigenvalues and $E_j^\perp$ to be the orthogonal complement of E_j in E. Let j and $\rho < R_{j+1}$ be free for the moment. By (p_2) there are constants a_5, a_6 such that

$$|P(x, \xi)| \le a_5 + a_6|\xi|^{s+1} \tag{4.26}$$

for all $\xi \in \mathbb{R}$ and $x \in \bar{\Omega}$. Hence (4.26) and the form of I show if $v \in \partial B_\rho$,

$$I(v) \ge \frac{1}{2}\rho^2 - a_6\|v\|^{s+1}_{L^{s+1}(\Omega)} - a_7 \tag{4.27}$$

for some constant a_7 independent of v. By the Gagliardo-Nirenberg inequality [31],

$$\|v\|_{L^{s+1}(\Omega)} \le c\|v\|^a\|v\|^{1-a}_{L^2(\Omega)} \tag{4.28}$$

for all $v \in E$ where c is a constant independent of v and

$$\frac{1}{s+1} = a(\frac{1}{2} - \frac{1}{n}) + (1-a)\frac{1}{2} \tag{4.29}$$

so $a < 1$. Substituting (4.28) into (4.27) yields

$$I(v) \ge \frac{1}{2}\rho^2 - a_8\rho^{(s+1)a}\|v\|^{(1-a)(s+1)}_{L^2(\Omega)} - a_7 \tag{4.3}$$

for $v \in \partial B_\rho$ and a_8 independent of v. Next if $v \in E_j$, by the Poincaré inequality,

$$(4.31) \qquad \|v\|_{L^2(\Omega)} \le \lambda_{j+1}^{-\frac{1}{2}}\|v\|$$

where λ_k is the kth eigenvalue of

$$(4.32) \qquad -\Delta w = \lambda w, \quad x \in \Omega; \quad w = 0, \quad x \in \partial\Omega.$$

Combining (4.30) - (4.31) shows

$$(4.33) \qquad I(v) \ge \frac{1}{2}(1-2a_8\rho^{s-1}\lambda_{j+1}^{\delta})\rho^2 - a_7$$

for $v \in \partial B_\rho \cap E_j^\perp$ where $\delta = -\frac{1}{2}(1-a)(s+1) < 0$ via (4.29). As is well known, $\lambda_k \to \infty$ as $k \to \infty$ [32]. Thus choose ρ such that $\rho^2 > 8a_7$ and choose j so that $1 - 2a_8\rho^{s-1}\lambda_{j+1}^{\delta} \ge \frac{1}{2}$ and $R_{j+1} > \rho$. Hence

$$(4.34) \qquad I(v) \ge \frac{1}{8}\rho^2 \equiv \alpha > 0$$

on $\partial B_\rho \cap E_j$, I satisfies (I_9), and Theorem 4.25 is proved.

Remark 4.35. One can use the ideas that go into the proofs of Theorems 4.23 and 4.25 to solve equations like (3.1) which are perturbed by an inhomogeneous term, e.g.

$$(4.36) \qquad -\Delta u \equiv p(x, u) + f(x), \ x \in \Omega; \ u = 0, \ x \in \partial\Omega$$

where p satisfies (p_1), (p_2), (p_4), and is odd in ξ provided that s in (p_2) is further restricted. There has been recent research in this direction by Struwe [37], Bahri and Berestycki [34], Dong and Li [35] and the author [36].

§5. Variational Methods in Bifurcation Problems

In this section we will indicate how the Mountain Pass Theorem can be used in a bifurcation setting. In particular we will sketch the proof of Theorem 1.13. Thus suppose now that E is a real Hilbert space with inner product $(.,.)$ and $L \in C^2(E, \mathbb{R})$ such that $I'(u) = Lu + H(u)$ with L linear and $H(u) = o(\|u\|)$ at $u = 0$. We are interested in solving

$$(5.1) \qquad I'(u) = \lambda u$$

for (λ, u) near $(\mu, 0)$ where μ is an isolated eigenvalue of L of finite multiplicity. This problem can be reduced to a finite dimensional one in a standard fashion. We can write $E = N(L-\mu\, id) \oplus N^{\perp}$ where $N = N(A)$ denotes the null space of a linear operator A and $N^{\perp}$ its orthogonal complement. Thus $u \in E$ implies $u = v+w \in N \oplus N^{\perp}$. Letting P and $P^{\perp}$ denote respectively the orthogonal projectors of E onto N and $N^{\perp}$, (5.1) can be rewritten as an equivalent pair of equations

$$(5.2) \qquad \begin{aligned} &\text{(i)} \quad \mu v + PH(v + w) = \lambda v \\ &\text{(ii)} \quad Lw + P^{\perp}H(v+w) = \lambda w \end{aligned}$$

since $Lv = \mu v$ for $v \in N$. Set

$$(5.3) \qquad F(\lambda, v, w) \equiv Lw - \lambda w + P^{\perp}H(v+w).$$

Note that $F: \mathbb{R} \times N \times N^{\perp} \to N$, F is C^1 near $(\mu, 0, 0)$, $F(\mu, 0, 0) = 0$, and $F_w(\mu, 0, 0) = L - \mu\, id$, an isomorphism of $N^{\perp}$ onto $N^{\perp}$. Hence by the Implicit Function Theorem, there is a neighborhood of $(\mu, 0)$ in $\mathbb{R} \times N$ in which the zeroes of F and therefore solutions of (5.2) (ii) are given by $w = \phi(\lambda, v)$ with $\phi \in C^1$ near $(0, 0)$. Note also that since $L - \mu\, id$ is an isomorphism of $N^{\perp}$ onto $N^{\perp}$, so is $L - \lambda\, id$ for all λ near μ. Thus

$$(5.4) \qquad \phi(\lambda, v) = -(L-\lambda\, id)^{-1}P^{\perp}H(v+\phi(\lambda, u)).$$

Since $H(u) = o(\|u\|)$ at $u = o$, (5.4) implies $\phi(\lambda, v) = o(\|v\|)$ at $v = o$ uniformly for λ near μ.

Thus to solve (5.1) near $(\mu, 0)$ in $\mathbb{R} \times E$, it suffices to solve the finite dimensional problem (5.2) (i) with $w = \phi(\lambda, v)$ for

(λ, v) near $(\mu, 0)$ in $\mathbb{R} \times N$. This reduction is called the *method of Lyapunov-Schmidt.* We can further convert (5.2) (i) to a finite dimensional variational problem. Set

$$f(\lambda, u) \equiv I(u) - \frac{\lambda}{2} \|u\|^2 \quad \text{and} \quad g(\lambda, v) \equiv f(\lambda, v+\phi(\lambda, v)).$$

Since $\phi \in C^1$, $g \in C^1$ near $(\mu, 0)$. Using (5.2) (ii), it is easy to see that for fixed λ near μ, critical points of $g(\lambda, .)$ near $v = 0$ are solutions of (5.2) (i). Note that $v = 0$ is a critical point of $g(\lambda, .)$ corresponding to the solution $(\lambda, 0)$ of (5.1). Either $v = 0$ is not an isolated critical point of $g(\mu, .)$ in which case alternative (i) of Theorem 1.13 prevails or it is isolated. In this latter case either (a) $v = 0$ is a strict local maximum or minimum of $g(\mu, .)$ or (b) $g(\mu, .)$ takes on both positive and negative values in every neighborhood of o. We will show (a) and (b) imply respectively alternatives (ii) and (iii) of Theorem 1.13.

To obtain (a) and (b) we study the form of g more closely:

$$(5.5) \quad g(\lambda, v) = \frac{1}{2} (\mu-\lambda)\|v\|^2 + \frac{1}{2} ((L - \lambda\, id)\phi(\lambda, v), \phi(\lambda, v)) + h(v + \phi(\lambda, v))$$

where $h'(u) = H(u)$ and $h(0) = 0$. As was noted above, $\phi(\lambda, v) = o(\|v\|)$ at $v = 0$ and $H(u) = o(\|u\|)$ at $u = o$. It follows that for $\lambda \neq \mu$, the dominating term in (5.5) near $v = 0$ is the first term on the right hand side of the equation, the other terms being $o(\|v\|^2)$ at $v = o$. Thus suppose that (a) holds and e.g. $v = 0$ is a strict local minimum of $g(\mu, .)$. Let $r > 0$ be small. Then there is a $\beta < 0$ such that $g(\mu, .) \leq 2\beta$ on ∂B_r (in N). Thus for λ near μ,

$$(5.6) \qquad g(\lambda, v) \leq \beta < 0, \quad v \in \partial B_r.$$

Now let $\lambda < \mu$ and λ near μ. Our above remarks imply there is a $\rho = \rho(\lambda) < r$ such that

$$(5.7) \qquad g(\lambda, v) \geq \frac{1}{4}(\mu-\lambda)\|v\|^2 = \frac{1}{4}(\mu-\lambda)\rho^2 \equiv \alpha(\lambda)$$

for $v \in \partial B_\rho$. Consider $g(\lambda, v)$ in B_r. It is clear from (5.6)-(5.7) that $g(\lambda, v)$ possesses a positive maximum $\bar{c}(\lambda)$ in B_r. Moreover $g(\lambda, v)$ satisfies (I_1) and (I_2) of the Mountain Pass Theorem. We cannot invoke that theorem directly since $g(\lambda, .)$ is only defined near 0 but due to (5.6) - (5.7), a mapping η can be constructed as in the proof of Theorem 2.1 such that $\eta = id$ on ∂B_r, $\eta \in C([0, 1] \times B_r, B_r)$, and η otherwise satisfies the conclusions of Theorem 2.1. The proof of Theorem 1.3 then yields a second critical value $\underline{c}(\lambda) \geq \alpha(\lambda) > 0$ together with the characterization (1.4) where we take Γ to be the collection of curves in B_r which join o to ∂B_r. If $\underline{c}(\lambda) < \bar{c}(\lambda)$, we have two distinct nontrivial critical points of $g(\lambda, .)$ while if

$$\underline{c}(\lambda) = \bar{c}(\lambda) = \max_{B_r} g(\lambda, .),$$

then $\bar{c}(\lambda)$ is the maximum of $g(\lambda, \cdot)$ over every curve in B_r joining 0 to ∂B_r. Therefore the sets of points at which $g(\lambda, \cdot)$ achieves its maximum separates o and B_r in N. In either event, we have established (ii) of Theorem 1.13. If o is a local maximum of $g(\mu, \cdot)$ we can take $\lambda > \mu$, replace g by -g and use the result we've just proved.

Now we turn to case (b). This is more delicate than (a) and we will only sketch the ideas here. See [23] for more details. Consider the negative gradiant flow corresponding to $g(\mu, .)$:

(5.8) $$\frac{d\psi}{dt} = -g_v(\mu, \psi), \quad \psi(0, v) = v$$

for $v \in \mathcal{B}$, a small neighborhood of o in $\mathbb{N}$. We have noted above that $g \in C^1$ and therefore g_v is merely continuous. However computing $g_v(\lambda, v)$ shows

(5.9) $$g_v(\lambda, v) = (\mu-\lambda)v+PH(v+\phi(\lambda, v))$$

which is C^1 (so in fact $g \in C^2$ with respect to v). Therefore there exists a unique local in t solution to (5.8). Since $g(\mu, v)$

takes on both positive and negative values near $v = o$, it is not too difficult to show that both S^+, S^- are nonempty where

$$S^+ = \left\{v \in \mathcal{B} \mid \psi(t, v) \in \mathcal{B} \text{ for all } t \geq 0\right\},$$

$$S^- = \left\{v \in \mathcal{B} \mid \psi(t, v) \in \mathcal{B} \text{ for all } t \leq 0\right\}.$$

Note that since $H(u) = o(\|u\|)$ at $u = o$, the matrix $g_{vv}(\mu, v)$ vanishes at $v = o$. If $g_{vv}(\mu, 0)$ were nonsingular, S^+, S^- would be respectively the stable and unstable manifolds of the isolated equilibrium point $v = o$ of $g_v(\mu, .)$.

We can now find a neighborhood Q of 0 in N, $Q \subset \mathcal{B}$ and constants $c^+ > 0 > c^-$ such that $v \in \partial Q$ implies (i) $g(\mu, v) = c^+$ or c^- or (ii) $\psi(t, v) \in \partial Q$ for t near o. Let $T^\pm \equiv g(\mu, \cdot)^{-1}(c^\pm) \cap S^\pm$ and let

$$\Gamma = \left\{\chi \in C([0, 1], Q) | \chi(0) = 0 \text{ and } \chi(1) \in T^-\right\}.$$

Define

$$c(\lambda) = \inf_{\chi \in \Gamma} \max_{t \in [0, 1]} g(\lambda, \chi(t)). \tag{5.10}$$

With ρ as in (5.7), we can assume $B_{\rho(\lambda)} \subset Q$ and (5.7) holds if $\lambda < \mu$ and λ is near μ. It follows that $c(\lambda) \geq \alpha(\lambda) > 0$. Moreover $c(\lambda)$ is a critical value of $g(\lambda, \cdot)$ since we can modify the proof of the Deformation Theorem so that there exists $\eta \in C([0, 1] \times Q, Q)$ satisfying 1° - 5° of Theorem 2.1. Then the proof of the Mountain Pass Theorem shows $c(\lambda, \cdot)$ is a critical value of $g(\lambda, \cdot)$. We get the case $\lambda > \mu$ by replacing g by $-g$. Thus (iii) of Theorem 1.13 follows and our sketch of the proof of the theorem is complete.

Remark 5.11. The proof of Corollary 1.14 paralles the above. By the Lyapunov-Schmidt reduction, $\phi(\lambda, v)$ is odd in v and therefore $g(\lambda, v)$ is even in v. The set Q constructed above can then be made symmetric with respect to the origin. A family of sets related to the Γ_m of Theorem 1.9 can be constructed and minimaxing $g(\lambda, \cdot)$ over these sets yields Corollary 1.14. The numbers k, m of the corollary are related to $\gamma(T^{\pm})$. See [24].

Remark 5.12. Theorem 1.13 can be applied to obtain bifurcation results for partial differential equations like

$$(5.13) \qquad -\Delta u = \nu(a(x)u+p(x, u)), \quad x \in \Omega \subset \mathbb{R}^n; \quad u = 0, \quad x \in \partial\Omega$$

where p satisfies (p_1) and (p_3) and $a(x) \in C(\bar{\Omega}, \mathbb{R})$ with $a > 0$ in $\bar{\Omega}$. In particular it is not difficult to show that every eigenvalue of

$$(5.14) \qquad -\Delta v = \beta a(x)v, \quad x \in \Omega; \quad v = 0, \quad x \in \partial\Omega$$

is a bifurcation point for (5.13).

REFERENCES

[1] Castro, A., Methodos de Reduccion via Minimax, this volume.

[2] Nirenberg, L., Variational and topological methods in nonlinear problems, Bull. A.M.S., 4, (1981), 267-302.

[3] Palais, R. S., Critical point theory and the minimax principle, Proc. Sym. Pure Math, *15*, Amer. Math. Soc. Providence, R.I. (1970), 185-212.

[4] Rabinowitz, P. H., Variational methods for nonlinear eigenvalue problems, in Eigenvalues of Nonlinear Problems, G. Prodi (ed.), C.I.M.E. Edizioni Cremonese, Rome, (1975), 141-195.

[5] Ambrosetti, A. and P. H. Rabinowitz, Dual variational methods in critical point theory and applications, J. Functional Anal., *14*, (1973), 349-381.

[6] Rabinowitz, P. H., Some minimax theorems and applications to nonlinear partial differential equations, Nonlinear Analysis: A Collection of Papers in Honor of Erich H. Rothe, edited by L. Cesari, R. Kannan, and H. F. Weinberger, Academic Press, (1978), 161-177.

[7] Benci, V. and P. H. Rabinowitz, Critical point theorems for indefinite functionals, Inv. Math. *52*, (1979), 241-273.

[8] Lazer, A. C., E. M. Landesman, and D. Meyers, On saddle point problems in the calculus of variations, the Ritz algorithm, and monotone convergence, J. Math. Anal. Appl. *52*, (1975), 594-614.

[9] Ahmad, S., A. C. Lazer, and J. Paul, Elementary critical point theory and perturbations of elliptic boundary value problems at resonance, Ind. Univ. Math. J. *25*, (1976) 933-944.

[10] Castro, A. and A. C. Lazer, Applications of a maximum principle, Rev. Colombiana Mat., *10*, (1976), 141-149.

[11] Amann, H. Saddle points and multiple solutions of differential equations, Math. Z., *196*, (1979), 127-166.

[12] Amann, H. and E. Zehnder, Nontrivial solutions for a class of nonresonance problems and applications to nonlinear differential equations, to appear.

[13] Coron, J. M., Resolution de l'équation Au + Bu = f ou A est lineare autoadjoint et B dédeirt d'un potential convexe, to appear Ann. Fac. Sc. Toulouse.

[14] Hofer, H., On strongly indefinite functionals with applications, to appear Trans. A. M. S..

[15] Ljusternik, L. A. and L. G. Schnirelman, Topological Methods in the Calculus of Variations, Hermann, Paris, 1934.

[16] Krasnoselski, M. A., Topological Methods in the Theory of Nonlinear Integral Equations, Macmillan, New York, 1964.

[17] Berger, M. S., Nonlinearity and Functional Analysis, Academic Press, New York, 1978.

[18] Browder, F. E., Nonlinear eigenvalue problems and group invariance, in Functional Analysis and Related Fields, F. E. Browder (ed.), Springer, 1970, 1-58.

[19] Amann, H., Ljusternik-Schnirelmann theory and nonlinear eigenvalue problems, Math. Ann. *199*, (1972) 55-72.

[20] Clark, D. C., A variant of the Lusternik-Schnirelmann theory, Indiana Univ. Math. J., *22*, (1972), 65-74.

[21] Böhme, R., Die Lösung der Verzweigungsgleichungen für nichtlineare Eigenwertproblem, Math. Z., *127*, (1972), 105-126.

[22] Marino, A.: La biforcazione nel caso variazionale, Conf. Sem. Math. dell'Univ Bari, *132* (1977).

[23] Rabinowitz, P. H., A bifurcation theorem for potential operators, J. Funct. Anal., *25*, 1977, 412-424.

[24] Fadell, E. R. and P. H. Rabinowitz, Bifurcation for odd potential operators and an alternative topological index. J. Funct. Anal., *26*, 1977, 48-67.

[25] Crandall, M. G. and P. H. Rabinowitz, Continuation and variational methods for the existence of positive solutions of nonlinear elliptic eigenvalue problems, Arch. Rat. Mec. Anal., *58*, (1975), 207-218.

[26] Fadell, E. R. and P. H. Rabinowitz, Generalized cohomological index theories for Lie group actions with an application to bifurcation questions for Hamiltonian systems, Inv. Math., *45*, (1978), 139-174.

[27] Benci, V., A geometrical index for the group S^1 and some applications to the research of periodic solutions of O.D.E.'s, to appear Comm. Pure Appl. Math.

[28] Fadell, E. R., S. Y. Husseini, and P. H. Rabinowitz, Borsuk-Ulam theorems for arbitrary S^1 actions and applications, to appear Trans. Amer. Math. Soc.

[29] Schwartz, J., Nonlinear Functional Analysis, lecture notes Courant Inst. of Math. Sc., New York Univ., 1965.

[30] Conner, P. E. and E. E. Floyd, Fixed point free involutions and equivariant maps, Bull. A. M. S., *66*, 1960, 416-441.

[31] Nirenberg, L., On elliptic partial differential equations, Ann. Scuola Norm. Sup. Pisa, Ser. 3, *13*, (1959), 115-162.

[32] Courant, R. and D. Hilbert, Methods of Mathematical Physics, VI, Interscience, New York, 1953.

[33] Struwe, M., Infinitely many critical points for functional which are not even and applications to superlinear boundary value problems, Manus. Math. *32*, (1980), 335-364.

[34] Bahri, A. and H. Berestycki, A perturbation method in critical point theory and applications, Trans. Amer. Math. Soc.

[35] Dong, G. -C. and S. Li, On the existence of infinitely many solutions of the Dirichlet problem for some nonlinear elliptic equations, Univ. of Wisconsin Math. Res. Center Tech. Rep. #2161, Dec. 1980.

[36] Rabinowitz, P. H., Multiple critical points of perturbed symmetric functionals, to appear Trans. Amer. Math. Soc.

OPTIMAL SPLINE SOLUTIONS OF SYSTEMS OF ORDINARY DIFFERENTIAL EQUATIONS

Larry L. Schumaker*

Department of Mathematics and Center for Approximation Theory
Texas A&M University
College Station, Texas 77843 USA

§1. Introduction

In 1959, Golomb & Weinberger [5] discussed the general problem of constructing an optimal approximation to an unknown function based on a finite amount of numerical data on the function. The technique has found many applications in approximation theory, leading to an extensive theory of spline functions (see eg. [1, 8] and references therein).

Later Golomb [4] elaborated on the application of the method to numerical solution of two-point boundary-value problems for ordinary differential equations, and gave error bounds and algorithms for constructing the spline solutions. Initial-value problems and an alternative numberical algorithm were discussed in Eidson & Schumaker [2], where it was found that the method is particularly effective on stiff problems.

The purpose of this paper is to take advantage of recent results

* Supported in part by a grant from Fundação de Amparo ã Pesquisa de Estado de São Paulo, Brazil.

on vector-valued splines (see Sidhu & Weinert [9]) to develop an analogous optimal spline method for systems. As with a single equation, the idea is to examine a large class of smooth functions which satisfy the initial conditions and the differential equations at a finite number of points, and then to select a "best" or "optimal" member of this class as our approximate solution.

The initial-value problem treated in this paper is presented in §2, and the extremal problem defining the vector-spline solution is discussed in §3. Sections 4 and 5 are devoted to developing the exact structural characterization of the spline and to numerical algorithms for constructing it. It is to be expected that as with a single equation, the method should perform especially well on stiff problems. In §6 we discuss a generalized Lg-spline method, while in §7 the problem of uniqueness is addressed. We close with a section containing notes and remarks.

§2. The initial-value problem

Throughout the remainder of this paper we shall suppose that d is a positive integer, and that all boldface letters represent vectors with d-components; for example, we write

$$\underset{\sim}{\phi}(t) = [\phi_1(t), \ldots, \phi_d(t)]^T$$

and use

$$\dot{\underset{\sim}{\phi}}(t) = [\dot{\phi}_1(t), \ldots, \dot{\phi}_d(t)]^T$$

for the derivative of ϕ with respect to t.

We are interested in the following *initial-value problem for a system of ordinary differential equations.*

Problem 1. Let $A(t) = (A_{ij}(t))_{i,j=1}^{d}$ and $\underset{\sim}{r}(t)$ be given with $A_{ij}(t)$ and $r_i(t)$ in $C[0, 1]$, all $1 \leq i,j \leq d$. Find $\underset{\sim}{\phi}(t)$ such that

(1) $$\dot{\underset{\sim}{\phi}}(t) = A(t)\underset{\sim}{\phi}(t) + \underset{\sim}{r}(t), \quad \text{for } 0 \leq t \leq 1$$

and

(2) $$\underset{\sim}{\phi}(0) = \underset{\sim}{0}.$$

We assume that Problem 1 has a unique solution $\underset{\sim}{\phi}$ for each $\underset{\sim}{r}$ with $\phi_j \in C^1[0, 1]$, $j = 1, 2, \ldots, d$. We have elected to work on $[0, 1]$ and with homogeneous boundary conditions for the sake of simplicity. This is no essential restriction since problems on general intervals $[a, b]$ with non-zero initial data can be reduced to Problem 1 by an appropriate change of variables.

§3. An extremal problem

We are going to find a numerical solution to Problem 1 by setting up and solving an extremal problem whose solution is the smoothest function satisfying the initial conditions along with the system of differential equations at a finite number of points in $[0, 1]$. To define the extremal problem, we need some notation.

For m a positive integer (which we shall choose greater than 1 here), let

$$L_2^m[0, 1] = \Big\{f\colon f^{(m-1)} \text{ is absolutely continuous on } [0, 1] \text{ and } f^{(m)} \in L_2[0, 1]\Big\}.$$

This is a well known Sobolev space. We shall need the vector version

(3) $$\underset{\sim}{L}{}_2^m[0, 1] = \Big\{\underset{\sim}{f} = (f_1, \ldots, f_d)^T\colon f_i \in L_2^m[0, 1],\ i = 1, 2, \ldots, d\Big\}.$$

We now define a space of smooth functions satisfying the initial conditions and the system of differential equations at points

$$0 < t_1 < t_2 < \ldots < t_N = 1.$$

Given $\underset{\sim}{r}$, let

(4) $$U(\underset{\sim}{r}) = \Big\{\underset{\sim}{f} \in \underset{\sim}{L}{}_2^m[0, 1]\colon \underset{\sim}{f}(0)=0 \text{ and } \dot{\underset{\sim}{f}}(t_i) = A\underset{\sim}{f}(t_i) + \underset{\sim}{r}(t_i),\quad i = 1, \ldots, N\Big\}.$$

Problem 2. Find $\underset{\sim}{s} \in U(\underset{\sim}{r})$ such that

$$(5) \qquad \int_0^1 (D^m \underset{\sim}{s}(t))^T D^m \underset{\sim}{s}(t)dt = \inf_{\underset{\sim}{u} \in U(\underset{\sim}{r})} \int_0^1 (D^m \underset{\sim}{u}(t))^T D^m \underset{\sim}{u}(t)dt.$$

We call a solution $\underset{\sim}{s}$ of Problem 2 a *spline interpolating the initial-value problem*. It is a function in $U(\underset{\sim}{r})$ which is the smoothest in the sense that the sum of the L_2-norms of the m-th derivatives of the components of $\underset{\sim}{s}$ is as small as possible. We discuss the problem in the following section -- for a more general version producing Lg-splines, see Section 6.

§4. The spline solution

Problems of the form of (5) have been considered by several authors -- for $d = 1$ see [1, 6] and references therein. The vector case was considered in [9] by Sidhu & Weinert. Applying their results, we have

Theorem 3. If $U(\underset{\sim}{r}) \neq \phi$, then there exists a solution $\underset{\sim}{s}$ of Problem 2. It is unique if and only if

$$(6) \qquad U(\underset{\sim}{0}) \cap P_m^d = (\underset{\sim}{0}),$$

where

$$(7) \qquad P_m^d = \left\{ \underset{\sim}{p} = (p_1, \ldots, p_d): p_j \in P_m, \quad j = 1, \ldots, d \right\}$$

and P_m is the set of polynomials of order m.

The uniqueness condition depends on the nature of $U(\underset{\sim}{0})$ -- in Section 7 we discuss conditions under which it is satisfied. In the remainder of this section we suppose there is a unique $\underset{\sim}{s}$ solving Problem 2, and turn to the question of characterizing it.

Theorem 4. Suppose condition (6) is satisfied. Then $\underset{\sim}{s} \in \underset{\sim}{L}_2^m[0, 1]$ is the unique solution of Problem 2 if and only if

(8a) $\underset{\sim}{s}(0) = \underset{\sim}{0}$

(8b) $\dot{\underset{\sim}{s}}(t_j) = A(t_j)\underset{\sim}{s}(t_j) + \underset{\sim}{r}(t_j), \quad j = 1, \ldots, N$

(8c) $\underset{\sim}{s}|_{(t_j, t_{j+1})} \in P^d_{2m}, \quad j = 0, \ldots, N-1, \quad (t_o = 0)$

(8d) $\underset{\sim}{s} \in C^{2m-3}[0, 1]$

(8e) $D^m\underset{\sim}{s}(0) = \ldots = D^{2m-2}\underset{\sim}{s}(0) = 0$

(8f) $D^m\underset{\sim}{s}(1) = \ldots = D^{2m-3}\underset{\sim}{s}(1) = \underset{\sim}{0}$

(8g) $R\underset{\sim}{s}(1) = \underset{\sim}{0}$

(8h) $\text{jump}[R\underset{\sim}{s}]_{t_j} = 0, \quad j = 1, 2, \ldots, N-1,$

where

$$R\underset{\sim}{s} = D^{2m-1}\underset{\sim}{s} + A^T D^{2m-2}\underset{\sim}{s}$$

and

$$\text{jump}[R\underset{\sim}{s}]_t = R\underset{\sim}{s}(t+) - R\underset{\sim}{s}(t-).$$

<u>Proof</u>. It is not hard to see (cf. [1, 6, 9] that $\underset{\sim}{s}$ is the solution iff

(9) $$\int_{-\infty}^{\infty} (D^m\underset{\sim}{s}(t))^T D^m\underset{\sim}{g}(t)dt = 0 \quad \text{for all} \quad \underset{\sim}{g} \in U(\underset{\sim}{0}),$$

where all functions are extended to $\underset{\sim}{L}^m_2(-\infty, \infty)$. Conditions (8a)-(8b) are necessary for $\underset{\sim}{s}$ to belong to $U(\underset{\sim}{r})$. To show the necessity of the remaining conditions we integrate (9) by parts. Choosing $g_i \in C^\infty(t_j, t_{j+1})$, $i = 1, \ldots, d$, a standard argument (cf. [1]) shows that

(10) $$D^m\underset{\sim}{s}(t) \equiv 0 \quad \text{on} \quad (-\infty, 0) \cup (1, \infty)$$

and

$$D^{2m}\underset{\sim}{s}(t) \equiv 0 \quad \text{on} \quad (t_j, t_{j+1}), \quad j = 0, \ldots, N-1.$$

This shows that (8c) must hold. The fact that $\underset{\sim}{s} \in U(\underset{\sim}{r})$ automatically implies $\underset{\sim}{s} \in C^{m-1}[0, 1]$.

We now examine the boundary terms arising in the integration by parts of (9). For any $\underset{\sim}{g} \in C^m(-\varepsilon, \varepsilon) \cap U(\underset{\sim}{0})$, we have

$$(11) \qquad 0 = \text{jump}[(D^{2m-1}\underset{\sim}{s})^T, \ldots, (D^m\underset{\sim}{s})^T]_0 \begin{bmatrix} \underset{\sim}{g}(0) \\ D\underset{\sim}{g}(0) \\ \vdots \\ D^{m-1}\underset{\sim}{g}(0) \end{bmatrix}$$

Now fix $1 \le j \le m-1$, and choose $\underset{\sim}{g}$ such that $D^i\underset{\sim}{g}(0) = 0$, $i=0, \ldots,$ $j-1, j+1, \ldots, m-1$. Then for each $1 \le \mu \le d$ we can set $D^j\underset{\sim}{g}_\nu(0)=\delta_{\nu\mu}$ for $1 \le \nu \le d$ to deduce that

$$\text{jump}[D^{2m-j-1}\underset{\sim}{s}]_0 = 0,$$

and coupling this with (10) yields (8e).

The argument at t_j is similar. For every $\underset{\sim}{g} \in C^m(t_j-\varepsilon, t_j+\varepsilon) \cap U(\underset{\sim}{0})$, we have

$$(12) \qquad 0 = \text{jump}[(D^{2m-1}\underset{\sim}{s})^T, \ldots, (D^m\underset{\sim}{s})^T]_{t_j} \begin{bmatrix} 0 & I_d & \bigcirc \\ I_d & A & \\ \bigcirc & & I_{(m-2)d} \end{bmatrix} \bullet$$

$$\bullet \begin{bmatrix} -A & I_d & \bigcirc \\ I_d & 0 & \\ \bigcirc & & I_{(m-2)d} \end{bmatrix} \begin{bmatrix} \underset{\sim}{g}(t_j) \\ D\underset{\sim}{g}(t_j) \\ \vdots \\ D^{m-1}\underset{\sim}{g}(t_j) \end{bmatrix}.$$

This condition is of the same nature as (11) except for the insertion of the matrices which are clearly inverses of each other (here I_n stands for the $n\times n$ indentity matrix). In terms of the operator R, we may rewrite this

$$0 = \text{jump}[(D^{2m-2}\underset{\sim}{s})^T, (R\underset{\sim}{s})^T, (D^{2m-3}\underset{\sim}{s})^T, \ldots, (D^m\underset{\sim}{s})^T]_{t_j} \begin{bmatrix} (\dot{\underset{\sim}{g}}-A\underset{\sim}{g})(t_j) \\ \underset{\sim}{g}(t_j) \\ D^2\underset{\sim}{g}(t_j) \\ \vdots \\ D^{m-1}\underset{\sim}{g}(t_j) \end{bmatrix}$$

Now let $\underset{\sim}{g} \in C^m(t_j-\varepsilon, t_j+\varepsilon)$ be such that $(\dot{\underset{\sim}{g}} - A\underset{\sim}{g})(t_j) = 0$ while

$D^i\underset{\sim}{g}(t_j) = 0$, $i = 2, \ldots, m-1$. For each $1 \le \mu \le d$, we can choose $\underset{\sim}{g}$ such that $\underset{\sim}{g}_\nu(t_j) = \delta_{\nu\mu}$, $1 \le \mu \le d$, and it follows that

$$\text{jump}[R\underset{\sim}{s}]_{t_j} = 0.$$

Similarly we can show

$$\text{jump}[D^{2m-\mu-1}\underset{\sim}{s}]_{t_j} = 0 \quad \text{if} \quad 2 \le \mu \le m-1.$$

This establishes the rest of (8d) as well as (8h). Coupling these equalities with (10) yields (8f) - (8g). We have shown that if $\underset{\sim}{s}$ is a solution of Problem 2 it must satisfy (8).

For the converse, we note that if $\underset{\sim}{s}$ satisfies (8), then (looking at each term obtained by integration by parts) the condition (9) must hold. It follows that $\underset{\sim}{s}$ is the unique solution of (5).

Theorem 4 shows clearly that $\underset{\sim}{s}$ is a vector-valued spline (i.e. each component s_i is a polynomial spline of order $2m$ with knots at $t_1, \ldots, t_{N-1}$). When $m = 2$ we have a cubic spline which is globally C^1. For larger m we have higher order splines with more smoothness.

§5. Computing the spline solution

There are several approaches to the numerical construction of the spline $\underset{\sim}{s}$ satisfying (5). It is clear from Theorem 4 that we may write

$$\underset{\sim}{s}(t) = \left\{\underset{\sim}{p}^j(t) \quad \text{for} \quad t_{j-1} \le t < t_j, \quad j=1,\ldots,N\right\} \tag{13}$$

where

$$\underset{\sim}{p}^j(t) = \begin{bmatrix} p_1^j(t) \\ \vdots \\ p_d^j(t) \end{bmatrix} = \begin{bmatrix} \sum_{i=0}^{2m-1} a_{1i}^j (t - t_{j-1})^i \\ \vdots \\ \sum_{i=0}^{2m-1} a_{di}^j (t - t_{j-1})^i \end{bmatrix},$$

$j = 1, \ldots, N$. This shows clearly that there are $N(2m)(d)$ coefficients to be determined.

Now we claim that the conditions (8) provide exactly $N(2m)(d)$ conditions on these coefficients. Indeed, we have

Condition	8a	8b	8d	8e	8f	8g	8h
Number of equations	d	Nd	(N-1)(2m-2)(d)	(m-1)d	(m-2)d	d	(N-1)d

One approach to constructing $\underset{\sim}{s}$ is to simply set up these $N(2m)(d)$ equations for the coefficients of the polynomial pieces. (It must be nonsingular by the uniqueness assumption).

A second possibility for the construction of $\underset{\sim}{s}$ is to set up a factorization procedure along the lines of that discussed in [3, 7] in the 1-dimensional case. The problem in this case can be treated as a sequence of problems involving numerical solution of linear systems.

Finally, we also mention the recursive algorithm of [9] for numerical construction of vector-valued Lg-splines. It is based on the identification of $\underset{\sim}{s}$ as the sample function for the linear least squares estimate for a certain lumped stochastic process.

§6. Lg-splines

In this section we discuss a generalization of Problem 2 whose solution is a so-called Lg-spline. To state the problem we need some additional notation. Let

$$L = (L_{ij})_{i,j=1}^{d}$$

be a matrix of differential operators; say

$$L_{ij} = \sum_{\nu=0}^{m_j} a_{ij\nu}(t)D^{\nu}, \quad i, j = 1, \ldots, d.$$

We assume that L_{jj} is of order m_j and that L_{ij} is of lesser order for $i = 1, \ldots, j-1, j+1, \ldots, d$; i.e., in particular,

$$a_{ij,m_j} = \delta_{ij}, \quad i, j = 1, \ldots, d.$$

Associated with the vector $\underset{\sim}{m} = (m_1, \ldots, m_d)^T$, we define

$$\underset{\sim}{L}_2^m[0, 1] = \left\{f: f_j \in L_2^{m_j}[0, 1], \quad j = 1, \ldots, d\right\}.$$

Suppose now that $m_j > 1$, $j = 1, \ldots, d$, and that

$$U(\underset{\sim}{r}) = \underset{\sim}{f} \in \underset{\sim}{L}_2^m[0, 1]: \underset{\sim}{f}(0) = \underset{\sim}{0} \text{ and } \dot{\underset{\sim}{f}}(t_j) = A(t_j)\underset{\sim}{f}(t_j) + \underset{\sim}{r}(t_j), \quad j = 1, \ldots, N\Big\}$$

where $0 = t_o < t_1 < \ldots < t_N = 1$ are given.

<u>Problem 5</u>. Find $\underset{\sim}{s} \in U(\underset{\sim}{r})$ such that

$$(15) \qquad \int_0^1 (L\underset{\sim}{s}(t))^T L\underset{\sim}{s}(t)dt = \inf_{\underset{\sim}{u} \in U(\underset{\sim}{r})} \int_0^1 (L\underset{\sim}{u}(t))^T L\underset{\sim}{u}(t)dt.$$

We call a function $\underset{\sim}{s}$ satisfying (15) an Lg-*spline interpolating the data* $U(\underset{\sim}{r})$. The polynomial splines discussed in Sections 3-5 clearly correspond to the case where $m = m_1 = \ldots = m_d$ and

$$L = \begin{bmatrix} D^m & & \bigcirc \\ & \ddots & \\ \bigcirc & & D^m \end{bmatrix}.$$

By the results of [9], we immediately have

<u>Theorem 6</u>. If $U(\underset{\sim}{r}) \neq \phi$, then there exists a solution of (15). It is unique if and only if

$$(16) \qquad N_L \cap U(\underset{\sim}{0}) = (\underset{\sim}{0}),$$

where

$$(17) \qquad N_L = \left\{\underset{\sim}{f} \in \underset{\sim}{L}_2^m[0, 1]: L\underset{\sim}{f}(t) = \underset{\sim}{0} \text{ for all } 0 \leq t \leq 1\right\}.$$

It is also possible to give a complete characterization of the interpolating Lg-spline. Indeed, (see [9]) it can be shown that

$$L^*L\underset{\sim}{s}(t) = 0 \qquad t_{j-1} < t < t_j, \quad j = 1, \ldots, N,$$

where

$$L^* = (L_{ij}^*)_{i,j=1}^d$$

$$L_{ij}^*\phi = \sum_{k=0}^{m_i} (-1)^k D^k [a_{ji,k}\phi].$$

In this case the smoothness conditions appearing in (8) must be replaced by similar conditions involving $L_{ij}^{*}\underset{\sim}{s}$ instead of simply derivatives of $\underset{\sim}{s}$. For details on the smoothness of vector valued Lg-splines see [9].

§7. Uniqueness

It is important for the numerical construction of the spline solution of Problem 2 that condition (6) guaranteeing uniqueness be satisfied. In this section we show that (6) holds whenever N is sufficiently large, provided that the t's are reasonably uniformly distributed throughout [0, 1].

Theorem 7. There exists $\varepsilon > 0$ such that whenever

(18) $$\bar{\Delta} = \max_{1 \leq i \leq N} (t_i - t_{i-1}) < \varepsilon,$$

then

(19) $$U(\underset{\sim}{0}) \cap P_m^d = (\underset{\sim}{0}).$$

Proof: Let

(20) $$\mathring{P}_m^d = \left\{ \underset{\sim}{p} \in P_m^d : \underset{\sim}{p}(0) = \underset{\sim}{0} \right\}.$$

Suppose there are sequences $\underset{\sim}{p}^{(\nu)} \in P_m^d$ and $\Delta^{(\nu)} = \{0 = t_0^{(\nu)} < t_1^{(\nu)} < \ldots < t_N^{(\nu)} = 1\}$ with $\bar{\Delta}_\nu \to 0$ such that

$$\dot{\underset{\sim}{p}}^{(\nu)}(t_i^{(\nu)}) = A(t_i^{(\nu)})\underset{\sim}{p}^{(\nu)}(t_i^{(\nu)}), \quad i=1, \ldots, N_\nu.$$

We may normalize these polynomials such that

$$\|\underset{\sim}{p}^{(\nu)}\| = \sum_{j=1}^{d} \|p_j^{(\nu)}\| = 1.$$

Then we may choose a subsequence (which we denote by ν again), such that $\underset{\sim}{p}^{(\nu)}$ converges to $\underset{\sim}{p} \in \mathring{P}_m^d$ with $\|\underset{\sim}{p}\| = 1$. It is easily seen that

$$\dot{\underset{\sim}{p}}(t) = A(t)\underset{\sim}{p}(t) \quad \text{for all} \quad 0 \leq t \leq 1.$$

But then by the uniqueness of solutions to the initial-value problem, we conclude that $\underset{\sim}{p} \equiv \underset{\sim}{0}$, contradicting the fact that $\|\underset{\sim}{p}\| = 1$. It follows that $U(\underset{\sim}{0}) \cap P_m^d = (\underset{\sim}{0})$ whenever $\bar{\Delta}$ is sufficiently small.

§8. Remarks

1. Although we have given no error bounds or convergence results in this paper, it is possible (cf. Golomb [4] for $d = 1$) to derive such results using the standard techniques of spline interpolation theory (with Extended-Hermite-Birkhoff linear functionals -- see eg. Varga [10]). The important fact is that spline interpolation is local so that error bounds in a given interval I depend only on the smoothness of the interpolated function in a neighborhood of I.

2. The local nature of convergence mentioned in Remark 1 has the practical consequence that if we are solving a stiff system where $\underset{\sim}{\phi}$ can misbehave badly for a substantial interval $[0, t_0]$, for intervals to the right of t_0 we may have excellent approximations even with $\bar{\Delta}$ relatively large.

3. While we have not yet performed numerical experiments on systems, our experience with a single stiff initial value-problem (cf. [2]) does bear out the behavior suggested in Remark 2. The spline method permits numerical integration with substantial $\bar{\Delta}$ (and thus moderate N) while discrete-variable methods have to march through the region of transients and hence need small step-size and encounter stability problems.

4. When A has constant coefficients, then the IVP can be solved exactly. This case provides a useful set of test examples.

5. The IVP for a single d-th order differential equation can, of course, always be converted to a system of d first order equations as in (1). Thus, the methods described here can be used to solve

initial-value problems of any order. An advantage over the spline method of [2, 4] may be that here we can use splines of order as low as 4 (cubic splines), whereas for a d-th order equation the methods of [2, 4] require splines of order 2(d+1) at least.

6. It is possible to give analogs for the theorems of §7 for the Lg-spline case. Since to do so requires special properties for the operators of L, and since in practice there would usually be no need for Lg-splines, we do not give the messy details here.

REFERENCES

[1] Copley, P. and L. L. Schumaker, On pLg-splines, J. Approximation Th. *23* (1978), 1-28.

[2] Eidson, H. D. and L. L. Schumaker, Spline solution of linear initial and boundary-value problems, in ISNM 32, Birkhäuser-Verlag, 1976, 67-80.

[3] Eidson, H. D. and L. L. Schumaker, Computation of g-splines via a factorization method, Comm. A.C.M. 17 (1974), 526-530.

[4] Golomb, M., Spline approximations to the solution of two-point boundary-value problems, MRC report 1066, Math. Research Center, U. of Wisconsin, 1970.

[5] Golomb, M. and H. F. Weinberger, Optimal approximations and error bounds, in *On Numerical Approximation*, R. E. Langer, ed., U. Wisconsin Press, 1959, 117-190.

[6] Jerome, J. W. and L. L. Schumaker, On Lg-splines, J. Approximation Th. *2* (1969), 29-49.

[7] Munteanu, M. J. and L. L. Schumaker, On a method of Carasso and Laurent for constructing interpolating splines, Math. Comp. *27* (1973), 317-325.

[8] Schumaker, L. L., *Spline Functions: Basic Theory*, Wiley Interscience, 1981.

[9] Sidhu, G. S. and H. L. Weinert, Vector-valued Lg-splines, J. Math. Anal. Appl. *70* (1979), 505-529.

[10] Varga, R. S., Error bounds for spline interpolation, in *Approximations with Special Emphasis on Spline Functions*, I. J. Schoenberg, ed., Academic Press, N.Y., 1969, 367-388.

STRUCTURALLY STABLE SECOND ORDER DIFFERENTIAL EQUATIONS

J. Sotomayor

Instituto de Matemática Pura e Aplicada
Rio de Janeiro, RJ - Brazil

1. Introduction

We consider second order differential equations of the form

$$(E_f) \qquad \ddot{x} = f(x, \dot{x}), \quad \dot{x} = \frac{dx}{dt}, \quad \ddot{x} = \frac{d^2x}{dt^2},$$

where f is a function of class C^k, $k \geq 1$, in $\mathbb{R}^2$.

Associated to (E_f) or rather to f itself, we consider the vector field $X(f) = y \frac{\partial}{\partial x} + f(x, y) \frac{\partial}{\partial y}$. Let M be a closed region in R^2; that is, M is the closure of its interior. If $p_o = (x_o, y_o)$ is point of M and $\phi(p_o, t)$, $t \in I(p_o, f)$, is the maximal integral curve of $X(f)$ through p_o. Call $\hat{I}(p_o, f)$ the connected component containing 0 of the closed set $\{t \in I(p_o, f), \ \phi(p_o, t) \in M\}$. Notice that $\hat{I}(p_o, f)$ is a closed interval which can be reduced to a point. The curve $\hat{\phi}(p_o, t) = \phi(p_o, t)$ restricted to $\hat{I}(p_o, f)$ is called the maximal integral curve through p_o of $X(f)$ in M. The maximal orbits of $X(f)$ in M are the images of its maximal integral curves.

Now we define C^k-structurally stable second order differential equations.

1.1 Definition. We say that (E_f) is C^k-structurally stable in M if there is an $\varepsilon > 0$ such that if g is of class C^k and

$$(P_\varepsilon) \qquad \|g-f\|_k^M = \sup_{\substack{(x,y)\in M \\ 0\le r+s\le k}} \left| \frac{\partial^{r+s} g(x,y)}{\partial^r x\, \partial^s y} - \frac{\partial^{r+s} f(x,y)}{\partial^r x\, \partial^s y} \right| < \varepsilon$$

then there is a homeomorphism $h = h_g : M \to M$ which maps maximal ortits of $X(g)$ in M onto maximal orbits of $X(f)$ in M.

A homeomorphism such as h_g is called a *topological equivalence* between E_g and E_f.

As far ao we know, the study of structurally stable second order differential equations was started in [2]. There was treated the case $k = 1$ and the restriction to functions f periodic in x was imposed; also M was the whole R^2 and the homeomorphism h was required to be also periodic in x. Finally, in the proximity relation imposed by (P_ε) it was required that ε instead of a constant be a positive continuous function (Whitney Topology).

In this paper we will prove a characterization theorem for C^k-structurally stable second order differential equations in compact regions M. We also establish that such equations, in accordance with [2, 3], form an open and dense set in the space of C^k systems in M and, in some cases, also a set of total Lebesgue measure. We will restrict ourselves to the following cases for M, the class of functions f and the homeomorphism h_g which realizes the topological equivalence between E_g and E_f:

A.- M is compact connected and has a boundary M of class C^2, transversal to the x-axis.

B.- $M = \{(x, y);\ a \le x \le b,\ \alpha(x) \le y \le \beta(x)\}$ where α and β are functions of class C^2 in $[a, b]$ such that $\alpha < 0 < \beta$ there.

In this case it is required that h_g map the segments $x = a$, and $x = b$ of M onto themselves.

C.- M is the same as in B except that $b = \infty$ and $a = -\infty$. In this case we restrict ourselves to the following situation:

1) α and β are periodic of period 1. That is $\alpha(x+1) = \alpha(x)$ for all x.

2) f, g etc are periodic in x of period 1. That is $f(x+1, y) = f(x,y)$ for all x and y.

3) h_g is periodic in x. That is $h_g(x+1, y) = h_g(x, y)$ for all x, y.

2. The Main Results

In order to formulate the main results of this paper we need to define the following class of functions $f \in C^k$, i.e., of class C^k in R^2:

(2.1) $$\Sigma^k(M) = \Sigma^k(1, M) \cap \Sigma^k(2, M) \cap \Sigma^k(3, M) \cap \Sigma^k(4, M)$$

where

(2.1.1) $\Sigma^k(1, M)$ denotes the set of functions $f \in C^k$ such that all the singularities of $X(f)$ in M are hyperbolic and contained in $\text{Int}\, M = M - \partial M$. Since all the singularities of $X(f) = y \frac{\partial}{\partial x} + f(x, y) \frac{\partial}{\partial y}$ are located in the x-axis and the Jacobian matrix has rows $(0, 1)$ and (f_x, f_y), it follows that $(x, 0)$ is a *saddle point* if and only if $f_x(x, 0) > 0$; it is a *focus* if $f_x(x, 0) < 0$ and $f_y^2 + 4f_x < 0$, it is a *node* if $f_x(x, 0) < 0$ and $f_y^2 + 4f_x \geq 0$.

(2.1.2) $\Sigma^k(2, M)$ denotes the set of functions $f \in C^k$ such that all the periodic orbits γ of $X(f)$ in M are hyperbolic and contained in $\text{Int}\, M$. In case C the orbits γ whose quotient $\tilde{\gamma}$ in $\tilde{\mathbb{R}}^2 = \mathbb{R}/\mathbb{Z} \times R$ are periodic orbits of the quotient field $\tilde{X}(f)$ on the manifold $\tilde{M}$ obtained projecting

M to $\tilde{R}^2$, will be also called periodic orbits of $X(f)$ non homotopic to zero. In either case we recall that a periodic orbit $\gamma = \{x(t), y(t)\}$ of $X(f)$ is hyperbolic if and only if

$$\int_0^\tau f_y(x(t), y(t))dt \neq 0,$$

where τ is the minimal period of γ in cases A, B, C or of $\tilde{\gamma}$ (i.e. $x(t+\tau) = x(t) + 1$, $y(t+\tau) = y(t)$), in case C.

(2.1.3) $\Sigma^k(3, M)$ denotes the set of $f \in C^k$ such that every point $p \in \partial M$, $X(f)$ and ∂M are transversal or are tangent and have parabolic contact. In case B if p is a corner of ∂M, for example $p = (a, \alpha(a))$, it is required that $X(f)$ be transverse to both edges, $\{x = a\}$ and $\{y = \alpha(x)\}$, of the corner.

Recall that if ∂M is defined locally by $b = 0$, where b is a C^2-function such that $b = 0$ on ∂M, $db_p \neq 0$ at $p \in \partial M$ and $b > 0$ on Int M, then $X(f)$ is transverse to ∂M if and only if $X(f).b = yb_x(x, y) + f(x, y)b_y(x, y) \neq 0$. If $X(f)\cdot b > 0$ (resp. < 0) $X(f)$ points inwards (resp. outwards) M.

If $[X(f)\cdot b] = 0$ (tangency) and $X^2(f)\cdot b = X(f)\cdot[X(f)\cdot b] \neq 0$, the contact is *parabolic*. The tangency is external (resp. internal) if $X^2(f)\cdot b < 0$ (resp. > 0).

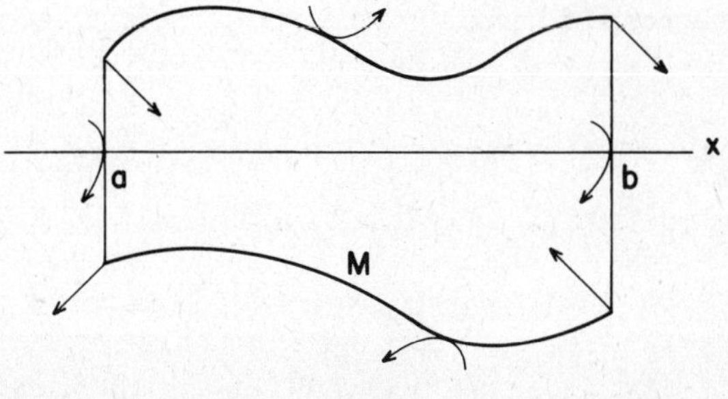

Fig. 1. Conditions for $\Sigma^k(3, M)$ in case B.

(2.1.4) $\Sigma^k(4, M)$ denotes the set of $f \in C^k$ such that $X(f)$ does not have

a) saddle connections in M,

b) saddle separatrizes in M either tangent to ∂M or, in case B, passing through a corner of M,

c) orbits tangent to ∂M at two different points or, in case B, tangent at one point and passing through a corner or, still in case B, passing through two corners.

Now we state the main conclusions of this paper of which only Theorems 2, 3, 4 and 5 are essentially new in the context of second order differential equations. For general systems of first order differential equations (i.e. vector fields of general form $X = g(x, y) \frac{\partial}{\partial x} + f(x, y) \frac{\partial}{\partial y}$) they have been proved in [5, 6]. However, the theorems of this paper, because of the restriction on the form of the fields, do not follow from the general case. Nevertheless, the idea of considering the two-parameter family of Theorem 2, which is the essential tool here, comes from the three parameter translated-rotated vector field family defined below, with $V = (u, v)$,

$R_\theta(X+V) = ((g+u)\sin\theta + (f+v)\cos\theta)\frac{\partial}{\partial x} + (-(g+u)\cos\theta + (f+v)\sin\theta)\frac{\partial}{\partial y}$,

this family was used in the general case for the proof of an analogous measure theoretical result.

<u>Theorem 1</u>. In either of the cases A, B and C we have that

a) $\Sigma^k(M)$ is open in $C^k(M)$, $k \geq 1$. That is, if $f \in \Sigma^k(M)$, there is an $\varepsilon > 0$ such that $\|g-f\|_k^M < \varepsilon$ implies that $g \in \Sigma^k(M)$.

b) Moreover, every $f \in \Sigma^k(M)$ is such that (E_f) is C^k-structurally stable.

The proof follows from [3].

Theorem 2. Let f be a function of class C^k, $k \geq 1$, in R^2. In either of the cases A, B and C

$$F_1(f) = \left\{(\lambda, \mu) \in R^2;\ \ f + \lambda p_2 + \mu \notin \Sigma^k(M)\right\},$$

where $p_2(x, y) = y$ in R^2, is a set of zero Lebesgue measure in R^2.

The proof is given in 3.2.

Theorem 3. In either of the cases A, B and C, the set $\Sigma^k(M)$ is dense in $C^k(M)$, $k \geq 1$. That is, for any f of class C^k in R^2 and $\varepsilon > 0$, given, there is $g \in \Sigma^k(M)$ such that $\|f-g\|_k^M < \varepsilon$.

The proof follows immediately from Theorem 2. Another application of this theorem is the following.

Theorem 4. a) In the vector space ω_m, $m \geq 1$, of polynomials

$$g = \sum_{0 \leq k+\ell \leq m} a_{k,\ell}\, x^k \dot{x}^\ell,$$

those which do not belong to $\Sigma^1(M)$, in cases A or B form a set of zero Lebesgue measure.

b) In the vector space $F_{m,n}$, $m \geq 1$, $n \geq 0$, of functions

$$g = \sum_{k=0}^{m} A_k(x)\dot{x}^k,$$

where

$$A_k(x) = \sum_{i=1}^{n} a_{k,i} \cos 2\pi ix + b_{k,i} \sin 2\pi ix,$$

those which do not belong to $\Sigma^1(M)$, in cases A, B or C form a set of zero Lebesgue measure.

Theorem 5. In either of the cases A, B and C if E_f is C^k-structurally stable in M, then $f \in \Sigma^k(M)$.

The proof is given in 3.

3. Proof of the Main Results

3.1 The proof of Theorem 1 follows from sections 4 and 5 of [3]. The only somewhat new situation appers at the corners in case B. This, however, does not imply any essential difference in the proof of the openness or in the construction of the homeomorphisms, by the method of "canonical regions", as developed in [3]. Actually the corners can be regarded as external tangencies, as in region C_1 of Fig. 2, or as separatrices as in regions C_2 and C_3 of Fig. 2.

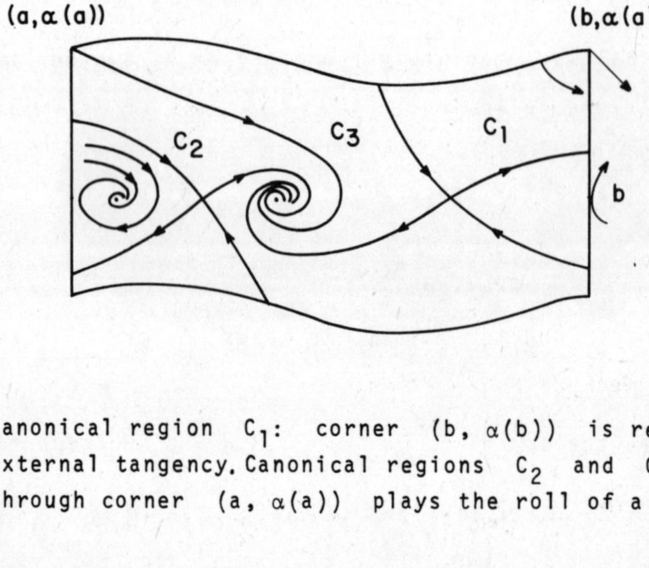

Fig. 2 - Canonical region C_1: corner $(b, \alpha(b))$ is regarded as external tangency. Canonical regions C_2 and C_3 orbit through corner $(a, \alpha(a))$ plays the roll of a separatrix.

3.2 The proof of Theorem 2 depends on serveral lemmas.

3.2.1 Lemma. Let $s: [a, b] \to R$ be a C^1 function. The set of critical values of s, defined by

$$\Gamma(s) = \{y = s(x); \quad s'(x) = 0\}$$

has zero Lebesgue measure in R.

This is the easiest case of Sard's Theorem [4].

3.2.2 <u>Lemma</u>. Let g be a function of class C^2 in $\mathbb{R}^2$.

Let $F_1(0, g)$ be the complement in R of the set of points $\mu \in R$ such that $X(g+\mu)$ has all its singular points in M with $-g_x \neq 0$ (non-zero Jacobian) and are contained in Int M. Then $F_1(0, g)$ has zero Lebesgue measure in R.

<u>Proof</u>. Follows from 3.2.1, in fact

$$F_1(0, g) = -g(\partial M \cap R\times\{0\}) \cup \Gamma(-g \mid R\times\{0\}).$$

3.2.3 <u>Lemma</u>. Let f be a C^1 function in $\mathbb{R}^2$ and $\mu \notin F_1(0, f)$ in 3.2.2. Then the set

$$F_1(1, f, \mu) = \left\{\lambda;\ f + \lambda p_2 + \mu \notin \Sigma^1(1,M)\right\}$$

has zero Lebesgue measure in R.

<u>Proof</u>. Let $x_i(\mu)$, $i = 1,2,\dots,m$, be the singular points of $X(f+\mu)$ in M

$$F_1(1, g, \mu) = \bigcup_{i=1}^{m} \Gamma\left[-(f+\mu) \mid \{x_i(\mu)\}\times R\right].$$

The proof follows from 3.2.1.

3.2.4 <u>Lemma</u>. Let g be a function of class C^1 in $\mathbb{R}^2$. The set $P(g)$ of points $(x, y; \lambda)$ in R^3 such that $X(g+\lambda p_2)$ has a periodic orbit through (x, y) is a surface of class C^1. The projection of $P(g)$ into the (x, y) plane is locally a diffeomorphism onto an open set of R^2.

<u>Proof</u>. Any periodic orbit of $X(g+\lambda p_2)$ homotopic to zero intersects the x-axis since it must contain a singularity in its interior. Therefore the intersection of $P(g)$ with the plane $\mathbb{R}\times\{0\}\times\mathbb{R}$ is

defined locally by

$$\pi(x, \lambda) - x = 0,$$

where π is the Poincaré first return map defined by the flow of $X(g+\lambda p_2)$ in a neighborhood of any point of this intersection.

By the formula derived in [1], pg. 384, it follows that

$$(1)\quad \frac{\partial\pi}{\partial\lambda}(x, \lambda) = \frac{\partial\pi}{\partial x}(x, \lambda) \int_0^{\tau(x, \lambda)} \exp\left[-\int_0^s (g_y(\phi(t, x, \lambda)) + \lambda)dt\right]\phi_2^2(t, x, \lambda)ds$$

there $\tau(x, \lambda)$ is the period of the orbit $\phi(t, x, \lambda) = (\phi_1(t, x, \lambda), \phi_2(t, x, \lambda))$ of $X(g+\lambda p_2)$ through $(x, 0)$, and

$$(2)\quad \frac{\partial\pi}{\partial x}(x, \lambda) = \exp\left[\int_0^{\tau(x, \lambda)} (g_y(\phi(t, x, \lambda)) + \lambda)dt\right].$$

This shows that $P(g)$ is a C^1 surface which is locally of the form $S^1 \times \{\lambda = \lambda(x)\}$ in cases A, B and partially in C. To complete the proof in case C we must also consider the periodic orbits of $\tilde{X}(g+ p_2)$ not homotopic to zero in $\tilde{M}$. These orbits meet the axis $\tilde{0} \times R$ and are disjoint of the $\tilde{x}$-axis. In this case $P(g) \cap \mathbb{R}\times\{0\}\times\mathbb{R}$ is given locally by

$$\pi(y, \lambda) - y = 0;$$

$\pi(y, \lambda) = \phi(1, y, \lambda)$ where $\phi(x, y, \lambda)$ is the solution of

$$\frac{d\phi}{dx} = G(x, \phi, \lambda),$$

with $G(x, \phi, \lambda) = \dfrac{g(x, \phi, \lambda) + \lambda\phi}{\phi}$, with initial condition

$$\phi(0, y, \lambda) = y.$$

Therefore, by the well known formula which gives the derivatives of solutions of differential equations with respect to parameters, as solutions of linear homogeneous equations, we get

$$(3)\quad \frac{\partial\pi}{\partial\lambda}(y, \lambda) = \frac{\partial\pi}{\partial x}(y, \lambda) \int_0^1 \exp\left[-\int_0^x G_2(x, \phi(x, y, \lambda), \lambda)du\right]dx$$

where

$$(4) \qquad \frac{\partial \pi}{\partial x}(y, \lambda) = \exp\left[\int_0^1 G_2(x, \phi(x, y, \lambda))dx\right]$$

This shows that the set of periodic orbits of $\tilde{X}(g+\lambda p_2)$, in case C, which are not homotopic to zero im $\tilde{M}$ are locally of the form $S^1 \times \{\lambda = \lambda(y)\}$.

This finishes the proof of the Lemma.

3.2.5 Remarks

a) From expressions (1) to (4) in 3.2.4, it follows that a periodic orbit γ of $X(g+\lambda p_2)$ is hyperbolic if and only if the projection p_3: $P(g) \to R$ on the axis λ is such that $dp_3(\gamma) \neq 0$ or equivalently if and only if $\frac{d\lambda(x_o)}{dx} \neq 0$, where λ is a local expression for a cross section of $P(g)$.

b) From 3.2.4, $P(g)$ is transversal to $\partial M \times R$. Therefore there are only finitely many λ's such that $X(g+\lambda p_q)$ has a periodic orbit in M tangent to ∂M.

c) It can be proved that the projection of $P(g)$ onto the (x, y)-space (resp. onto $\tilde{R}^2$, in case C) is a globally a diffeomorphism onto an open set of R^2 (resp. $\tilde{R}^2$). This property will not be used in the proof of Theorem 2.

It is enough to verify that any two periodic orbits, γ_1 of $X(g+\lambda_1 p_2)$ and γ_2 of $X(g+\lambda_2 p_2)$, are disjoint if $\lambda_1 \neq \lambda_2$. For instance in case C for curves $\tilde{\gamma}_1$, $\tilde{\gamma}_2$ non homotopic to zero this is obvious since $X(g+\lambda_1 p_2)$ is transversal to γ_2. See Fig. 3

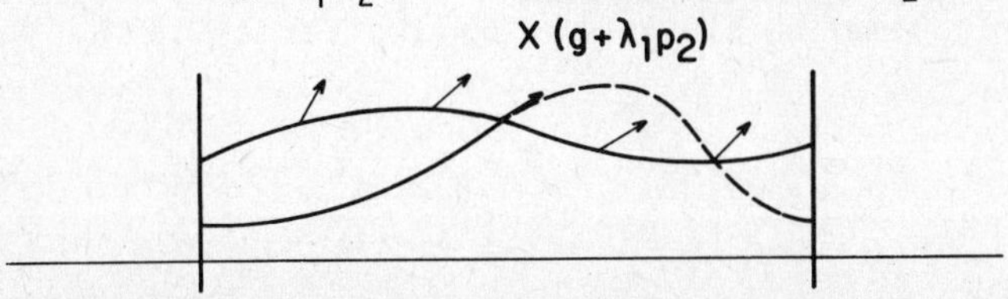

Fig. 3. Dotted line means impossibility

Also a curve γ_1 such that $\tilde{\gamma}_1$ is not homotopic to zero can not meet a curve γ_2 homotopic to zero. See Fig. 4.

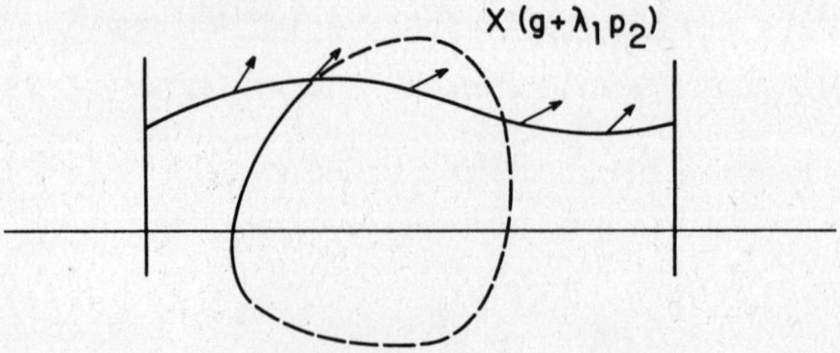

Fig. 4. Dotted line means impossible intersection

The case of γ_1 and γ_2 homotopic to zero is illustrated in Fig. 5.

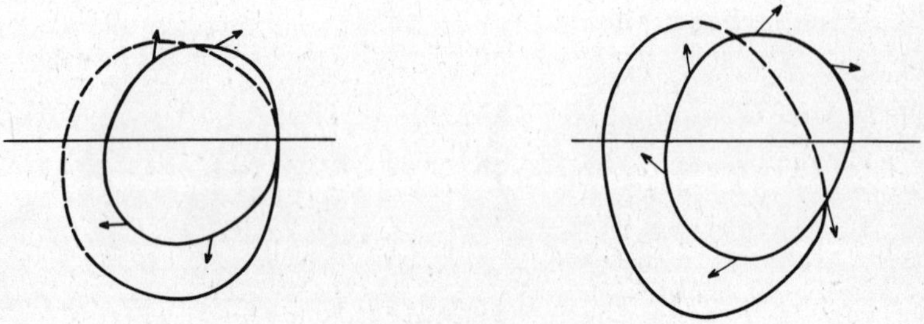

Fig. 5. Impossible intersection

3.2.6 Corollary. Let g be of class C^1 in $\mathbb{R}^2$. For every $\mu \in R$, the set

$$F_1(2, g, \mu) = \left\{\lambda;\ g + \lambda p_2 + \mu \notin \Sigma^1(2, M)\right\}$$

has zero Lebesgue measure in R.

Proof. Follows from Remarks 3.2.5, a), b) and Lemma 3.2.1

3.2.7 Lemma. Let μ be such that $\mu \notin F_1(0, f) \cup F_1'(0, f)$ where $F_1'(0, f) = \{\mu;\ b_x + b_{yy}(f+\mu) \mid \partial M \cap \{y=0\} = 0\}$. Then, the set $C(f, \mu)$ of points $(q, \lambda) \in \partial M \times R$ such that $X(f+\lambda p_2+\mu)$ is tangent at q

to ∂M is a C^1 curve. The projection $p_3: C(f, \mu) \to R$ onto the λ-axis is such that (q, λ) is regular if and only if q is a point of parabolic contact of $X(f+\lambda p_2+\mu)$ with ∂M.

Proof. Let b be a local expression for ∂M. We have that $C(f, \mu) = \{(q, \lambda); b(q) = 0, X(f+\lambda p_2+\mu).b = b_x y+b_y(f+\lambda p_2+\mu)(q) = 0\}$.
If $q_o = (x_o, y_o)$ with
If $y_o \neq 0$ and $b_y(q_o) \neq 0$, the Jacobian matrix of $(b, X(f+\lambda p_2+\mu)b)$ at (q_o, λ_o) is

$$(1) \qquad \begin{pmatrix} b_x(q_o) & b_y(q_o) & 0 \\ \alpha & \beta & b_y(q_o)y_o \end{pmatrix}$$

and has rank 2 since $b_y^2(q_o)y_o \neq 0$. Therefore $C(f, \mu)$ near (q_o, λ_o) is of the form $\{(x, u(x), \lambda(x))\}$ for x in a neighborhood of x_o.

If the contact is parabolic, i.e., $X^2(f+\lambda p_2+\mu)b(q_o, \lambda_o) \neq 0$ clearly $C(f, \mu)$ is locally of the form $\{(x(\lambda), y(\lambda), \lambda)\}$, since $X^2(f+\lambda p_2+\mu)b(q_o, \lambda_o) = d[X(f+\lambda p_2+\mu)b]X(f+\lambda p_2+\mu)$ with $X(f+\lambda p_2+\mu)(q_o)$ tangent to ∂M at q_o.

If (q_o, λ_o) is a regular point of p_3 then $\frac{d\lambda}{dx}(x_o) \neq 0$, but this, by the implicit function theorem, is given by the left square minor δ of (1). But δ not zero is equivalent to parabolic contact.

If $y_o = 0$ and $b_y(q_o) = 0$, the condition for parabolic contact between $X(f+\lambda p_2+\mu)$ and ∂M is $[b_x + b_{yy}(f+\mu)](q_o) \neq 0$. By the very hypothesis $\mu \notin F_1'(0, f)$ follows in this case the conclusion.

3.2.8 Corollary. Let f be of class C^1 in R^2. Let $\mu \notin F_1'(0, f) \cup F_1(0, f)$ of 3.2.2, 3.2.7. Then

$$F_1(3, f, \mu) = \left\{\lambda;\ X(f+\lambda p_2+\mu) \notin \Sigma^1(3, M)\right\}$$

has zero Lebesgue measure.

3.2.9 Lemma. Let $\mu \notin F_1(0, f) \cup F_1'(0, f)$. See 3.2.2 and 3.2.7. Then the set $F_1(4, f, \mu) = \{\lambda;\ X(f+\lambda p_2+\mu) \notin \Sigma^1(4, M)\}$ has zero Lebesgue measure in R.

Proof: Denote by $F_1'(4, f, \mu)$ the set of points λ for which $X(f+\lambda p_2+\mu)$ has either

A) at least a saddle connection or

B) at least a saddle separatrix parabolically tangent to ∂M, or passing through a corner of ∂M or

C) at least an orbit with two parabolic tangencies with ∂M or

D) at least an orbit with one parabolic tangency with ∂M and passing through a corner in ∂M.

We prove that the λ's in each of the cases A) to D) are countable. In fact, the set of points λ which correspond to A is the union of finitely many discrete subsets. Each such subset corresponds to a pair (s_p, u_q), where s_p a stable separatrix of a saddle p and u_q an unstable separatrix of a saddle q. See Fig. 6

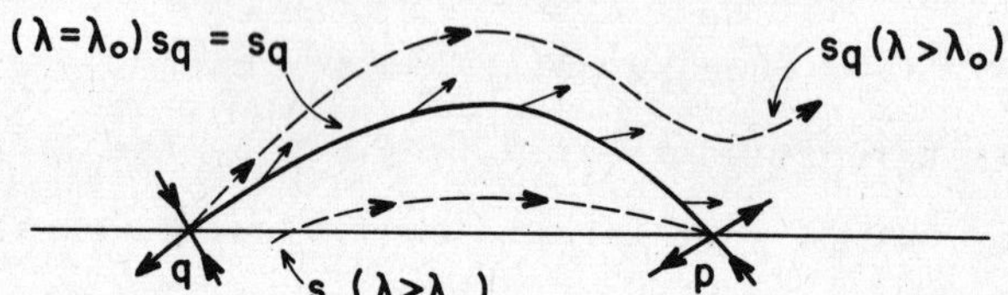

Fig. 6

which illustrates the fact that no saddle connection is possible between $s_p(\lambda)$ and $u_q(\lambda)$ for $\lambda > \lambda_0$ small. The same works for $\lambda < \lambda_0$.

Similar analysis works for the other cases. See Fig. 7 for cases B and C.

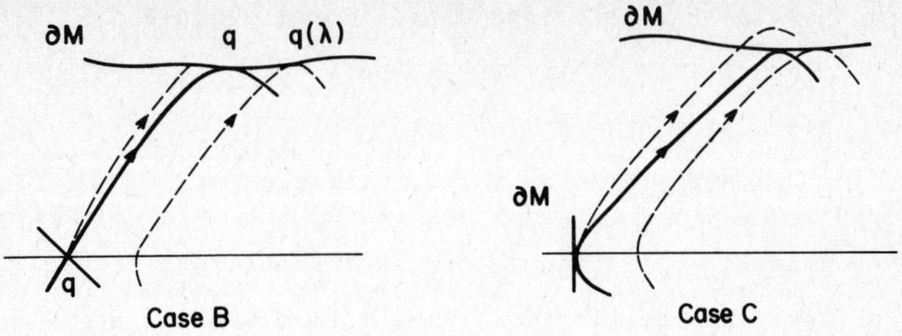

Fig. 7

The proof of the Lemma follows from the fact that $F_1(4, f, \mu) = F_1(3, f, \mu) \cup F_1'(4, f, \mu)$.

Proof of Theorem 2: The set

$$F_1(f) = \left\{(\lambda, \mu);\ f+\lambda p_2+\mu \notin \Sigma^1(M)\right\}$$

is contained in

$$(F_1(0, f) \cup F_1'(0, f)) \times R \cup (R-F_1(0, f) \cup F_1'(0, f)) \times R.$$

The first set has zero Lebesgue measure in $\mathbb{R}^2$, by 3.2.2 and 3.2.7. The Part of $F_1(f)$ contained in the second set is of the form

$$\bigcup_{\mu \notin F_1(0, f) \cup F_1'(0, f)} \{\mu\} \times \left[\bigcup_{i=1}^{4} F_1(i, f, \mu))\right].$$

This set has zero Lebesgue measure in $\mathbb{R}^2$, by Fubini's Theorem and the assertions of 3.2.3; 3.2.6; 3.2.8; 3.2.9.

This ends the proof of Theorem 2.

3.3 The proof of Theorem 3 is immediate from Theorem 2 since for any f of class C^k in $\mathbb{R}^2$

$$\| f+\lambda p_2+\mu-f\|_k^M \leq L|\lambda| + |\mu|,$$

where $L = \sup\{y;\ (x, y) \in M\}$.

3.4 The proof of Theorem 4 follows from Theorem 2 and Fubini's Theorem. In case a), for instance, apply Fubini's Theorem to the product space

$$\mathbb{R}(a_{o,o};\ a_{o,1}) \times R(a_{1,o},\dots,a_{m,m}).$$

For each element f in the second factor the function $g(f)$ which defines the equation is of the form given in Theorem 2 with $\lambda = a_{o,1}$, $\mu = a_{o,o}$ and $f = a_{1,o}x + \sum_{2\leq i+j\leq m} a_{ij}x^i\dot{x}^j$. From Theorem 2,

$$A_f = (\mathbb{R}(a_{o,o};\ a_{o1}) \times \{f\}) \cap \left[C^1(M) - \Sigma^1(M)\right]$$

has zero Lebesgue measure in $\mathbb{R}^2$. Therefore the union in f of the sets A_f above has zero Lebesgue measure in $\omega_m = \mathbb{R}^2 \times R(a_{1,o};\dots;a_{m,m})$.

3.5 <u>Proof of Theorem 5</u>: Let f be of class C^k in $\mathbb{R}^2$. We will prove that if (E_f) is C^k-structurally stable on M then $f \in \Sigma^k(M)$.

By Theorem 3, $X(f)$ is topologically equivalent to $X(g)$, with $g \in \Sigma^k(M)$. Then we can assert that

a) $X(f)$ has finitely many singular points all topologically equivalent to saddles or nodes and contained in Int M.

b) $X(f)$ (resp. $\tilde{X}(f)$, in case C) has finitely many periodic orbits, all attractors or repellors and contained in Int M (resp. Int $\tilde{M}$, in case C).

c) $X(f)$ has finitely many points of non transversal contact with ∂M which are locally internal or external. A priori topologically transversal contacts with ∂M are not excluded by Theorem 3.

d) $X(f)$ satisfies 2.1.4 with saddles replaced by topological saddles and parabolic tangencies by external and internal tangencies.

a_1) Now we prove that $f \in \Sigma^k(1, M)$. In fact, at a singular point

$p_0 = (x_0, 0)$ we must have that $f_x(x_0, 0) \neq 0$. If it were $f_x(x_0, 0) = 0$ we could consider the perturbation $f_\varepsilon = f + \varepsilon\beta(x)(x-x_0)$, where β is a bump function with $\beta(x_0) = 1$ and whose support containts $(x_0, 0)$ as unique singularity. Since $f(x, 0)$ has isolated zeros it must be of one of the following sorts

Fig. 8

By properly chosing the sign of ε, small, we get $X(f_\varepsilon)$ with two singularities more than $X(f)$. This contradicts the hypothesis of structural stability of (E_f).

b_1) Now we prove that $f \in \sum^k(2, M)$.

If $X(f)$ (resp. $\tilde{X}(f)$, in case C) had a non hyperbolic periodic orbit γ contained in M. That is $\int_\gamma f_y = 0$. We could consider the following perturbation

$$f_\varepsilon = f + \varepsilon\beta g g_y,$$

where g is a function of class C^{k+1} which vanishes on γ and $dg(q) \neq 0$ for each q in γ; β is a bump function such that $\beta(\gamma) = 1$ and its support is disjoint from other periodic orbits of $X(f)$. In case C, g and β can be taken periodic in x, of period 1.

We have that

$$\int_\gamma \frac{\partial f_\varepsilon}{\partial y} = \varepsilon\int_\gamma g_y^2,$$

with $\int_\gamma g_y^2 > 0$, since g_y^2 is positive at least at points of γ

maximal y-coordinate.

Therefore, if γ were an attractor of $X(f)$ (resp. $\tilde{X}(f)$, in case C) we could obtain, by chosing $\varepsilon > 0$, small, that $X(f_\varepsilon)$ (resp. $\tilde{X}(f_\varepsilon)$) would have at least two more periodic orbits than $X(f)$ (resp. $\tilde{X}(f)$). Which contradicts the hypothesis of structural stability of (E_f).

c_1) Now we prove that $f \in \sum^k(3, M)$.

If $X(f)$ had a point $p_o = (x_o, y_o)$ of non parabolic contact with ∂M, we could consider the following perturbation $f_{\varepsilon,\ell,m} = f + (\varepsilon + \ell(x-x_o) + m(y-y_o))\beta$, where β is a bump function whose support is disjoint from points of external or internal tangency of $X(f)$ with ∂M other than p_o, in case p_o is a point of internal or external tangency.

Let b be a C^2 function which defines locally M. We have

$$X^2(f_{\varepsilon,\ell,m})\cdot b(p_o) = y_o[b_{xx}+b_{xy}+(f+\varepsilon)+b_y(f_x+\ell)]$$

(*)

$$+ [b_{xy}y_o+b_x+b_{yy}(f+\varepsilon)+b_y(f_y+m)](f+\varepsilon).$$

If $b_y(p_o) \neq 0$, take $\varepsilon = 0$ and (ℓ, m) small so that $X^2(f_{o,\ell,m})b(p_o) = b_y[\ell y_o + mf(p_o)]$ is positive (resp. negative) if the contact between $X(f)$ and ∂M was supposed external (resp. internal). In such case $X(f_{o,\ell,m})$ would have at least one more point of internal tangency and one more point of external tangency with ∂M. This would contradict the hypothesis of structural stability of (E_f).

If $b_y(p_o) = 0$, the argument is similar. In this case it must be $y_o = 0$ and $b_x(x_o, 0) \neq 0$.

We take $(m, \ell) = (0, 0)$ and (*) reduces to

$$X^2(f_{\varepsilon,o,o})\cdot b(p_o) = \varepsilon[b_x + b_{yy}(f+\varepsilon)],$$

whose sign can be made positive or negative by chosing ε appropriate.

By a_1), b_1), c_1) and d) it follows that $f \in \sum^k(M)$. This ends the proof of Theorem 5.

REFERENCES

[1] Andronov, A., Leontovich, E., et al. - Theory of Bifurcations of Dynamical Systems on a Plane, PIST, Jerusalen, 1971.

[2] Barreto, A. - Estabilidade Estrutural das Equações Diferenciais da forma $\ddot{x} = f(x, \dot{x})$. Tese. IMPA, 1964.

[3] Peixoto, M. C., Peixoto, M. M. - Structural Stability in the plane with enlarged boundary conditions, Anais Acad. Bras. Cien., 31, 1959.

[4] Sard, A. - The measure of the critical set of differentiable maps, B.A.M.S., 48, 1942.

[5] Sotomayor, J. - La mesure de l'ensemble de bifurcation des champs de vecteurs dans le plan, C. R. Acad. Sc. Paris, t. 294 (2 nov 1981) Serie I.

[6] Sotomayor, J. - Curvas definidas por equações diferenciais no Plano, 13th Coloq. Bras. Mat. IMPA, 1981.

Vol. 817: L. Gerritzen, M. van der Put, Schottky Groups and Mumford Curves. VIII, 317 pages. 1980.

Vol. 818: S. Montgomery, Fixed Rings of Finite Automorphism Groups of Associative Rings. VII, 126 pages. 1980.

Vol. 819: Global Theory of Dynamical Systems. Proceedings, 1979. Edited by Z. Nitecki and C. Robinson. IX, 499 pages. 1980.

Vol. 820: W. Abikoff, The Real Analytic Theory of Teichmüller Space. VII, 144 pages. 1980.

Vol. 821: Statistique non Paramétrique Asymptotique. Proceedings, 1979. Edited by J.-P. Raoult. VII, 175 pages. 1980.

Vol. 822: Séminaire Pierre Lelong–Henri Skoda, (Analyse) Annees 1978/79. Proceedings. Edited by P. Lelong et H. Skoda. VIII, 356 pages, 1980.

Vol. 823: J. Král, Integral Operators in Potential Theory. III, 171 pages. 1980.

Vol. 824: D. Frank Hsu, Cyclic Neofields and Combinatorial Designs. VI, 230 pages. 1980.

Vol. 825: Ring Theory, Antwerp 1980. Proceedings. Edited by F. van Oystaeyen. VII, 209 pages. 1980.

Vol. 826: Ph. G. Ciarlet et P. Rabier, Les Equations de von Kármán. VI, 181 pages. 1980.

Vol. 827: Ordinary and Partial Differential Equations. Proceedings, 1978. Edited by W. N. Everitt. XVI, 271 pages. 1980.

Vol. 828: Probability Theory on Vector Spaces II. Proceedings, 1979. Edited by A. Weron. XIII, 324 pages. 1980.

Vol. 829: Combinatorial Mathematics VII. Proceedings, 1979. Edited by R. W. Robinson et al.. X, 256 pages. 1980.

Vol. 830: J. A. Green, Polynomial Representations of GL_n. VI, 118 pages. 1980.

Vol. 831: Representation Theory I. Proceedings, 1979. Edited by V. Dlab and P. Gabriel. XIV, 373 pages. 1980.

Vol. 832: Representation Theory II. Proceedings, 1979. Edited by V. Dlab and P. Gabriel. XIV, 673 pages. 1980.

Vol. 833: Th. Jeulin, Semi-Martingales et Grossissement d'une Filtration. IX, 142 Seiten. 1980.

Vol. 834: Model Theory of Algebra and Arithmetic. Proceedings, 1979. Edited by L. Pacholski, J. Wierzejewski, and A. J. Wilkie. VI, 410 pages. 1980.

Vol. 835: H. Zieschang, E. Vogt and H.-D. Coldewey, Surfaces and Planar Discontinuous Groups. X, 334 pages. 1980.

Vol. 836: Differential Geometrical Methods in Mathematical Physics. Proceedings, 1979. Edited by P. L. García, A. Pérez-Rendón, and J. M. Souriau. XII, 538 pages. 1980.

Vol. 837: J. Meixner, F. W. Schäfke and G. Wolf, Mathieu Functions and Spheroidal Functions and their Mathematical Foundations Further Studies. VII, 126 pages. 1980.

Vol. 838: Global Differential Geometry and Global Analysis. Proceedings 1979. Edited by D. Ferus et al. XI, 299 pages. 1981.

Vol. 839: Cabal Seminar 77 – 79. Proceedings. Edited by A. S. Kechris, D. A. Martin and Y. N. Moschovakis. V, 274 pages. 1981.

Vol. 840: D. Henry, Geometric Theory of Semilinear Parabolic Equations. IV, 348 pages. 1981.

Vol. 841: A. Haraux, Nonlinear Evolution Equations- Global Behaviour of Solutions. XII, 313 pages. 1981.

Vol. 842: Séminaire Bourbaki vol. 1979/80. Exposés 543–560. IV, 317 pages. 1981.

Vol. 843: Functional Analysis, Holomorphy, and Approximation Theory. Proceedings. Edited by S. Machado. VI, 636 pages. 1981.

Vol. 844: Groupe de Brauer. Proceedings. Edited by M. Kervaire and M. Ojanguren. VII, 274 pages. 1981.

Vol. 845: A. Tannenbaum, Invariance and System Theory: Algebraic and Geometric Aspects. X, 161 pages. 1981.

Vol. 846: Ordinary and Partial Differential Equations, Proceedings. Edited by W. N. Everitt and B. D. Sleeman. XIV, 384 pages. 1981.

Vol. 847: U. Koschorke, Vector Fields and Other Vector Bundle Morphisms – A Singularity Approach. IV, 304 pages. 1981.

Vol. 848: Algebra, Carbondale 1980. Proceedings. Ed. by R. K. Amayo. VI, 298 pages. 1981.

Vol. 849: P. Major, Multiple Wiener-Itô Integrals. VII, 127 pages. 1981.

Vol. 850: Séminaire de Probabilités XV. 1979/80. Avec table générale des exposés de 1966/67 à 1978/79. Edited by J. Azéma and M. Yor. IV, 704 pages. 1981.

Vol. 851: Stochastic Integrals. Proceedings, 1980. Edited by D. Williams. IX, 540 pages. 1981.

Vol. 852: L. Schwartz, Geometry and Probability in Banach Spaces. X, 101 pages. 1981.

Vol. 853: N. Boboc, G. Bucur, A. Cornea, Order and Convexity in Potential Theory: H-Cones. IV, 286 pages. 1981.

Vol. 854: Algebraic K-Theory. Evanston 1980. Proceedings. Edited by E. M. Friedlander and M. R. Stein. V, 517 pages. 1981.

Vol. 855: Semigroups. Proceedings 1978. Edited by H. Jürgensen, M. Petrich and H. J. Weinert. V, 221 pages. 1981.

Vol. 856: R. Lascar, Propagation des Singularités des Solutions d'Equations Pseudo-Différentielles à Caractéristiques de Multiplicités Variables. VIII, 237 pages. 1981.

Vol. 857: M. Miyanishi. Non-complete Algebraic Surfaces. XVIII, 244 pages. 1981.

Vol. 858: E. A. Coddington, H. S. V. de Snoo: Regular Boundary Value Problems Associated with Pairs of Ordinary Differential Expressions. V, 225 pages. 1981.

Vol. 859: Logic Year 1979–80. Proceedings. Edited by M. Lerman, J. Schmerl and R. Soare. VIII, 326 pages. 1981.

Vol. 860: Probability in Banach Spaces III. Proceedings, 1980. Edited by A. Beck. VI, 329 pages. 1981.

Vol. 861: Analytical Methods in Probability Theory. Proceedings 1980. Edited by D. Dugué, E. Lukacs, V. K. Rohatgi. X, 183 pages. 1981.

Vol. 862: Algebraic Geometry. Proceedings 1980. Edited by A. Libgober and P. Wagreich. V, 281 pages. 1981.

Vol. 863: Processus Aléatoires à Deux Indices. Proceedings, 1980. Edited by H. Korezlioglu, G. Mazziotto and J. Szpirglas. V, 274 pages. 1981.

Vol. 864: Complex Analysis and Spectral Theory. Proceedings, 1979/80. Edited by V. P. Havin and N. K. Nikol'skii, VI, 480 pages. 1981.

Vol. 865: R. W. Bruggeman, Fourier Coefficients of Automorphic Forms. III, 201 pages. 1981.

Vol. 866: J.-M. Bismut, Mécanique Aléatoire. XVI, 563 pages. 1981.

Vol. 867: Séminaire d'Algèbre Paul Dubreil et Marie-Paule Malliavin. Proceedings, 1980. Edited by M.-P. Malliavin. V, 476 pages. 1981.

Vol. 868: Surfaces Algébriques. Proceedings 1976–78. Edited by J. Giraud, L. Illusie et M. Raynaud. V, 314 pages. 1981.

Vol. 869: A. V. Zelevinsky, Representations of Finite Classical Groups. IV, 184 pages. 1981.

Vol. 870: Shape Theory and Geometric Topology. Proceedings, 1981. Edited by S. Mardešić and J. Segal. V, 265 pages. 1981.

Vol. 871: Continuous Lattices. Proceedings, 1979. Edited by B. Banaschewski and R.-E. Hoffmann. X, 413 pages. 1981.

Vol. 872: Set Theory and Model Theory. Proceedings, 1979. Edited by R. B. Jensen and A. Prestel. V, 174 pages. 1981.

Vol. 873: Constructive Mathematics, Proceedings, 1980. Edited by F. Richman. VII, 347 pages. 1981.

Vol. 874: Abelian Group Theory. Proceedings, 1981. Edited by R. Göbel and E. Walker. XXI, 447 pages. 1981.

Vol. 875: H. Zieschang, Finite Groups of Mapping Classes of Surfaces. VIII, 340 pages. 1981.

Vol. 876: J. P. Bickel, N. El Karoui and M. Yor. Ecole d'Eté de Probabilités de Saint-Flour IX – 1979. Edited by P. L. Hennequin. XI, 280 pages. 1981.

Vol. 877: J. Erven, B.-J. Falkowski, Low Order Cohomology and Applications. VI, 126 pages. 1981.

Vol. 878: Numerical Solution of Nonlinear Equations. Proceedings, 1980. Edited by E. L. Allgower, K. Glashoff, and H.-O. Peitgen. XIV, 440 pages. 1981.

Vol. 879: V. V. Sazonov, Normal Approximation – Some Recent Advances. VII, 105 pages. 1981.

Vol. 880: Non Commutative Harmonic Analysis and Lie Groups. Proceedings, 1980. Edited by J. Carmona and M. Vergne. IV, 553 pages. 1981.

Vol. 881: R. Lutz, M. Goze, Nonstandard Analysis. XIV, 261 pages. 1981.

Vol. 882: Integral Representations and Applications. Proceedings, 1980. Edited by K. Roggenkamp. XII, 479 pages. 1981.

Vol. 883: Cylindric Set Algebras. By L. Henkin, J. D. Monk, A. Tarski, H. Andréka, and I. Németi. VII, 323 pages. 1981.

Vol. 884: Combinatorial Mathematics VIII. Proceedings, 1980. Edited by K. L. McAvaney. XIII, 359 pages. 1981.

Vol. 885: Combinatorics and Graph Theory. Edited by S. B. Rao. Proceedings, 1980. VII, 500 pages. 1981.

Vol. 886: Fixed Point Theory. Proceedings, 1980. Edited by E. Fadell and G. Fournier. XII, 511 pages. 1981.

Vol. 887: F. van Oystaeyen, A. Verschoren, Non-commutative Algebraic Geometry, VI, 404 pages. 1981.

Vol. 888: Padé Approximation and its Applications. Proceedings, 1980. Edited by M. G. de Bruin and H. van Rossum. VI, 383 pages. 1981.

Vol. 889: J. Bourgain, New Classes of $\mathcal{L}^p$-Spaces. V, 143 pages. 1981.

Vol. 890: Model Theory and Arithmetic. Proceedings, 1979/80. Edited by C. Berline, K. McAloon, and J.-P. Ressayre. VI, 306 pages. 1981.

Vol. 891: Logic Symposia, Hakone, 1979, 1980. Proceedings, 1979, 1980. Edited by G. H. Müller, G. Takeuti, and T. Tugué. XI, 394 pages. 1981.

Vol. 892: H. Cajar, Billingsley Dimension in Probability Spaces. III, 106 pages. 1981.

Vol. 893: Geometries and Groups. Proceedings. Edited by M. Aigner and D. Jungnickel. X, 250 pages. 1981.

Vol. 894: Geometry Symposium. Utrecht 1980, Proceedings. Edited by E. Looijenga, D. Siersma, and F. Takens. V, 153 pages. 1981.

Vol. 895: J.A. Hillman, Alexander Ideals of Links. V, 178 pages. 1981.

Vol. 896: B. Angéniol, Familles de Cycles Algébriques – Schéma de Chow. VI, 140 pages. 1981.

Vol. 897: W. Buchholz, S. Feferman, W. Pohlers, W. Sieg, Iterated Inductive Definitions and Subsystems of Analysis: Recent Proof-Theoretical Studies. V, 383 pages. 1981.

Vol. 898: Dynamical Systems and Turbulence, Warwick, 1980. Proceedings. Edited by D. Rand and L.-S. Young. VI, 390 pages. 1981.

Vol. 899: Analytic Number Theory. Proceedings, 1980. Edited by M.I. Knopp. X, 478 pages. 1981.

Vol. 900: P. Deligne, J. S. Milne, A. Ogus, and K.-Y. Shih, Hodge Cycles, Motives, and Shimura Varieties. V, 414 pages. 1982.

Vol. 901: Séminaire Bourbaki vol. 1980/81 Exposés 561–578. III, 299 pages. 1981.

Vol. 902: F. Dumortier, P.R. Rodrigues, and R. Roussarie, Germs of Diffeomorphisms in the Plane. IV, 197 pages. 1981.

Vol. 903: Representations of Algebras. Proceedings, 1980. Edited by M. Auslander and E. Lluis. XV, 371 pages. 1981.

Vol. 904: K. Donner, Extension of Positive Operators and Korovkin Theorems. XII, 182 pages. 1982.

Vol. 905: Differential Geometric Methods in Mathematical Physics. Proceedings, 1980. Edited by H.-D. Doebner, S.J. Andersson, and H.R. Petry. VI, 309 pages. 1982.

Vol. 906: Séminaire de Théorie du Potentiel, Paris, No. 6. Proceedings. Edité par F. Hirsch et G. Mokobodzki. IV, 328 pages. 1982.

Vol. 907: P. Schenzel, Dualisierende Komplexe in der lokalen Algebra und Buchsbaum-Ringe. VII, 161 Seiten. 1982.

Vol. 908: Harmonic Analysis. Proceedings, 1981. Edited by F. Ricci and G. Weiss. V, 325 pages. 1982.

Vol. 909: Numerical Analysis. Proceedings, 1981. Edited by J.P. Hennart. VII, 247 pages. 1982.

Vol. 910: S.S. Abhyankar, Weighted Expansions for Canonical Desingularization. VII, 236 pages. 1982.

Vol. 911: O.G. Jørsboe, L. Mejlbro, The Carleson-Hunt Theorem on Fourier Series. IV, 123 pages. 1982.

Vol. 912: Numerical Analysis. Proceedings, 1981. Edited by G. A. Watson. XIII, 245 pages. 1982.

Vol. 913: O. Tammi, Extremum Problems for Bounded Univalent Functions II. VI, 168 pages. 1982.

Vol. 914: M. L. Warshauer, The Witt Group of Degree k Maps and Asymmetric Inner Product Spaces. IV, 269 pages. 1982.

Vol. 915: Categorical Aspects of Topology and Analysis. Proceedings, 1981. Edited by B. Banaschewski. XI, 385 pages. 1982.

Vol. 916: K.-U. Grusa, Zweidimensionale, interpolierende Lg-Splines und ihre Anwendungen. VIII, 238 Seiten. 1982.

Vol. 917: Brauer Groups in Ring Theory and Algebraic Geometry. Proceedings, 1981. Edited by F. van Oystaeyen and A. Verschoren. VIII, 300 pages. 1982.

Vol. 918: Z. Semadeni, Schauder Bases in Banach Spaces of Continuous Functions. V, 136 pages. 1982.

Vol. 919: Séminaire Pierre Lelong – Henri Skoda (Analyse) Années 1980/81 et Colloque de Wimereux, Mai 1981. Proceedings. Edité par P. Lelong et H. Skoda. VII, 383 pages. 1982.

Vol. 920: Séminaire de Probabilités XVI, 1980/81. Proceedings. Edité par J. Azéma et M. Yor. V, 622 pages. 1982.

Vol. 921: Séminaire de Probabilités XVI, 1980/81. Supplément: Géométrie Différentielle Stochastique. Proceedings. Edité par J. Azéma et M. Yor. III, 285 pages. 1982.

Vol. 922: B. Dacorogna, Weak Continuity and Weak Lower Semicontinuity of Non-Linear Functionals. V, 120 pages. 1982.

Vol. 923: Functional Analysis in Markov Processes. Proceedings, 1981. Edited by M. Fukushima. V, 307 pages. 1982.

Vol. 924: Séminaire d'Algèbre Paul Dubreil et Marie-Paule Malliavin. Proceedings, 1981. Edité par M.-P. Malliavin. V, 461 pages. 1982.

Vol. 925: The Riemann Problem, Complete Integrability and Arithmetic Applications. Proceedings, 1979-1980. Edited by D. Chudnovsky and G. Chudnovsky. VI, 373 pages. 1982.

Vol. 926: Geometric Techniques in Gauge Theories. Proceedings, 1981. Edited by R. Martini and E.M.de Jager. IX, 219 pages. 1982.